AF353110

Land Use Conflict Detection and Multi-Objective Optimization Based on the Productivity, Sustainability, and Livability Perspective

Land Use Conflict Detection and Multi-Objective Optimization Based on the Productivity, Sustainability, and Livability Perspective

Editors

Dong Jiang
Jinwei Dong
Gang Lin

MDPI • Basel • Beijing • Wuhan • Barcelona • Belgrade • Manchester • Tokyo • Cluj • Tianjin

Editors

Dong Jiang
Institute of Geographic
Sciences and Natural
Resources Research, Chinese
Academy of Sciences
China

Jinwei Dong
Institute of Geographic
Sciences and Natural
Resources Research, Chinese
Academy of Sciences
China

Gang Lin
Institute of Geographic
Sciences and Natural
Resources Research, Chinese
Academy of Sciences
China

Editorial Office
MDPI
St. Alban-Anlage 66
4052 Basel, Switzerland

This is a reprint of articles from the Special Issue published online in the open access journal *Land* (ISSN 2073-445X) (available at: https://www.mdpi.com/journal/land/special_issues/PLE).

For citation purposes, cite each article independently as indicated on the article page online and as indicated below:

LastName, A.A.; LastName, B.B.; LastName, C.C. Article Title. *Journal Name* **Year**, *Volume Number*, Page Range.

ISBN 978-3-0365-4621-6 (Hbk)
ISBN 978-3-0365-4622-3 (PDF)

Contents

About the Editors

Dong Jiang

Dong Jiang, Professor at the Institute of Geographic Sciences and Natural Resource Research (IGSNRR), the Chinese Academy of Sciences (CAS), Beijing, China. His current research interests include renewable energy, energy–food–water nexus, and applications of remote sensing and geographical information system. He received his MS degree in Hydrological Geology from China University of Mining and Technology in 1997 and PhD degree in Ecology from the IGSNRR, CAS, in 2000. He has published more than 150 journal papers (>60 papers in English and in SCI-indexed journals) and is coauthor of 11 books (4 in English). He is an Editorial Board Member of Scientific Reports, Associate Chief Editor of Journal of Global Change Data & Discovery, and the leading Guest Editor of the Special Issue "Monitoring and Modeling Terrestrial Ecosystems' Response to Climate Change" of Advances in Meteorology (2013 and 2015).

Jinwei Dong

Jinwei Dong, Professor at the Institute of Geographic Sciences and Natural Resource Research (IGSNRR), the Chinese Academy of Sciences (CAS), Beijing, China. His research focuses on land cover and land use change (LCLUC) and environmental remote sensing, especially the application of advanced cloud-based geospatial processing technology. He has developed new algorithms to map rubber plantations and improve paddy rice mapping using phenology-based approaches. Dr. Dong has also investigated the uncertainties of remote sensing data and remote sensing-based models while addressing ecological scientific questions. His current research interests center on the monitoring of two typical land use change processes in monsoon Asia: paddy rice expansion in high latitude regions and the conversion of natural forests to plantations in tropical monsoon Asia. He investigates changes in these landscapes and their regional climatic and ecological effects using multi-source remote sensing, in situ observations, and modeling strategies.

Gang Lin

Gang Lin, Research Assistant at the Institute of Geographic Sciences and Natural Resource Research (IGSNRR), the Chinese Academy of Sciences (CAS), Beijing, China. His current research interests include land use classification and evolvement based on the production–living–ecological (PLE) perspective; resource utilization and its environmental impact; coupling simulation of water–energy–land systems.

Preface to "Land Use Conflict Detection and Multi-Objective Optimization Based on the Productivity, Sustainability, and Livability Perspective"

Land use affects many aspects of regional sustainable development, so insight into its influence is of great importance for the optimization of national space. With the rapid economic growth and the accelerating industrialization, there has been a noticeable increase in the contention and conflict between various land uses. For example, the expansion of urban and industrial land has occupied a large amount of high-quality farmland, leading to serious contradictions between socioeconomic development and the protection of cultivated land resources. Solving these problems has thus become an important scientific proposition for regional sustainable development in the field of human-economic geography. Based on the widely recognized development goals based on productivity, sustainability, and livability perspectives, a scientific classification system, spatial conflict detection, and multi-objective optimization of land use functions (LUFs) have provided an efficient means for land use planning and management, attracting the extensive attention of researchers and policymakers. For example, production–living–ecological space (PLE), the shortened form for productive space (PS), living space (LS), and ecological space (ES), is reclassified considering both land use functions and utilization types based on productivity, sustainability, and livability perspectives. First proposed by the Chinese Government Report in 2012, it has made prominent contributions to improving the protection system of spatial development while implementing major function-oriented zoning at all scales. Successful implementation in China proves that PLE holds promotional value to provide support for space use allocation, land use conflict management, and sustainable development. However, compared to the literature available on functional classification and identification based on productivity, sustainability, and livability perspectives, it is surprising that very few among the increasing number of papers dealing with LUFs are devoted to its conflict detection and multi-objective optimization with land use planning, which has left many significant questions unanswered. There are few reports on integrated research, and international practical application is especially lacking. Thus, the theoretical research and practical applications of the multi-functional complexity of land use still need to be further refined.

This Special Issue entitled "Land Use Conflict Detection and Multi-Objective Optimization Based on Productivity, Sustainability, and Livability Perspectives" mainly focuses on functional classification, spatial conflict detection, and spatial development pattern optimization based on productivity, sustainability, and livability perspectives, presenting a relevant opportunity for all scholars to share their knowledge from the multidisciplinary community across the world that includes landscape ecologists, social scientists, and geographers. Further progress in theoretical research and practical applications on the scientific classification system toward productivity, sustainability, and livability, such as PLE classification of China, is covered. Integrative studies regarding suitability evaluation and optimized modes of e multi-functional land use systems to meet the target of geographical space optimization and regional sustainable development are also presented.

Dong Jiang , Jinwei Dong, and Gang Lin

Editors

Review

A Review on the Overall Optimization of Production–Living–Ecological Space: Theoretical Basis and Conceptual Framework

Gang Lin [1,2], Dong Jiang [1,2,3], Jingying Fu [1,2,3,*] and Yi Zhao [4]

[1] Institute of Geographic Sciences and Natural Resources Research, Chinese Academy of Sciences, No. 11A Datun Road, Chaoyang District, Beijing 100101, China; ling@lreis.ac.cn (G.L.); jiangd@igsnrr.ac.cn (D.J.)

[2] College of Resources and Environment, University of Chinese Academy of Sciences, No. 19A Yuquan Road, Haidian District, Beijing 100049, China

[3] Key Laboratory of Carrying Capacity Assessment for Resource and Environment, Ministry of Natural Resources, No. 46 Fuchengmen Road, Xicheng District, Beijing 100812, China

[4] College of Geoscience and Surveying Engineering, China University of Mining and Technology, No. 11 Xueyuan Road, Haidian District, Beijing 100083, China; zy@student.cumtb.edu.cn

* Correspondence: fujy@igsnrr.ac.cn; Tel.: +86-10-6488-9221

Abstract: The 18th National Congress of the Communist Party of China put forward the optimization of territorial space development patterns as the primary measure of ecological civilization construction, and put forward the goal of "promoting intensive and efficient production space, livable and moderate living space, and beautiful (picturesque scenery) ecological space". Through literature research and summing induction, this paper combs the research progress of the overall optimization of "Production–Living–Ecological" space (PLES) systematically. It is found that the existing work mainly focuses on the overall optimization of PLES from the perspectives of land-use quality, land-use suitability evaluation, resource and environmental carrying capacity, and comparative advantages. However, due to the lack of understanding of the scientific connotation of PLES, and the imperfect construction of quantitative identification methods and classification system, there are many problems in the technical approaches of the overall optimization of PLES, which remain to be clarified. In the future, the technological approach to the overall optimization of PLES should be guided by the vision of building a beautiful China, with the theory of a human–Earth coupling system as the core, and systematically build a theoretical system and technical framework to identify and optimize territorial space.

Keywords: Production–Living–Ecological space; overall optimization; beautiful China; ecological civilization

Citation: Lin, G.; Jiang, D.; Fu, J.; Zhao, Y. A Review on the Overall Optimization of Production–Living –Ecological Space: Theoretical Basis and Conceptual Framework. *Land* **2022**, *11*, 345. https://doi.org/10.3390/land11030345

Academic Editors: Benedetto Rugani and Hossein Azadi

Received: 31 December 2021
Accepted: 24 February 2022
Published: 25 February 2022

Publisher's Note: MDPI stays neutral with regard to jurisdictional claims in published maps and institutional affiliations.

1. Introduction

Since China's reform and opening up, with the acceleration of industrialization, large-scale land resources have been continuously developed. Urban construction land is also faced with environmental problems, such as occupying ecological space, air pollution, water pollution, and ecological imbalance. How to coordinate social and economic development with ecological and environmental protection has become the core issue of sustainable development research in China. At the 18th National Congress of the Communist Party, the five-sphere integrated strategy of giving prominence to ecological progress was put forward from an overall and strategic perspective. At the same time, it pointed out that optimizing the spatial development pattern of the national territory should be taken as the primary measure of ecological civilization construction, so as to promote intensive and efficient production space, livable and appropriate living space, and beautiful (picturesque scenery) ecological space [1–3].The "Production–Living–Ecological" space (production space, living space, and ecological space, PLES), as the main body of the spatial pattern optimization, has become an important foundation for the planning and implementation of functional zones at all levels, the establishment of the spatial planning system, and the improvement of the system for the development and protection of territorial space. At the third Plenary Session of the 18th CPC Central Committee in November 2013, a number of major issues concerning comprehensively deepening reform were discussed in depth, and the Decision of the CPC Central Committee on Some Major Issues Concerning Comprehensively Deepening Reform was adopted. The decision calls for accelerating the construction of an ecological civilization system, establishing a spatial planning system, delimiting the limits of space development control, and implementing use control. In May 2019, the Opinions of the CPC Central Committee and the State Council on Establishing a Territorial Space Planning System and Supervising its Implementation identified the main objectives of territorial space planning, which included that "By 2035, China will comprehensively modernize its governance system and capacity of land and space, and basically form the national space pattern featuring intensive and efficient production space, livable and appropriate living space, beautiful ecological space, secure, harmonious, competitive and sustainable."

PLES basically covers the space activity scope of human social life, and is the basic carrier of human economic and social development. The three are independent and inter-related, with a symbiotic integration and restriction effect. The synergistic effect of "Production–Living–Ecological" (PLE) will be greater than the sum of its parts [4]. Coordinating the spatial function and land-use structure under the linkage of PLES; promoting the coordinated development of the quantity structure and spatial layout of PLES; and taking into account factors such as population distribution, economic development, territorial space utilization, and ecological and environmental protection to improve the layout of PLES in a scientific and reasonable way is the key measure to promote the building of a beautiful China, and accelerate the transformation of the way of production and living to "green" under the overall layout of the national ecological civilization construction [5]. It is also an important means to promote high-quality development and quality of life with the center of people, which is both necessary in reality, and urgent of the times. Since the report of the 18th CPC National Congress first clarified the development requirements of PLES from the height of national strategy, PLES has become the practice subject and research focus of territorial space planning and urban planning [6]. At present, the relevant theoretical research and local practice of PLES are still in the exploratory stage [7]. Related research and work mostly focused on the formulation of policies and management methods, or the evaluation and analysis of a single "space", such as demarcating the red line for basic farmland protection, the red line for ecological protection, and the boundary of urban development. These studies lack comprehensive and integrated studies of PLES in the whole domain, and lack systematic and comprehensive technical ideas and application practices of overall optimization of PLES.

Based on the classification, identification, evaluation, and optimization of PLES as the theme, through filtering and sorting the published literature since 2012 (the 18th CPC National Congress), the optimization theory, method, and practice of PLES was combed systematically, which oriented the national spatial planning. In order to provide theoretical support for the optimization of territorial space development, protection, and planning, this study also tries to put forward the development direction of the overall optimization of PLES in the future.

2. Connotation and Development of PLES

2.1. Definition of PLES

As early as 1984, some scholars in China put forward the dialectical relationship of PLE in the process of material exchange between human beings and nature [8]. Subsequently, the concept of PLE began to appear in research on sociology and ecology, and it was not until 2002 that PLE was gradually applied to research on geography [9]. At that time, research was guided only by the literal meaning of PLE, rather than the core of the research, and did not give further explanations for its scientific connotations. Different scholars have discussed the definition of the basic connotation of PLES from different perspectives. Some scholars started from the top-level design of national planning and correlated PLES with main function zoning. These authors argued that ecological space mainly has ecological functions, accumulates ecological capital, and relates to the functions of production and living; production space focuses on the accumulation of production capital, and relates to living functions, which is equivalent to the key development zone and optimized development zone in the main functional area of the country; and living space mainly plays the service function of living, accumulates living capital, and gives play to the functions of both production and ecology. This understanding is equivalent to restricted development zones in the national main function area [10]. However, main function zoning is only stipulated at the national and provincial levels, whereas the basic evaluation unit is the county administrative region, which is unable to control smaller spatial areas, and lacks overall planning capabilities [11].

Wu [12] was the first to clearly define PLES from the perspective of the PLE function. He argued that ecological space refers to an area featuring important ecological functions, with ecological products and services as its main function; production space is an area with the main function of providing industrial products, agricultural products, and service products, primarily including industrial and mining construction space and agricultural production space; and living space is a region that provides the main function of human habitation and public activity, primarily including the urban and town settlement space and rural living space. Together, the three dimensions of production, living, and ecology constitute the whole national space. Zhu [13,14] enriched the above definition while also emphasizing the dominant function of land-use, arguing that production space is mainly used for production and operation activities. Living space provides places for people to live, consume, and engage in recreation, whereas ecological space is the sum of the environment needed or occupied by a species in a macroscopic stable state, and provides ecological products necessary for human beings.

Territorial space is a complex geographical system, and includes land, water, mineral resources, ecology, social economy, and other different resource elements; moreover, there are extremely complex interactions among the elements [15]. From this perspective, PLES covers biophysical processes; direct and indirect production; as well as spiritual, cultural, leisure, and aesthetic needs, and represents the product of the synergy and coupling of multiple systems of nature, society, and economy [16–18]. Among them, ecological space is the foundation that enables production space and living space to realize their own functions, and is key to coordinating the relationship between humans and land, and achieving regional sustainable development [19,20].

2.2. Classification of PLES

There are two types of studies on the classification of PLES. One type classifies different land uses based on the single functions of production space, living space, and ecological space. For example, Hu [21] and others established a classification system for PLES at the scales of urban and rural regions and urban built-up areas by referring to *the Standard of Urban Land Classification and Planning and Construction Land*. According to the dominant function of a single land category, Ma [22] proposed a one-to-one correspondence between the secondary land-use types and PLES, and conducted a classification study on PLES. This classification method simplified the complex relationship between these three spaces, but ignored the complex multiple functions of land. For example, arable land (paddy field and dry land) was uniformly divided into production space, but its ecological function characteristics were not fully considered. The other classification method is mainly based on the multiple functions of land-use. This method extends the function of PLE, lists the land by each separate compound function type, and classifies the land according to the land-use standard on this basis [23]. Starting from the main function of land, and taking other functions into account, Zhang et al. [24] incorporated the concept of ecological land,

and constructed a classification system of Production–Living–Ecological land (PLEL) use at a national scale to coordinate the space of PLE land. This method compensates for the shortcoming that land ecological function is not sufficiently considered in land-use classification, and realizes the connection between land function classification, land-use classification, and urban land classification. Thus, this method is widely used in production practice and scientific research related to land function regulations, ecological political achievement assessments, and ecological environmental effects [25–27]. In addition, based on the theoretical connotations of PLEL, Liu et al. [28] analyzed the dialectical relationship between land-use functions and land-use types from the perspective of the complexity of the land PLE multifunctional complex, and constructed a classification standard system for PLEL according to the national classification standard of land-use status. To some extent, this classification method also reflects that PLES is, in essence, a result of the evolution and regional differentiation of the human–Earth relationship system.

Existing research on the identification of mainstream PLES methods can be divided into research using the merge classification method and the quantitative measurement method. The former involves qualitative research, which mainly merges and classifies land-use data based on yearbook data, national land surveys, and remote-sensing image data to identify PLES [29–33]. This method facilitates the connection between land-use function and classification standards, and compensates for the deficiency of ecological functions in land-use classification to some extent. Thus, this method has been widely used in practical fields [34–36]. However, this method produces errors in the recognition results, as it ignores the composite function of space. Moreover, these results will be different under different classification systems. Most of the latter systems focus on quantitative analysis, and mainly use the calculation function group of the spatial function value to establish a measurement system for land-use function, and identify PLES through a quantitative measure of the leading function of land-use [37]. For example, Li et al. [16] integrated the calculation function group of spatial function value based on ecosystem-service value assessment, and quantitatively identified the PLE functions of different value quantities, as well as the distribution of dominant function space. By processing and analyzing POI data, Cao et al. [38] identified, in more detail, the dominant function of land-use at a micro scale, and then, differentiated and delineated the PLES with single and mixed functions in the city. The advantage of the quantitative measurement method is that it can accurately identify the dominant function of PLES. However, it remains difficult to carry out multi-subject integration and multi-scale integrated expression using this method. Moreover, this method is difficult to apply in practice.

3. Research Progress on the Overall Optimization of PLES

3.1. Theoretical Basis for Overall Optimization of the PLES

The goal of PLES optimization is to achieve a "win–win" PLE situation, which is the key to ecological civilization construction [39]. Achieving a win–win PLE situation is the basic principle of environmental management put forward by Ye [40], based on China's basic situation. The core idea of this principle is that environmental governance should not involve disruptive, passive, or excessive governance, or damage the economy. Tian Daqing et al. argued that a win–win PLE situation should not only be the goal criterion of sustainable development, but also the behavioral judgment criterion of sustainable development. Based on this concept, Gao [41] constructed an evaluation system for comprehensive management of the water environment in small watersheds, and endowed this system with the scientific connotation of a win–win PLE situation.

The optimization of PLES should be based on the carrying capacity of resources and the environment, and be carried out in accordance with the principle of balance between the population, resources, and the environment, in order to coordinate the development of the land-use structure and scale of PLES [7]. Scientific evaluation of the carrying capacity of resources and the environment, and the suitability of territorial space development, are key to scientifically develop a PLES [42]. Therefore, in order to achieve the goal of a win–win PLE situation, the overall optimization of PLES should be supported by regional sustainable development theory, human–Earth system coupling theory, system science theory, spatial equilibrium theory, and community theory.

3.1.1. Theory of Sustainable Development

The theory of sustainable development refers to development that meets the needs of the present without jeopardizing the ability of future generations to also meet their own needs. Sustainable development includes common development, coordinated development, fair development, efficient development, and multidimensional development; the core theories of sustainable development are the theory of sustainable utilization of resources, externality theory, and three production theories [43,44]. In PLES, the intensive and efficient production space emphasizes improving the intensive utilization level and output efficiency of production space, that is, the economic goal; appropriate living space refers to improving the quality of life and livable level of residents, that is, the environmental goal; the beautiful ecological space attaches importance to improving the ecological service function and the quality of ecological environment, that is, the environmental goal; the overall planning of PLES is the optimization goal of coordinating different functional spaces to maximize the comprehensive benefits of the economy, society, and environment. Therefore, the overall planning and optimization of PLES is an important practical way for the goal of a "beautiful China" and China's SDGs. To meet the strategic needs of territorial space optimization, the coordinated optimization and overall development of PLES should be guided by the sustainable development concept of "people-oriented" development to develop scientific cognitive methods, and clarify the basic logic problems relevant to the situation.

3.1.2. Human–Earth System Coupling Theory

The human–Earth system is a dynamic, open, large, and complex system composed of subsystems related to population, resources, ecology, the environment, the economy, and society. Each subsystem has a relationship of mutual influence, mutual promotion, and restrictions. Furthermore, there are frequent exchanges of personnel, materials, energy, capital, technology, and information inside and outside the system, and the complex feedback structure inside the system presents obvious non-linear and dissipative characteristics [45,46]. Human–Earth coupling refers to the dynamic correlation between humanity and nature through the interactions and complex feedback mechanisms between human economic and social activities, resources, ecology, and the environment [47,48]. The human–Earth coupling system emphasizes the multi-dimensional coupling in organization, space, and time. Through the complex interaction between one element and many elements, as well as the interaction and coupling between many elements, this system embodies the characteristics of comprehensiveness, complexity, and nonlinearity at a high level [49,50]. Optimization of the human–Earth system refers to the reasonable combination and matching of sub-systems and components of the regional human–Earth system in the space–time process [51]. The PLES system involves different resource elements and their combinations, such as water resources, land resources, and energy resources. These elements have extremely complicated mutually influential relationships. The overall optimization of PLES should take the human–Earth coupling theory as its core, and effectively measure the nonlinear effects among the subsystems and elements in the system; scientifically clarify the ordered structure of matter, energy, and information in the system; and emphasize the multi-dimensional coupling in organization, space, and time.

3.1.3. System Science Theory

According to the perspective of system theory, a system refers to an agglomeration of many elements with specific functions and organic connections [52]. From this perspective, "space" is the collection of all material flow, energy flow, and information flow generated by the interaction between man and nature, including natural resource elements (water resources, land resources, energy, etc.), the artificial environment, and other material space.

"Space" also includes attribute space, which is formed from changes to the distribution, structure, and function characteristics of material space, along with the formation and development of technology and information [53]. Territorial space is a large, dynamic, complex system involving the interaction of multiple factors. Territorial space is a dynamic, multi-dimensional, and complex human–Earth relationship space–time system developed along the time axis under the participation of human activities, with time–space–human is its core element [54]. According to the theory of "element–structure–function" in system theory, system structure is the foundation of system–function realization. However, the system structure depends on the organizational form and action mode of the system elements. Only by systematically splitting the territorial space structure, and analyzing the interactions between space and function, can we model comprehensive zoning, and explain the geographical decision mechanism of territorial space optimization using a quantitative decision analysis model [55].

In terms of the "territorial space" system, the research topics cover land resources, water resources, mineral resources, the ecological environment, economic and social development, and other multi-dimensional and multifaceted factors. Only the comprehensive integration and overall optimization of these spatial elements can maximize the PLE functions, and achieve the ultimate goal of the optimal allocation of territorial space [11,56–58]. Therefore, the identification of key elements is the basis of structural optimization and functional realization. As the main carrier of future spatial planning, land resources represent the core of PLES optimization. The optimization of PLES is driven and guided by optimizing the quantity ratio and spatial allocation of land.

3.1.4. Spatial Equilibrium Theory

Spatial equilibrium is a spatial "Pareto efficiency" state based on the coordination of the population, economy, resources, and environment [59]. State equilibrium considers not only economic development (that is, the development of production space), but also the development of living space and ecological space, as well as the spatial allocation of land resources in the region. On the one hand, the balanced development of PLES should optimize the allocation of various elements; give full play to the potential and advantages of various spaces; and, at the same time, make the social, economic, resource, environmental, and other elements harmonious and orderly, in order to maximize the overall benefits.

3.1.5. Community Theory

The concept of "community" emerges from Aristotle's definition of the relationship between a combination of various elements to achieve a common "good". With the rapid development of urbanization, community has yielded the concept of "union" for each region, which not only corresponds to the combination of people in regional space, but also refers to the community at different levels [60]. In addition, the idea of a community with a shared future for humans means that, on the whole, humans have become an increasingly close community in the era of globalization and information. Only by establishing the central position of a community with a shared future for mankind can we truly grasp the essence and future of the world. The essence of territorial space optimization is pursuing the sustainable development of PLES, and represents a systematic project to optimize PLES. For the community of PLES optimization, the ultimate goals are to realize the optimization of PLEL; achieve the mutual restriction, adaptation, promotion, and coordination of the subsystems of ecology, production, and living; and achieve a benign impact in a win–win PLES situation [61–63].

3.2. Technical Methods for the Overall Optimization of PLES

3.2.1. Spatial Optimization Based on Territorial Space Utilization Quality

The expression of various functions of territorial space directly reflects the sustainability of territorial space utilization, and the utilization quality of these functions reflects the demand of PLES [64,65]. From the perspective of PLES, it is an important element of territorial space planning to evaluate the internal mechanism and coupling coordination degree of territorial space utilization quality development. This task can realize the balanced development of regional PLES, and promote the coordination and coupling of economic and social resources, and the ecological environment [66–68]. Li et al. [69] took the utilization quality of PLES as an assessment index, evaluated the utilization quality index of PLES in different provinces, and coordinated construction of the land-use mechanism of "population–land–industry" as a goal; the authors also proposed countermeasures to optimize construction land patterns from the perspective of industrial land-use structures. Zhang [66] constructed an evaluation index system of territorial space utilization quality from a realistic state of territorial space and the guaranteed function and support ability of economic development and social harmony. The author evaluated and analyzed the spatio-temporal differentiation characteristics of utilization quality and the coupling coordination degree of PLES in combination with the entropy method and comprehensive evaluation method, providing effective guidance for the rational development and utilization of PLES in regional land. The concepts and connotations of the utilization quality evaluation of national territorial space from the perspective of PLES emphasize the comprehensive diagnosis and evaluation of the utilization status, efficiency, and guarantees related to the sustainable development of human society. Using the landscape ecological index and GIS spatial analysis methods to identify the spatio-temporal competition and game processes of resource elements (with land as the core in the same spatial location), and carrying out the optimal allocation of PLES, also represent important means of spatial optimization in the early stage. For example, with the goal of minimizing land-use conflict and maximizing industrial suitability, Zhang et al. [70] proposed a spatial allocation scheme of industrial land based on the conflict identification of PLES by evaluating the conflict level of PLES and the suitability of industrial production space. With the help of the landscape ecological index method, Lin et al. [71] constructed a spatial conflict index, measured the conflict level of regional PLES, and constructed a spatial allocation model based on multi-objective constraints and scenario setting, thereby realizing the optimization and simulation of a spatial layout for PLEL. However, the conflict of PLES covers many aspects such as society, economy, and ecology [72,73], and is a result of the comprehensive effect of the regional human–Earth relationship system. It can only judge the rationality of the distribution pattern and coordination of the utilization quality using the theory of land-use function and landscape ecological patterns, and requires further verification and improvement.

3.2.2. Spatial Optimization Based on Suitability Evaluation

The suitability evaluation of PLES is an important basis for optimizing the quantity ratio and spatial layout of PLES. Here, suitability is defined as the suitability of land for PLE functions in a specific area under specific conditions, and focuses on comprehensive consideration of territorial space development, protection, and spatial carrying capacity [6]. Research on the optimization of PLES based on suitability evaluation can be roughly divided into two types. One type involves the comprehensive evaluation of multi-element superposition based on spatial superposition analysis. For example, Jin [30] applied the BP neural network model to evaluate the PLE functions of the standard evaluation units of territorial space, and completed research on the comprehensive functional zoning of territorial space based on double-constraint clustering. Starting from the concept of PLES suitability, Yang [74] used GIS spatial analysis tools to construct an evaluation system for PLES suitability that integrates natural, cultural, and ecological elements, and employed spatial superposition to divide the evaluation grades of PLES suitability within the whole domain. He analyzed the problems existing in the layout of PLES, and proposed an optimization path for PLES, along with relevant policy suggestions. The second method involves undertaking suitability evaluations using landscape ecological pattern analysis, the cumulative resistance model, or the gravity model. For example, by introducing the competition perspective of PLES, Chen [75] established an evolutionary game model of PLES, and proposed an optimal decision scheme for the allocation of PLES to optimize the allocation of resilient ecological space under the guidance of the action mechanism of the evolution of ecological space through the competition of PLES. Chen [75] introduced a competition perspective of "production–ecological space" to optimize the allocation of national resilience ecological space. Under the guidance of the action mechanism of ecological space evolution under the competition of PLES, the evolutionary game model of PLES was established, and an optimal decision scheme of allocation for PLES was put forward.

3.2.3. Spatial Optimization of Bearing Capacity Based on PLE

PLE capacity refers to the capacity of the ecosystem to provide resources and environmental capacity (ecological capacity), the capacity of economic development activities (productive capacity), and the capacity of social development under certain living standards (living capacity) [76]. Among them, ecological-space-bearing capacity refers to the bearing capacity of regional water and soil energy resources to production space and living space, as well as the strength of system elasticity and self-repair ability; production space bearing capacity refers to the intensity and scale of economic activities that can be achieved by the existing economic and technological level within the elastic limits of the system itself; and living-space bearing capacity refers to the bearing capacity of natural conditions, infrastructure, public transportation, medical and health care, culture and education, and other resources of a city provided to the population under a certain living standard, which can reflect the quality of living in a region. Some scholars believe that the essence of PLES is the dynamic mapping of social, economic, and ecological processes in land-use space, and that the key to its optimization is the PLE bearing capacity [76]. Wang [77] evaluated the bearing capacity and development potential of PLES by constructing an evaluation index system and development potential evaluation index system. He also put forward suggestions for optimizing the spatial development pattern of national land based on the evaluation results. Zhou [51] used the state-space method to construct a three-dimensional state model of PLE, evaluated the bearing capacity of the PLE composite system, detailed the problems and shortcomings of each research unit in development, and proposed the direction for spatial layout optimization of the PLE system.

3.2.4. Spatial Optimization Based on Comparative Advantage

The theory of comparative advantage holds that there are differences in resource endowment between different regions that determine the different efficiency levels in the utilization of different commodities within those regions. Comparative advantage can be gained through exchange [78]. The standard Normalized Revealed Comparative Advantage index (NRCA), which is not constrained by time and space, can be used to evaluate the dominant function of each city, and determine the comparative advantage function of the city scientifically and effectively [79]. Therefore, under the guidance of PLES function theory, the spatial pattern of urban agglomeration can be optimized by constructing a normalized revealed comparative advantage index combined with system clustering and GIS technology. For example, taking the PLE function as a breakthrough point, Xu [79] introduced the NRCA based on the comparative advantage theory, determined the distribution pattern in the dominant functions of land space in the middle reaches of the Yangtze River urban agglomeration, and put forward a realization path for the optimal utilization of territorial space; Wei et al. [80] analyzed the land spatial characteristics of urban agglomerations in the upper reaches of the Yangtze River via the entropy weight method and function evaluation method from the perspective of the PLE function, and built a spatial-function comparative-advantage index to explore the optimization path of the land-space optimization scheme.

The optimization of PLES belongs to the research category working on the optimal allocation of land and resources, and the optimization of its quantitative structure and spatial layout is an important part of previous research [6]. From a theoretical perspective, the related theories of regional resources and environmental carrying capacity, urbanization, and ecological environment coupling play a major supporting role. From the perspective of research methods, the traditional technology for the optimal allocation of land resources is often used to construct economic models or landscape ecological models for optimizing the quantity ratio or land allocation in order to realize the optimization of land spatial patterns based on PLES. Research can, moreover, explore different scales, such as provinces, cities (urban agglomerations), counties, towns, and villages, covering different areas, such as cities and villages; and urban–rural ecotones, such as coastal zones, mining reclamation areas, and island fishing villages [81–84]. However, due to the lack of a unified technical system for the division of PLES at different scales and for different geographical types, along with the systematic combing and summarization of concepts for optimizing PLES patterns, the optimization of PLES requires further research.

4. Optimization Approach for PLES Considering the Vision of "Beautiful China"

Forty years after China's reform and opening up, China's land space has basically formed a relatively stable pattern [14]. The corresponding zoning research work also has a relatively mature framework of support, which is constantly being enriched and improved upon. Since 1950, China's regionalization has entered a widescale developmental period. However, no in-depth theoretical or methodological support for the zoning work was provided at this stage. At the same time, due to the limitations of the objective conditions and basic data, most zoning plans were relatively simple. At the end of the 20th century, zoning entered an ongoing stage of comprehensive zoning research, during which, many zoning schemes in China were developed with profound historical backgrounds closely related to the levels and needs of national economic development in the same period. The study of regionalization accordingly changed from serving mainly agricultural production to considering both agricultural production and economic development, and then, serving sustainable development. However, due to some contents being quite different in regional division, there is also no institutionalized guarantee for the identification of natural regionalization, which leads to the fact that it is not really absorbed by the economic construction planning of local governments, resulting in the failure to combine natural regionalization with economic regionalization, which entails certain restrictions in supporting regional sustainable development [85].

Planning for the national main functional area should represent the strategic background of basic planning in the future. Indeed, as the core means of spatial optimization at present, territorial spatial planning is an important part of the future national planning system. Moreover, PLES embodies the Chinese people's vision for "future" development. This vision represents the final and most direct spatial carrier for the national strategy to penetrate into people's livelihoods, and the ultimate optimization goal of planning implementation [86]. However, under the current strategic background of ecological civilization construction, there are many problems in the layout of "main function zones", "territorial space planning", and PLES. "Beautiful China" was a major strategic goal proposed at the 18th National Congress of the CPC, and represents a spatial carrier for realizing sustainable development of the Chinese nation, and enabling the Chinese people to enjoy better lives. To achieve the vision of "beautiful China", the territorial and spatial patterns in the new era should be scientific and orderly, and remain in line with the processes of both modern and sustainable development. Based on analyzing the internal relationship between ecological civilization, beautiful China, and PLES, as well as the scientific connotations of PLES, the overall optimization system of PLES should be guided by the vision of building a beautiful China supported by the needs of national strategic applications, and centered on the human–Earth coupling system theory. By comprehensively considering the multi-dimensional coupling relationships between subsystems of the population, society, economy, resources, and the environment, as well as various elements within the system, a large, complex, dynamic, and open geographical system can be formed (Figure 1).

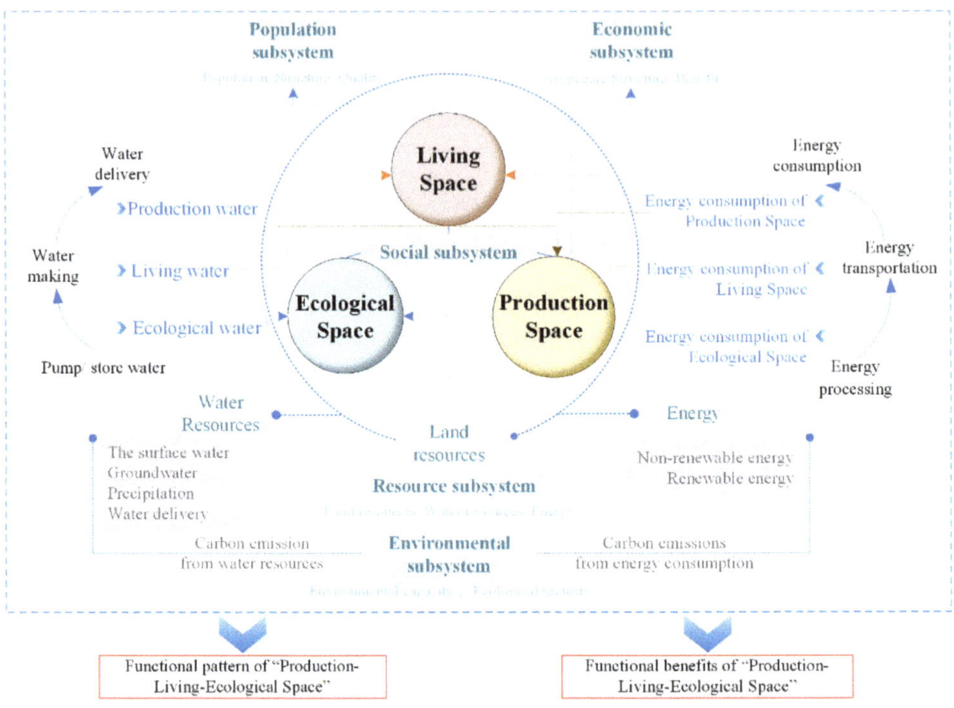

Figure 1. Framework system for the overall optimization of PLES.

The overall optimization system of PLES includes the resource subsystem, population subsystem, social subsystem, economic subsystem, and environmental subsystem. Among them, the resource subsystem includes land resources, water resources, and energy. As the main carrier of future spatial planning, land resources are the core of PLES optimization. From the perspective of function, land resources can be divided into PLES. PLES has the characteristics of differences in spatial scale, functional complexity, and dynamic scope. Under different spatial scales and time nodes, the same territorial space can feature different, or even overlapping, properties of PLES [86]. Living space mainly functions as living land, providing resource support for population subsystems, and guaranteeing living needs. Production space mainly functions as production land, providing resource input for the economic subsystem, and meeting regional production demands. Ecological space mainly functions as ecological land, which provides ecological environment security maintenance capabilities for the environmental subsystem. As a compound land with multiple functions, public resources provide services for the resource supply, management, substitution, and compensation of social subsystems. In the process of land resource utilization, the population, society, and economic subsystem are also accompanied by the utilization of water resources and energy. The utilization of water resources includes production water for the agricultural and service industries, domestic water for urban and rural areas, and ecological water. Energy utilization includes production space energy consumption, living-space energy consumption, and ecological-space energy consumption. In the resource subsystem, land resources, water resources, and energy are interrelated through different avenues that, together, constitute the resource basis for the utilization and development of PLES. The different utilization modes, industrial structures, and development degrees of PLES determine the intensity of regional carbon emissions caused by the different quantitative methods of mutual demand between water, soil, and energy elements [87]. Therefore, constraints such as boundaries, red lines, and standards of land resources, water resources, energy, and carbon emissions should be taken into consideration in the overall optimization system for PLES.

Optimizing the quantitative structure and spatial layout of PLES based on water–land–energy–carbon constraints solves the problem of the pattern level of PLES. However, this optimization only reflects the application of the national strategy in territorial space optimization, and does not express people's demands for high standards of living, production, and ecological and environmental safety, which would involve solving the qualitative level of PLES. In the Evaluation Index System and Implementation Plan for the Construction of a Beautiful China issued by the National Development and Reform Commission on 28 February 2020 [88], 22 indicators in the five categories of fresh air, clean water, safe soil, good ecology, and clean living were used to evaluate the construction of a beautiful China. These indicators are a direct reflection of the production environment, ecological environment, and living environment in the overall optimization system of PLES. Therefore, the evaluation results for the construction of a beautiful China will directly test the qualities and benefits of the overall optimization of PLES in order to realize the overall optimization of PLES under the dual constraints of patterns and benefits.

5. Conclusions and Future Prospects

By analyzing the internal relationship between ecological civilization, beautiful China, and PLES, as well as the scientific connotations of PLES, this paper systematically organized the optimization theory, methods, and practices for PLES around territorial spatial planning, with the overall planning of PLES as the goal, and the promotion of ecological civilization construction as the starting point. We also proposed an optimization approach for PLES. Based on the analysis and summary, the main research directions for the overall optimization of PLES in the future should include the following:

(1) Research on the main regional types and differentiation rules of PLES: This paper studied the distribution characteristics of the functional spatial patterns of PLES at multiple scales. In this way, the scale differences and functional complexities of PLES under different spatial scales were clarified. Moreover, classification and identification systems for PLES under different scales and different regional types were established.

(2) Analysis and research on the evolution process, structural characteristics, and development trends of PLES: The dynamic spatial and temporal evolution characteristics of the structural proportion and spatial scale of PLES were analyzed, including regional suitability and carrying capacity evaluation, regional differences in space, and development over time. Moreover, the coupling coordination degree and conflict mechanism in the evolutionary process of PLES were revealed.

(3) Analysis of the internal nonlinear mechanism for the PLES System: Guided by the theory of the human–Earth system, we revealed the competition and synergistic interactions and positive feedback mechanisms between population, resources, ecology, environment, and economic and social development in the PLES system, and identified the nonlinear dynamic effects among subsystems and elements in the system.

(4) Analysis of the mechanism for matter and energy transfer in the overall optimization system of PLES: Based on the regional resource metabolism theory and the geographical patterns of PLES, we deeply analyzed the transfer path, flow process, and metabolic mechanisms of key elements, such as water, soil, energy, and carbon, in the co-evolution process of PLES.

(5) Development and application of a simulation model and multi-objective optimization model for the overall optimization system of PLES: Studies should make full use of the research results for the nonlinear dynamic mechanisms, transmission paths, and flow processes of key elements in the PLES system, and carry out research on overall optimization strategies and scenario predictions for PLES in different regions, at different scales, and for different types.

(6) Research, development, and application of the overall optimization and decision support platform for PLES: A visual support platform was developed with integrated

functions for data processing, spatio-temporal analysis, scenario simulation, result output, problem diagnosis, and early warning and control. This platform can help coordinate the government at all levels, and between different departments and different areas of the target demand, to facilitate population migration, urban and rural construction and industrial development, resource development, ecological construction, environmental protection, infrastructure construction, public services and disaster prevention, and mitigation-system spatial deployment as components of unified and consensus development goals and plans.

Author Contributions: G.L. and J.F. contributed to all aspects of this work; D.J. conducted data analysis; and G.L. and Y.Z. wrote the main manuscript text. All authors reviewed the manuscript. All authors have read and agreed to the published version of the manuscript.

Funding: This work was supported by grants from the Strategic Priority Research Program of the Chinese Academy of Sciences (Grant No. XDA19040305), Youth Innovation Promotion Association (Grant No. 2018068), and Institute of Geographical Sciences and Natural Resources Research, Chinese Academy of Sciences (Grant No. E0V00112YZ).

Institutional Review Board Statement: Not applicable.

Informed Consent Statement: Not applicable.

Data Availability Statement: The data presented in this study are available on request from the author.

References

1. Li, S.D.; Liu, M.C.; Chen, Y.F. *Beautiful Ecology: Theoretical Exploration Index Evaluation and Development Strategy*, 3rd ed.; Beijing Science Press: Beijing, China, 2017.
2. Wang, Q.M. The development path of ecological civilization. *QIUSHI* **2008**, *3*, 43–44.
3. Xue, X.F. Accelerating ecological progress. *West J.* **2013**, *15–16*, 26.
4. Chen, Y.W.; Mai, J.M. Adhere to the harmonious coexistence of human and nature, accelerate the construction of green modern landscape city. *China Ecol. Civiliz.* **2017**, *S1*, 6–9.
5. Gao, F.; Zhao, X.Y.; Song, X.Y.; Wang, B.; Wang, P.L.; Niu, Y.B.; Wang, W.J.; Huang, C.L. Connotation and Evaluation Index System of Beautiful China for SDGs. *Adv. Earth Sci.* **2019**, *34*, 75–85.
6. Huang, J.C.; Lin, H.X.; Qi, X.X. A literature review on optimization of spatial development pattern based on ecological-production-living space. *Prog. Geogr.* **2017**, *36*, 378–391.
7. Lu, D.F.; Jiang, M.Q. Features, logical relations and optimizing strategies of urban "Production-living-ecological space". *Hebei Acad. J.* **2019**, *39*, 156–166.
8. Xu, D.X. Social production and ecological environment in human life. *J. Guangxi Norm. Univ. Philos. Soc. Sci. Ed.* **1984**, *4*, 3–11.
9. Zhang, C.G.; Fang, C.L. Driving mechanism analysis of ecological-economic-social capacity interactions in oasis systems of arid lands. *J. Nat. Resour.* **2002**, *17*, 181–187.
10. Fang, C.L. The Scientific Basis and Systematic Framework of the Optimization of Chinese Urban Development Pattern. *Econ. Geogr.* **2013**, *33*, 1–9.
11. Li, W. A Study on Theoretic Methods and Application of the Evaluation and Optimization of County Land Space. Ph.D. Dissertation, Fujian Normal University, Fuzhou, China, 2018.
12. Wu, Z.Y. Space optimization of "Production-living-ecological" and ecological environment protection of Beijing, Tianjin and Hebei. *City* **2014**, *12*, 26–29.
13. Zhu, Y.Y.; Yu, B.; Zeng, J.X.; Han, Y. Spatial Optimization from Three Spaces of Production, Living and Ecology in National Restricted Zones—A Case Study of Wufeng County in Hubei Province. *Econ. Geogr.* **2015**, *35*, 26–32.
14. Fu, X.; Wang, X.; Zhou, J.; Ma, J. Optimizing the Production-Living-Ecological Space for Reducing the Ecosystem Services Deficit. *Land* **2021**, *10*, 1001. [CrossRef]
15. Cao, X.S. Geogovernance of national land use based on coupled human and natural systems. *J. Nat. Resour.* **2019**, *34*, 2051–2059. [CrossRef]
16. LI, G.D.; Fang, C.L. Quantitative function identification and analysis of urban ecological-production-living spaces. *Acta Geogr. Sin.* **2016**, *71*, 49–65.
17. Liu, D.P. Thinking on the demarcation of "Production-living-ecological space" in urban and rural planning. *Anhui Archit.* **2021**, *28*, 38–39.
18. Yang, X.J.; Zhao, B.Y.; Li, M.L. Rural construction planning at county level based on "Production-living-ecological space" coordinated development: A case study of Pinglu County, Shanxi Province. *China Anc. City* **2021**, *35*, 66–74.
19. Kates, R.W.; Clark, W.C.; Corell, R.; Hall, J.M.; Svedlin, U. Environment and development—Sustainability science. *Science* **2001**, *292*, 641–642. [CrossRef]

20. Ge, Y.C.; Ge, D.B. Ecological spatial distribution based on ecological suitability: A case study of Xiangyin County, Yueyang City. *J. Northeast Agric. Sci.* **2021**, *46*, 120–124.
21. Hu, W.T.; Wang, L.G.; Shu, M.H. Reflections on Delimiting the Three Basic Spaces in the Compilation of Urban and Plans. *City Plan. Rev.* **2016**, *40*, 21–26.
22. Ma, Y.Y. The Study on the Optimization of "Production-Living-Ecological" Space in Puge County. Master's Dissertation, Sichuan Normal University, Chengdu, China, 2017.
23. Liao, G.T.; He, P.; Gao, X.S.; Deng, L.J.; Zhang, H.; Feng, N.N.; Zhou, W.; Deng, O.P. The Production–Living–Ecological Land Classification System and Its Characteristics in the Hilly Area of Sichuan Province, Southwest China Based on Identification of the Main Functions. *Sustainability* **2019**, *11*, 1600. [CrossRef]
24. Zhang, H.Q.; Xu, E.Q.; Zhu, H.Y. An ecological-living-industrial land classification system and its spatial distribution in China. *Resour. Sci.* **2015**, *37*, 1332–1338.
25. Gou, M.M.; Liu, C.F.; Li, L.; Xiao, W.F.; Wang, N.; Hu, J.W. Ecosystem service value effects of the Three Gorges Reservoir Area land use transformation under the perspective of "production-living-ecological space". *Chin. J. Appl. Ecol.* **2021**, *32*, 3933–3941.
26. Zhou, J.X.; Zhang, Q.J. Study on the Relationship between Land Use Transformation and Ecological Environment Change in Tianjin Based on "Ecological-production-living Spaces". *J. Tianjin Chengjian Univ.* **2021**, *27*, 333–339.
27. Wang, R.J.; Zhao, X.M.; Guo, X.; Ye, Y.C.; Li, Y.; Zhou, Y. A Study on Land Use Transformation and Ecological Environment Effects fromthe Perspective of "Production-Living-Ecological Space": A Case Study of Yingtan City, Jiangxi Province. *Acta Agric. Univ. Jiangxiensis* **2021**, *43*, 681–693.
28. Liu, J.L.; Liu, Y.S.; Li, Y.R. Classification evaluation and spatial-temporal analysis of "production-living-ecological" spaces in China. *Acta Geogr. Sin.* **2017**, *72*, 1290–1304.
29. Cheng, T.; Zhao, R.; Liang, Y. Production- living -ecological Space Classification and Its Functionzl Evaluation. *Remote Sens. Inf.* **2018**, *33*, 118–125.
30. Jin, G. Study on Comprehensive Function Regionalization of National Spatial Territory. Ph.D. Dissertation, China University of Geosciences, Wuhan, China, 2014.
31. Chen, L.; Zhou, S.L.; Zhou, B.B.; Gang, L.L.; Chang, T. Characteristics and Driving Forces of Regional Land Use Transition Based on the Leading Function Classification: A Case Study of Jiangsu Province. *Econ. Geogr.* **2015**, *35*, 155–162.
32. Liu, Y.; Shi, D.; Wang, J. Recognition and analysis of Sansheng space function in tourist towns—Taking Erdaobaihe Town Changbai Mountain, Jilin Province as an example. *Jiangsu Agric. Sci.* **2021**, *49*, 201–206.
33. Zhang, C.; Zhang, Y.B.; Chen, S.; Yang, L.H. Identification and Evaluation of Development Patterns of Small Towns in Hebei Province: A Functional Perspective of Production, Living and Ecology. *J. Hebei Univ. Technol. Soc. Sci. Ed.* **2021**, *13*, 16–24.
34. Huang, S.Q.; Xi, F.R.; Chen, Y.M.; Gao, M.; Pan, X.; Ren, C. Land Use Optimization and Simulation of Low-Carbon-Oriented—A Case Study of Jinhua, China. *Land* **2021**, *10*, 1020. [CrossRef]
35. Zhang, J.; Chen, Y.; Zhu, C.M.; Huang, B.B.; Gan, M.Y. Identification of Potential Land-Use Conflicts between Agricultural and Ecological Space in an Ecologically Fragile Area of Southeastern China. *Land* **2021**, *10*, 1011. [CrossRef]
36. Li, B.; Yang, C.X.; Xie, D.L.; Luo, Y.Z.; Li, Y.T. Analysis of Spatiotemporal Evolution Characteristics of 'Sansheng' Function in Hilly Areas—A Case Study of Jiangjin District of Chongqing. *Res. Soil Water Conserv.* **2021**, *28*, 316–323.
37. Guo, R.; Zhao, Y.W.; Song, X.Y.; Aihemaiti, N.M.T. Identification and layout of Production-living-ecological Space based on POI data: A case study of Qingdao city. In Proceedings of the 2020/2021 China Urban Planning Annual Conference and 2021 China Urban Planning Academic Season Chengdu 2021, Chengdu, China, 25 September 2021; pp. 881–888. (In Chinese).
38. Cao, G.R.; Gu, C.L.; Zhang, Q.Y. Recognition of "Ecological Space, Living Space, and Production Space" in Urban Central Area Based on POI Data: The Case of Shanghai. *Production* **2019**, *2*, 44–53.
39. Ye, W.H. Building a healthy society by coordinating ecology, life and production for ecological civilization. *J. Wuhan Univ. Sci. Technol. Soc. Sci. Ed.* **2010**, *12*, 7–10.
40. Tian, D.Q.; Wang, Q.; Ye, W.H. Common development of life, production and environment: Basic goal and behavior norms of sustainable development. China population. *Resour. Environ.* **2004**, *14*, 8–11.
41. Gao, S.; Zhu, D.L.; Hu, H.L. Water Environment Comprehensive Management of Small Watershed Based on Common Development of Life, Production and Environment. *Chin. J. Environ. Manag.* **2017**, *9*, 52–56.
42. Gu, C.L. Scientific Planning of "Production-Living-Ecological Space" Elements under the Land Space Planning System. CIYEW.COM. Available online: https://www.ciyew.com/policy/0637-4238.html (accessed on 7 October 2019).
43. Bossel, H. *Indicators for Sustainable Development: Theory, Method, Applications: A Report to the Balaton Group*; International Institute for Sustainable Development (IISD): Winnipeg, MB, Canada; Institut International du Développment Durable: Vernier, Switzerland, 1999.
44. Wang, F. Philosophical Thinking on sustainable development strategy. *Theory Horiz.* **2002**, *5*, 7–8.
45. Wu, C.J. On the research core of geography—The regional system of man earth relationship. *Econ. Geogr.* **1991**, *3*, 7–12.
46. Lu, D.D. Theoretical studies of man-land system as the core of geographical science. *Geogr. Res.* **2002**, *21*, 135–145.
47. Liu, J.G.; Dietz, T.; Carpenter, S.R.; Alberti, M.; Folke, C.; Moran, E.; Pell, A.N.; Deadman, P.; Kratz, T.; Lubchenco, J. Complexity of coupled human and natural systems. *Science* **2007**, *317*, 1513–1516. [CrossRef]
48. Liu, J.G.; Dietz, T.; Carpenter, S.R.; Folke, C.; Trade, S. Coupled Human and Natural Systems. *AMBIO J. Hum. Environ.* **2007**, *36*, 639–649. [CrossRef]

49. Liu, T.; Hou, L.G. Spatial-temporal evolution characteristics of the production-living-ecological spatial function coupling coordination in the western Sichuan urban agglomeration. *Hubei Agric. Sci.* **2021**, *60*, 76–83.

50. Niu, Y.X.; Wu, S.X.; Guo, C.Y.; Zhuang, Q.W.; Xie, C.H.; Zhang, Z.H.; Luo, G.P. Analysis on the spatio-temporal changes and coupling coordination of the function of "production-living-ecological" in Xinjiang. *Arid Land Geogr.* **2021**, *44*, 1–21. Available online: http://kns.cnki.net/kcms/detail/65.1103.X.20210622.1109.002.html (accessed on 20 October 2021).

51. Mao, H.Y. Theories and methods of optimal control of human-earth system: Commemoration of 100th anniversary of Academician Wu Chuanjun's birth. *Acta Geogr. Sin.* **2018**, *73*, 608–619.

52. Von Bertalanffy, L. The History and Status of General Systems Theory. *Acad. Manag. J.* **1972**, *15*, 407–426.

53. Zhang, X.J. The Elements, Structure and Model of Smart City System. Ph.D. Dissertation, South China University of Technology, Guangzhou, China, 2015.

54. Zhang, Y.Y.; Cheng, M.J. Spatial Systematic Cognition and Ideas on Spatial Planning System Reform. *China Land Sci.* **2016**, *30*, 11–21.

55. Ma, S.F.; Huang, H.Y.; Cai, Y.M.; Nian, P.H. Theoretical Framework with Regard to Comprehensive Sub-Areas of China's Land Spaces Based on the Functional Optimization of Production, Life and Ecology. *Nat. Resour. Econ. China* **2014**, *27*, 31–34.

56. Chen, H.; Yang, Q.; Su, K.; Zhang, H.; Lu, D.; Xiang, H.; Zhou, L. Identification and Optimization of Production-Living-Ecological Space in an Ecological Foundation Area in the Upper Reaches of the Yangtze River: A Case Study of Jiangjin District of Chongqing, China. *Land* **2021**, *10*, 863. [CrossRef]

57. Li, J.S.; Sun, W.; Yu, J.H. Change and regional differences of production-living-ecological space in the Yellow River Basin: Based on comparative analysis of resource-based and non-resource-based cities. *Resour. Sci.* **2020**, *42*, 2285–2299. [CrossRef]

58. Du, L.Z.; Yang, W.W. Multi-functional Evaluation and Zoning Optimization in Rural Areas under the Perspective of "Sansheng Spaces": Taking Heilongjiang Province as an Example. *Archit. Cult.* **2021**, *8*, 258–261.

59. Zhang, Y.Z.; Zhang, J.L.; Cheng, Y.; Ren, J.L. Study on the Meaning of Space Balance and Condition Assessment from the Perspective of Supply and Demand Driven—A Case of Shandong Province. *Soft Sci.* **2016**, *30*, 54–58.

60. Bauman, Z. Community: Seeking Safety in an Insecure World. *Contemp. Sociol.* **2001**, *31*, 442. [CrossRef]

61. Hu, X.D. Production, Living and Ecological Space Based on Human Settlements in Reclation Area of Coal Min. Master's Dissertation, China University of Geosciences, Beijing, China, 2016.

62. Ding, C.Y.; Tang, G.N.; Ji, Y.N.; Sun, Y.P. Study on the integrated development of "Production-living-Ecological Space" in beautiful countryside – A case study of Zhejiang Province. *Rural Sci. Technol.* **2021**, *12*, 99–103.

63. Zhang, C.W.; Chen, K.Q.; Guo, Y.W.; Wang, D.; Zhao, B.B.; Wu, Y. Evaluating livability of urban space with consideration of the interactive characteristics of ecological-living-productive spaces: A case of Wuhan city. *Bull. Surv. Mapp.* **2021**, *1*, 124–129.

64. Zhang, X.S.; Xu, Z.J. Spatial Temporal Evolution of Functional Coupling Coordination Degree of Production-Living-Ecological Space and Its Relationship with Human Activity Intensity in Ethnic Minority Areas—Taking Minority Autonomous Prefecture of Guizhou as an Example. *Res. Soil Water Conserv.* **2021**, *28*, 268–273.

65. Li, Z.Y.; Li, Y.Y.; Wang, L.; Pei, Y.L. Study on the Functional Characteristics and Division Optimization of "Production-Living-Ecological" of Geographical Space in Yunnan Province. *Ecol. Econ.* **2021**, *37*, 94–101.

66. Zhang, J.X. Assessment of land space utilization quality and its coupling and coordination based on producing, living and ecological—A case study of the southern Jiangsu Region. *J. Agric. Sci.* **2017**, *38*, 57–63.

67. Wang, D.; Jiang, D.; Fu, J.Y.; Lin, G.; Zhang, J.L. Comprehensive Assessment of Production–Living–Ecological Space Based on the Coupling Coordination Degree Model. *Sustainability* **2020**, *12*, 2009. [CrossRef]

68. Dong, G.; Ge, Y.; Jia, H.; Sun, C.Z.; Pan, S.Y. Land Use Multi-Suitability, Land Resource Scarcity and Diversity of Human Needs: A New Framework for Land Use Conflict Identification. *Land* **2021**, *10*, 1003. [CrossRef]

69. Li, Q.Y.; Fang, C.L.; Wang, S.J. Evaluation of Territorial Utilization Quality in China: Based on the Aspect of Production-Living-Ecological Space. *Areal Res. Dev.* **2016**, *35*, 163–169.

70. Zhang, L.; Cheng, X.Q.; Dong, X.C.; Ma, C.Q.; Wang, Y.X. Research on Spatial Layout Optimization of Industrial Land Based on Mutual Exclusion of Ecological-Production-Living Spaces in Tianjin. *Geogr. Geo-Inf. Sci.* **2019**, *35*, 112–119.

71. Lin, G.; Jiang, D.; Fu, J.Y.; Cao, C.L.; Zhang, D.W. Spatial Conflict of Production-Living-Ecological Space and Sustainable-Development Scenario Simulation in Yangtze River Delta Agglomerations. *Sustainability* **2020**, *12*, 2175. [CrossRef]

72. Yang, C.M.; Xu, X.F.; Zhang, H.; Hu, Y.G. Evolution and Optimization Features of Rural Residential Areas Based on 'the Functions of Production-living-ecological Spaces' in Shanghai. *Resour. Environ. Yangtze Basin* **2021**, *30*, 2392–2404.

73. Lin, X.H.; Wang, Y. Study on the Strategy of Village Space Optimization in Xunwu County of Jiangxi Province against the Background of Rural Vitalization Strategy. *Subtrop. Soil Water Conserv.* **2021**, *33*, 5–10.

74. Yang, H. Study on the Suitability Evaluation and the Optimization of "Production-Living-Ecological" Space-Taking Yangzhong for Example. Master's Dissertation, Nanjing Normal University, Nanjing, China, 2018.

75. Cheng, X.Q. A Study on the Delimitation of Resilient Ecological Space of Land Based on the Game of "Production-Living-Ecological". Master's Dissertation, Tianjin Polytechnic University, Tianjin, China, 2019.

76. Zhou, L.Q. The Study on the Optimization and Function Improvement of "Ecological-Production-Living" Space in Zhengzhou. Master's Dissertation, Zhengzhou University, Zhengzhou, China, 2018.

77. Wang, S. Optimization of Territorial Development Patterns Based on Production-Living-Ecological Space of Xi'an City. Master's Dissertation, Chang'an University, Xi'an, China, 2018.

78. Tang, J.H.; Liu, C.W.; Wu, Y.X. A Study on Evaluation of Energy Consumption in China Based on NRCA Model. *Econ. Geogr.* **2011**, *31*, 1313–1318.

79. Xu, L. Research on the Optimization of Geographical Spatial Pattern of Urban Agglomeration in the Middle Reaches of the Yangtze River Base on the Production-Living-Ecological Function. Ph.D. Dissertation, Huazhong Agricultural University, Wuhan, China, 2017.

80. Wei, X.F.; Zhao, Y.L.; Li, X.B.; Xue, C.L.; Xia, S.Y. Characteristics and Optimization of Geographical Space in Urban Agglomeration in the Upper Reaches of the Yangtze River Based on the Function of "Production-Living-Ecological". *Resour. Environ. Yangtze Basin* **2019**, *28*, 70–79.

81. Lin, G.; Fu, J.; Jiang, D. Production–Living–Ecological Conflict Identification Using a Multiscale Integration Model Based on Spatial Suitability Analysis and Sustainable Development Evaluation: A Case Study of Ningbo, China. *Land* **2021**, *10*, 383. [CrossRef]

82. Yuan, G.; Chen, W.B.; Yu, S.K.; Tang, S.L. A Study of Multi-scale Demarcation and Function Domination of the Production, Living and Ecological Space at County Level. *Acta Agric. Univ. Jiangxiensis* **2021**, *43*, 931–941.

83. Xiao, R.; Shao, H.Y.; Li, F.; Xie, H.B. Classification evaluation and spatial-temporal pattern analysis of the "production-living-ecological spaces" in Sichuan province. *Hubei Agric. Sci.* **2021**, *60*, 146–152.

84. Li, Q.; Luo, H.M.; Liu, C.X.; Yu, X.F.; Liu, Z.X. Study on Characteristics Evolution and Optimal Allocation of "Production-Living-Ecological Spaces" in Pearl River Delta Urban Agglomeration. *Land Resour. Inf.* **2021**, *298*, 126803. Available online: http://kns.cnki.net/kcms/detail/11.4479.n.20210918.0918.012.html (accessed on 2 November 2021).

85. Zhang, D.; Ge, Q.S.; Zhang, X.Q.; Wu, S.H.; Yang, Q.Y. Regionalization in China: Retrospect and prospect. *Geogr. Res.* **2005**, *24*, 330–344.

86. Wei, W.; Zhang, R. Exploration on the Optimization Path of Land Space Based on Main Functional Areas, Spatial Planning, and Three Living Spaces. *Urban. Archit.* **2019**, *16*, 45–51.

87. Lai, X.J. Research on "production-living-ecological" spatial governance mode of Guangdong Province for "carbon peak and carbon neutral". *Real Estate World* **2021**, *13*, 25–27.

88. National Development and Reform Commission. Circular of the National Development and Reform Commission on Printing and Distributing Evaluation Index System and Implementation Plan of Beautiful China Construction. Fghz [2020] No. 296. 2020. Available online: https://www.ndrc.gov.cn/xxgk/zcfb/tz/202003/t20200306_1222531.html (accessed on 3 July 2020).

Article

Spatio-Temporal Pattern and Conflict Identification of Production–Living–Ecological Space in the Yellow River Basin

Furui Xi [1,2], Runping Wang [3], Jusong Shi [1,2,*], Jinde Zhang [1,2], Yang Yu [1,2], Na Wang [1,2] and Zhiyi Wang [1,2]

[1]   China Institute of Geo-Environment Monitoring, Beijing 100081, China; xifurui@mail.cgs.gov.cn (F.X.); zhangjinde@mail.cgs.gov.cn (J.Z.); jcyyuyang@mail.cgs.gov.cn (Y.Y.); jcywangna@mail.cgs.gov.cn (N.W.); jcywangzhiyi@mail.cgs.gov.cn (Z.W.)
[2]   Key Laboratory of Mine Ecological Effects and Systematic Restoration, Ministry of Natural Resources, Beijing 100081, China
[3]   School of Geography and Information Engineering, China University of Geosciences, Wuhan 430074, China; wangrp@cug.edu.cn
*   Correspondence: sjusong@mail.cgs.gov.cn; Tel.: +86-10-83473309

Abstract: Production–living–ecological space (PLES) is the main body of the optimization of the development and protection pattern of territorial space, and the spatial conflict in PLES reflects a struggle for ecological protection and socio-economic development in the process of spatial development and utilization. The Yellow River Basin is one of the most concentrated and prominent areas of spatial conflict of PLES in China. Therefore, clarifying the spatio-temporal pattern of PLES of the region and scientifically identifying the characteristics of its spatial conflict will significantly improve the efficiency of comprehensive utilization of spatial resources, promote the integrated and orderly development of resource elements in the basin, and eventually achieve the strategic goals of ecological protection and high-quality development of the Yellow River Basin. In this research, the CA–Markov model was applied to simulate the spatio-temporal pattern of PLES in the Yellow River Basin from 2010 to 2025, and the landscape ecology method was adopted to construct the spatial conflict of the PLES measurement model for identifying the spatio-temporal trends of conflicts and their intensity. The results reveal that, from 2010 to 2025, ecological–production space (EPS) dominates the PLES in the Yellow River Basin, as its total area remains stable amid fluctuations; living–production space (LPS) shows the most notable change, as it grows yearly along with urbanization and industrialization process of the region; the transition between ecological–production space (EPS) and production–ecological space (PES) is the most frequent, and the two also account for the largest area. Spatial conflict of PLES in the Yellow River Basin is mainly reflected in the encroachment of LPS on other PLES, concentrated in the regions from Hekou Town to the left bank of Longmen, Fen River, Shizuishan to the southern bank of Hekou Town, and Daxia River and Tao River in the Yellow River Basin. From 2010 to 2025, the space conflict composite index of PLES (*SCCI*) of most regions in the basin lies within 0.7, which is a stable or basically controllable level. Among the 29 tertiary water resource divisions in the Yellow River Basin, the *SCCI* of 15 indicate a major, decreasing trend.

Keywords: Yellow River Basin; production–living–ecological space; spatio-temporal pattern; conflict identification

Citation: Xi, F.; Wang, R.; Shi, J.; Zhang, J.; Yu, Y.; Wang, N.; Wang, Z. Spatio-Temporal Pattern and Conflict Identification of Production–Living–Ecological Space in the Yellow River Basin. *Land* **2022**, *11*, 744. https://doi.org/10.3390/land11050744

Academic Editor: Dong Jiang

Received: 15 April 2022
Accepted: 15 May 2022
Published: 18 May 2022

Publisher's Note: MDPI stays neutral with regard to jurisdictional claims in published maps and institutional affiliations.

1. Introduction

1.1. Motivation and Literature Review

Due to the rapid development of the economy and the advancement of urbanization, the highly intensified exploitation of spatial resources has become a distinctive feature of urban and rural spatial development processes. As the number of spatial resources is limited, while their functions are highly adjustable, different groups utilize these resources in various intensities to meet their own interests, and therefore, the use of spatial resources is at times not in line with the ecological environment protection. As a result, a series of spatial conflicts have emerged, such as the uncontrolled expansion of urban space, the imbalance between agricultural space and ecological space, the degradation of ecosystems due to the encroachment of ecological space, and the unreasonable layout and function of various spaces within cities.

The human–land space competition and conflict of interests caused by land use have gradually become a hot issue in the international community [1], receiving close attention from global stakeholders such as NGOs, the United Nations, and governments [2,3]. To properly alleviate land-use conflicts, the Food and Agriculture Organization of the United Nations (FAO) formulated and promulgated the Land Evaluation Outline, proposing that land-use planning should be carried out scientifically based on land suitability. Accordingly, countries around the world established their own land-use suitability evaluation systems on the basis of this outline, for the purposes of coordinating the relationship between land resource supply and human demand and realizing the sustainable use of land resources [4]. At the same time, studies on land-use conflicts, concerning the areas of society, economy, geography, and environment are increasing in academia [5–9]. Research in this area is mainly focused on the sources, types, identification, evolution, and management of spatial conflicts [10–18].

Spatial resources can be functionally divided into three types: production space, living space, and ecological space [19]. Production–living–ecological space (PLES) basically covers the scope of spatial activities of human work and life and is the basic carrier of human socio-economic development. As the main body of the optimization of the spatial pattern of territorial space, PLES becomes an important basis for the implementation of the main functional area planning at all levels, the construction of the spatial planning system, and the improvement of the spatial development and protection system of the territorial space [20].

Spatial conflict in PLES is mainly manifested by the imbalance of structure and function of PLES, inappropriate territorial combination, and uncontrolled transformation of spatial types. In particular, it reveals the unreasonable occupation of living and ecological spaces by production space and the destruction of ecological space by spaces of living and production. The identification of spatial conflict of PLES, simulation of spatial conflict pattern in PLES, and analysis of its development and evolutionary characteristics can effectively reveal the complexity and vulnerability of the human–land relationship, fully reflecting the results and characteristics of spatial resource competition in the process of human–land interaction, and provide basic support for the optimization of regional territorial development and protection pattern [21].

The spatial conflict of PLES, in essence, belongs to the category of land-use conflicts. The study of land-use conflicts dates back to the 18th century and was initially focused on the conflict between the added economic value of land, human demand, and the land system [22]. Since the 1960s and 1970s, land-use conflicts have been characterized by interdisciplinary and diversified integration, and scholars' research perspectives have been enriched, revealing the causes, forms, and characteristics of land-use conflicts in relation to different dimensions such as regional deprivation [23–25], spatial competition [26–28], spatial integration [29–31], spatial control [32–36], ecological security [37], non-cooperative games [38], energy security and climate change [39], etc., as well as their impacts on socio-economic development and resources. In terms of research content, the spatial spillover effects of talent, policy, capital, technology, and resources [40–42] have been explored, the conflicts in spatial resource utilization between different interest groups and conflicts between spatial utilization and regional ecological environmental protection [43,44] have been analyzed, the conflicts in land-use subjects, land planning, and land systems [45–47] have been evaluated, and the spatial conflicts have been measured from the perspectives of economics and ecology, respectively. This body of research provides a basic framework for exploring the process of urbanization to promote the stability and harmony of human–land relationships and optimize the regional ecological security pattern. In terms of evaluation methods, scholars have mainly adopted the participatory survey method [48],

PSR model and fuzzy evaluation method [49], multiobjective planning method [50], landscape pattern analysis method [21,37], coupled coordination degree method [51], suitability evaluation [52,53], and actor–network analysis method [54], focusing on the scale of administrative regions such as urban agglomerations, provinces, and cities or special regions such as mining areas [45,55], and initially built a theoretical framework and methodological system for spatial conflict research.

To ensure that production space is used intensively and efficiently, living space is pleasant and proper in size, and ecological space is unspoiled and beautiful are important goals to realize the construction of ecological civilization in China [56]. From 2012 to 2017, the central working conference of urbanization, the 13th Five-Year Plan, and the report of the 19th National Congress of China set the coordinated development of the PLES as an essential strategic initiative to enhance the modernization of the spatial governance system and governance capacity of the country. How to alleviate the spatial conflicts among different PLES systems has become a primary issue that needs to be solved [57].

Although scholars have made some progress in spatial conflict identification, there are still certain shortcomings. First of all, at the spatial scale, previous studies mainly focus on administrative units such as urban agglomerations, provinces, and cities, and there are fewer studies at the watershed scale; in terms of research content, previous studies on spatial conflicts mostly concerned the space of land-use types, and there are fewer studies on the analysis of conflicts within the PLES system; at the temporal scale, most of the research considered the current situation of spatial conflicts, but less attention has been given to the evolution of future spatial conflicts. This becomes particularly important in the context of the recent emphasis on nature-based solutions for climate change mitigation [58].

1.2. Objective and Contribution

Based on the abovementioned state of research, in this paper, the Yellow River Basin was taken as the study area of spatial conflict of PLES, and the CA–Markov model was used to predict the future land-use pattern of the basin (until 2025). On account of the multifunctionality and composite nature of land use, four types of PLES were classified and analyzed. Using the grid as the basic evaluation unit and the 29 tertiary water resource divisions in the Yellow River Basin as the basic study unit for spatial conflicts, the spatial conflict of the PLES measurement model was constructed by adopting the landscape ecological index to evaluate the current and future spatial patterns of the spatial conflicts (until 2025), providing a basis for mitigating the spatial conflicts and optimizing the spatial development and protection pattern. It also provides a scientific reference to support and serve the major national strategies for ecological protection and high-quality development in the Yellow River Basin and helps to achieve sustainable development in relation to the economic, social, and ecological dimensions of the environment.

2. Research Methodology

2.1. Study Area

Yellow River, the second-largest river in China, originates in the Yueguzonglie Basin at the northern foot of the Bayan Khara Mountains in Qinghai. Its main stream is 5464 km long, flowing through nine provinces and regions in China—namely, Qinghai, Sichuan, Gansu, Ningxia, Inner Mongolia, Shaanxi, Shanxi, Henan, and Shandong—and finally running into the Bohai Sea. The Yellow River is regarded as the mother river of China, as Chinese civilization was born in the Yellow River Basin, which is also an important ecological barrier and economic zone in China. The Yellow River Basin covers an area of about 795,000 km^2, located between 95°59′–118°58′ E and 31°56′–42°03′ N (Figure 1). It amounts to 8.3% of China's land area, and the total population of provinces in which the Yellow River Basin is located was 420 million in 2018, accounting for 30.3% of China's population, with a regional gross national product of over 23.9 trillion yuan, making up 26.5% of China's GDP in 2018. Known as the "energy basin", the Yellow River Basin is rich in coal, oil, natural gas, and non-ferrous metals, among which coal reserves account for more than half of China's total amount, making it an important base for energy, chemical, raw material, and basic industrial production in China. At the same time, the Yellow River Basin connects the Qinghai–Tibet Plateau, the Loess Plateau, and the North China Plain, and has many national parks and national key ecological function areas such as the Sanjiangyuan and Qilian Mountains, making it an important ecological security barrier in northern China.

In recent years, the rapid industrialization and urbanization of the Yellow River Basin have accelerated the evolution of its natural geographic pattern, and the unreasonable human development and utilization activities have aggravated the deterioration of the ecological environment, resulting in the tightening of resource and environmental constraints in the basin, and the intensity of territorial space development is on the verge of overload. Since industries in the Yellow River Basin rely heavily on energy, and economic zones and urban agglomerations generally use land carelessly, the encroachment of production and living spaces on ecological space is serious, and the problem of spatial conflict of PLES is very prominent, which seriously restricts the high-quality socio-economic development. To achieve the strategic goal of ecological priority and green development, scientific implementation of territorial planning and land-use control and optimization of spatial development and protection pattern are first required to identify the spatio-temporal pattern of spatial conflict of PLES in the Yellow River Basin.

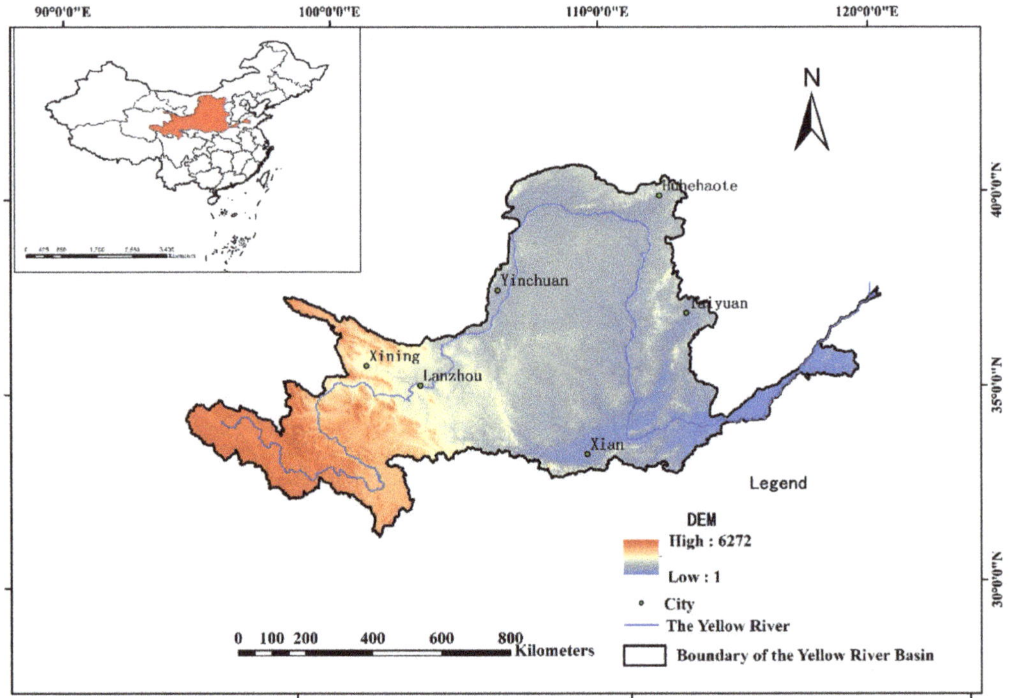

Figure 1. Location of the Yellow River Basin.

2.2. Research Framework

The framework of PLES simulation and identification of spatial conflict of PLES in the Yellow River Basin is as follows:

Step 1: Simulation of PLES distribution pattern in the Yellow River Basin in 2025 based on the CA–Markov model;

Step 2: Analysis of the spatio-temporal patterns of PLES in the Yellow River Basin from 2010 to 2025;

Step 3: Quantification of the spatial conflict of PLES in the Yellow River Basin from 2010 to 2025 using the Spatial Conflict Composite Index evaluation method of landscape ecology and evaluation of the degree of spatial conflict of PLES and its spatio-temporal pattern.

2.3. Data Sources and PLES Classification

The vector data of the boundaries of the Yellow River, Yellow River Basin, tertiary water resource divisions, and the land-use raster data of the Yellow River Basin in 2010, 2015, and 2018 were provided by the Resource and Environment Science and Data Center of the Chinese Academy of Sciences (https://www.resdc.cn/, accessed on 10 March 2022). The land-use data of the Yellow River Basin have a spatial resolution of 30 m × 30 m. Land-use types included 6 primary types—namely, arable land, forest land, grassland, waters, residential land, and unused land—and 25 secondary types.

Considering the multifunctional complexity of spatial resources, the same space resource may have one or more complex functions, including functions related to production, living, and ecology. In this research, the PLES classification of the Yellow River Basin was formed by referring to the current research results of other scholars on PLES classification [44,59,60], integrating and reclassifying the existing land-use types based on the actual situation of the Yellow River Basin, and forming the PLES classification in accordance with the classification principles of dominant and secondary functions, which contains living–production space (LPS), production–ecological space (PES), ecological–production space (EPS), and ecological space (ES), as shown in Table 1.

Table 1. PLES classification.

Ecological–Living–Production Space (PLES) Classification	Land-Use Types
Living–production space (LPS)	Urban and rural land, Industrial and mining land, Residential land (urban land, rural residential land, other types of construction land)
Production–ecological space (PES)	Cultivated land (paddy field, dry land)
Ecological–production space (EPS)	Forest land, grassland
Ecological space (ES)	Unused land (sand land, Gobi desert land, saline–alkali land, marshland, bare land, bare rock land, oceans, other types of unused land), water area

2.4. Space Conflict Composite Index (SCCI) of PLES

Based on the theory of landscape ecology, the characteristics of spatial complexity, spatial vulnerability, and spatial stability were used to determine the spatial conflict of the PLES index and quantitatively evaluate its intensity in the Yellow River Basin. The evaluation method using the space conflict composite index of PLES (*SCCI*) is described in [61], and the calculated *SCCI* values are normalized within the 0–1 interval. *SCCI* can be expressed as follows:

$$SCCI = CI + FI - SI$$

where *CI* is the spatial complexity index, which is quantified by using the area-weighted average patchwork fractal index (*AWMPFD*) in landscape ecology. With the rapid socio-economic development, land development and utilization activities gradually intensify; as a result, the shapes of patches tend to be complex, and spatial utilization conflicts grow accordingly. Therefore, the area-weighted average patch fractal index (*AWMPFD*) can better characterize the degree of interference of neighboring patches to the measured patches, which reflects the degree of influence of human development and utilization activities on the space. The higher the value, the greater the external force on the patches. The *AWMPFD* can be calculated as follows:

$$AWMPFD = \sum_{i=1}^{m} \sum_{j=1}^{n} \left[\frac{2\ln(0.25 P_{ij})}{\ln(a_{ij})} \left(\frac{a_{ij}}{A} \right) \right]$$

In this formula, P_{ij} is the perimeter of the patch, a_{ij} is the area of the patch, A is the total area of the spatial type, i and j are the j-th spatial type in the i-th spatial unit; m is the total number of units involved in the evaluation in the study area, and n is the number of PLES.

FI is the spatial vulnerability index, which is measured by using the vulnerability of each landscape type within the study area in landscape ecology. *FI* characterizes the ability of spatial patches to resist external pressure, which directly affects the degree of spatial vulnerability. The weaker the resistance, the more vulnerable to external influence, the stronger the spatial vulnerability, and the higher the level of spatial conflict. PLES is a redistribution of the landscape types, referring to the related literature [21,22,37,44,45,62,63], the vulnerability of each type of PLES is assigned as LPS −0.1, PES −0.44, EPS −0.3, and ES −0.75. The *FI* calculation equation is as follows:

$$FI = \sum_{i=1}^{n} F_i \times \frac{a_i}{S}$$

In the above formula, F_i is the vulnerability index of class i spatial type, n is the total number of PLES classifications, and a_i is the area of patches of various landscape types within the unit; S is the total spatial area.

SI is the spatial stability index, which is measured through the landscape fragmentation index in landscape ecology. The main effect of spatial conflict on the regional spatial pattern can lead to landscape patch fragmentation. The more fragmented the spatial pattern, the more homogeneous the type, the less spatial stability, and the higher the intensity of the spatial conflict. The *SI* value is calculated by the following formula:

$$SI = 1 - PD$$

$$PD = \frac{n_i}{A}$$

where *PD* is the patch density; the larger the *PD* value, the higher the fragmentation of the space, the lower the spatial stability, and the lower the stability of the corresponding spatial ecosystem. n_i is the number of patches of type i spatial type in each spatial unit, and A is the area of each spatial unit.

2.5. CA–Markov Scenario Simulation

The CA–Markov model predicts land-use change by combining the principles of cellular automata (CA), Markov chains, and multiobjective land allocation [64]. It also has the ability to predict and model spatial changes in complex systems over time. The CA–Markov model integrates spatio-temporal factors in a land-use raster map, treats the land-use type represented by each raster as a metacell state, and uses a land-use transfer area matrix and probability matrix to determine the transfer of metacell states and simulate the change in land-use pattern in a certain region in a specific time.

The simulation process for the PLES distribution of the Yellow River Basin in 2025 was as follows:

The spatial overlay analysis of the land-use data was first processed in ArcGIS and imported into IDRISI software; then, the probability matrix of PLES shift in the Yellow River Basin from 2010 to 2015 was calculated using the Markov model. Considering the data of terrain slope, elevation, and road, the MCE module was used to construct the land-use transfer suitability atlas, and the CA–Markov model was applied to simulate and generate the PLES distribution in 2018. Finally, using the PLES classification data in 2018 as the benchmark, the number of CA cycles was set to 7, based on which the CA–Markov model was used to simulate the PLES distribution of the Yellow River Basin in 2025.

3. Results

3.1. Spatio-Temporal Pattern of PLES

As shown in Table 1, land use in the Yellow River Basin was reclassified into living–production space (LPS), production–ecological space (PES), ecological–production space (EPS), and ecological space (ES) by using LUCC data. Additionally, the PLES distribution patterns in 2010, 2015, and 2018 were obtained (Figures 2–4). The CA–Markov model was applied to predict and analyze the distribution pattern of PLES in the Yellow River Basin in 2025, using the kappa coefficient to test the consistency between the simulation results and the current distribution of land-use types in 2018; the results suggested that the kappa value was greater than 0.85, indicating that the simulation results of the CA–Markov model were more satisfactory. The spatial pattern of PLES in the Yellow River Basin in 2025 is shown in Figure 5.

From 2010 to 2025, EPS remains the most prominent among PLES categories in the Yellow River Basin, which is concentrated and widely distributed in the middle and upper reaches of the Yellow River Basin, with an annual average area that exceeds 50% of the total area of the basin. PES in the Yellow River Basin is widely distributed as well, with an annual mean area of about 36%, concentrated in the Yellow Huaihai Plain, Fenwei Plain, Ning-Meng Plain, and other major agricultural production areas, and also widely scattered in the Loess Plateau and other areas. ES is spread mostly in the upper reaches of the Yellow River Basin and the source region of the Yellow River, mainly in form of desert, sand, Gobi, bare land, and other unused land types, with an average area of about 9% for many years. LPS is found mostly in Jinan, Zhengzhou, Xi'an, Taiyuan, Hohhot, Yinchuan, Lanzhou, Xining, and their surrounding areas, with a cluster-like concentrated distribution, sharing only a minimum area of about 4%.

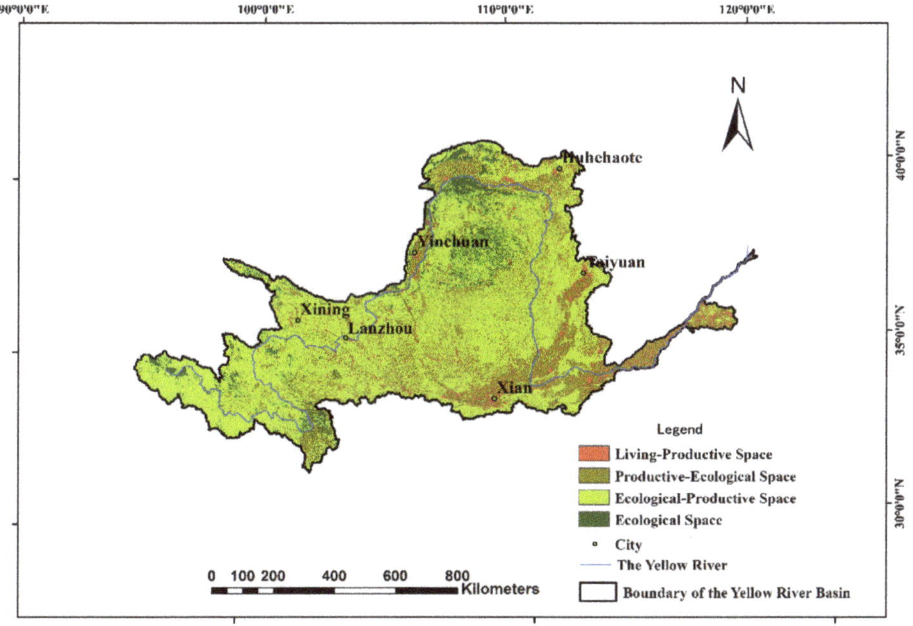

Figure 2. Spatial pattern of PLES in the Yellow River Basin in 2010.

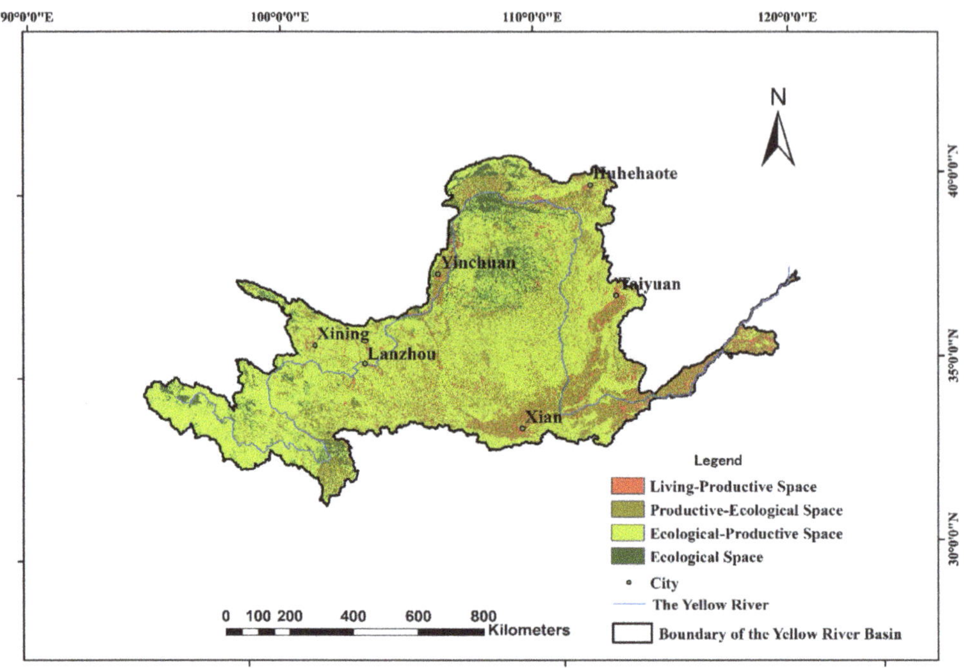

Figure 3. Spatial pattern of PLES in the Yellow River Basin in 2015.

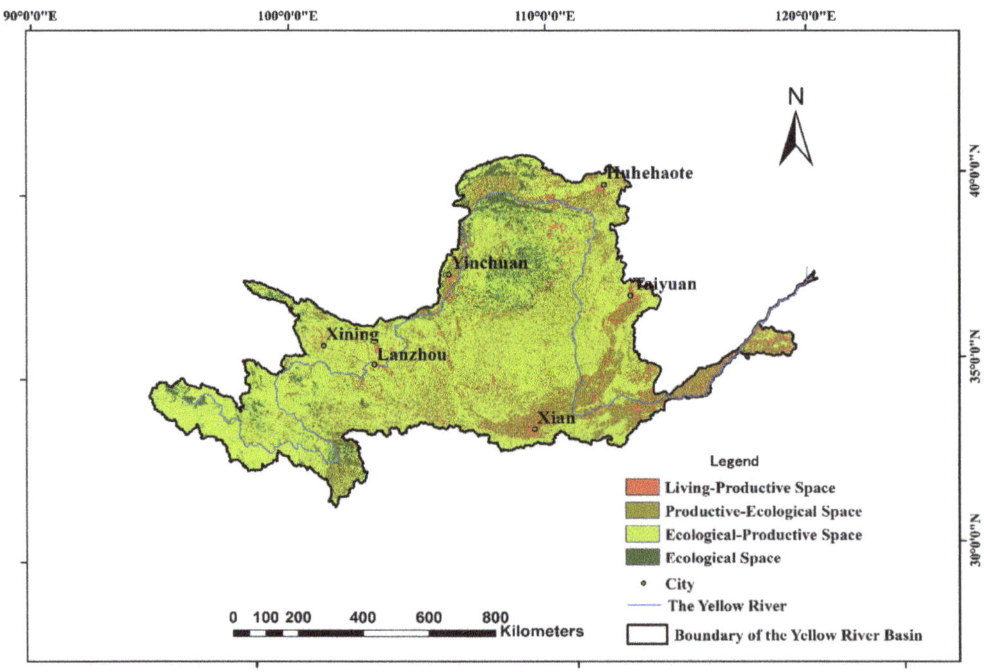

Figure 4. Spatial pattern of PLES in the Yellow River Basin in 2018.

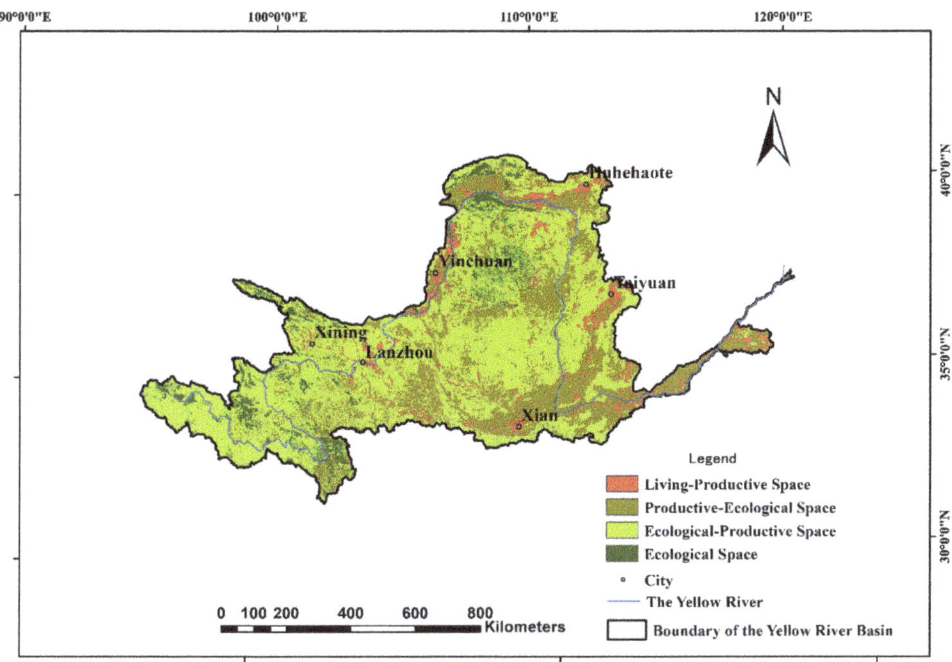

Figure 5. Predicted spatial pattern of PLES in the Yellow River Basin in 2025.

Table 2 shows the changes in the PLES area and their proportion in the Yellow River Basin from 2010 to 2025. Overall, EPS and PES are still the dominant PLES types in this period—the combined area of both exceeds 85% of the total area of the basin. The spatial area of the two does not vary significantly with time, and the year-to-year variation is mostly less than 0.1%, which is basically a stable state. In contrast, the LPS area of the Yellow River Basin shows a yearly growth trend, expected to increase from 2.51×10^4 km^2 in 2010 to 4.28×10^4 km^2 in 2025—an expansion of nearly 70% in 15 years, and the growth rate is increasing year by year. In addition, the ES area of the Yellow River Basin is gradually decreasing, from 7.59×10^4 km^2 in 2010 to 6.48×10^4 km^2 in 2025, and the decreasing trend is expected to gradually intensify from 2018 to 2025.

Table 2. Area statistics of PLES from 2010 to 2025 (area in 10^4 km^2, rate in %).

Ecological–Living–Production Space (PLES) Classification	2010		2015		2018		2025	
	Area	Rate	Area	Rate	Area	Rate	Area	Rate
Living–production space (LPS)	2.51	3.16	2.71	3.41	3.08	3.88	4.28	5.38
Production–ecological space (PES)	28.74	36.12	28.65	36.04	28.21	35.48	28.32	35.62
Ecological–production space (EPS)	40.66	51.14	40.58	51.04	40.82	51.34	40.42	50.84
Ecological space (ES)	7.59	9.58	7.56	9.51	7.39	9.30	6.48	8.16

3.2. PLES Transfer Matrix Analysis

According to Table 3, all four types of PLES have different degrees of transfer in and out of each other. The area where PLES type conversion occurs is 10,686.02 km^2, accounting for about 1.34% of the total area. The amounts of EPS and PES are larger than the amounts of the other three types of spatial transformation, with a total of 4481.32 and 4276.73 km^2, respectively. Among them, the type shift between EPS and PES is especially drastic. The transfer volume of EPS to PES is about 2910.06 km^2, amounting to 64.94% of its total transfer volume. The transfer from PES to EPS is also greater, about 2489.50 km^2, accounting for about 58.21% of its total spatial transfer. In comparison, the transfer between LPS and ES is more stable. The transfer from ES to LPS is 215.79 km^2, while the transfer from LPS to ES is the smallest among all PLES types, with an area of 11.15 km^2.

Table 3. PLES transfer matrix for 2010–2015 (km^2).

2010	2015			
	LPS	**PES**	**EPS**	**ES**
LPS		248.39	157.58	11.15
PES	1434.18		2489.50	353.05
EPS	786.05	2910.06		785.21
ES	215.79	377.99	917.07	

Note: PES: production–ecological space; LPS: living–production space; ES: ecological space; EPS: ecological–production space.

According to Table 4, the area where PLES type transfer occurred is 64,586.46 km^2, accounting for about 8.12% of the total area, which shows a substantial growth trend, compared with the PLES transfer from 2010 to 2015. In terms of the proportion of PLES type transfer, it is roughly similar to the percentage in 2010–2015. PES and EPS are less stable, as their transfer to other PLES types has the largest areas of 29,715.94 and 25,050.59 km, respectively. Among them, the transfer from EPS to PES is about 20,547.12 km^2, occupying more than 80% of their total transfer. The area where PES transferred to EPS remains the largest among all PLES types, about 23,211.14 km^2, taking up about 78.11% of its total transfer. In comparison, the transfer volume between LPS and ES is smaller, with 664.27 km^2 transferred from ES to LPS, while the transfer volume from LPS to ES is only 212.07 km^2, which is the smallest area among all PLES types.

Table 4. PLES transfer matrix for 2015–2018 (km^2).

2015	2018			
	LPS	**PES**	**EPS**	**ES**
LPS		2591.67	1017.29	212.07
PES	4687.27		23,211.14	1817.53
EPS	2188.10	20,547.12		2315.37
ES	664.27	2060.41	3274.22	

The PLES distribution of the Yellow River Basin in 2025 was predicted with the CA–Markov model. From the PLES area transfer matrix of the Yellow River Basin for 2018–2025 shown in Table 5, it is evident that the PLES transfer area will continue to grow substantially, to about 205,740.87 km^2, and the average annual PLES transfer area is 29,391.43 km^2, which is about 1.36 times that of 2015–2018, but the PLES transfer growth rate indicates a decreasing trend. With the same pattern of PLES transfer changes in the cycles of 2010–2015 and 2015–2018, EPS and PES are less stable, and the spatial type transfers in and out are relatively large. Among them, EPS and PES transfer areas are 87,864.46 and 83,044.38 km^2, respectively, both of which account for more than 40% of the total transfer volume. In comparison, LPS and ES are relatively stable, for which the area transferred from LPS to ES is the smallest, only 272.25 km^2, accounting for about 0.13% of the total transferred area.

Table 5. PLES transfer matrix for 2018–2025 (km^2).

2018	2025			
	LPS	**PES**	**EPS**	**ES**
LPS		9552.25	2421.51	272.25
PES	14,136.75		62,574.32	6333.31
EPS	7886.52	67,904.43		12,073.51
ES	2084.75	7363.75	13,137.52	

3.3. Spatial Conflict of PLES

The space conflict composite index values (*SCCI*) of the 29 tertiary water resource divisions in the Yellow River Basin were calculated in 2010, 2015, 2018, and 2025 (Table 6), respectively, based on the *SCCI* calculation method. According to the statistical distribution characteristics of the *SCCI*, the *SCCI* values were standardized and then classified into four levels: stably controllable, basically controllable, basically out of control, and seriously out of control. The specific division intervals are as follows: level 1: "stably controllable" [0.00, 0.30); level 2: "basically controllable" [0.30, 0.70); level 3: "basically out of control" [0.70, 0.90); and level 4: "seriously out of control" [0.90, 1.00].

Table 6. *SCCI* of the tertiary water resource division from 2010 to 2025.

NO.	Tertiary Water Resource Division	*SCCI* (2010)	*SCCI* (2015)	*SCCI* (2018)	*SCCI* (2025)
1	Heyuan to Maqu	0.08	0.36	0.07	0.62
2	Maqu to Longyangxia	0.02	0.07	0.06	0.56
3	Daxia River and Tao River	0.37	0.36	0.39	1
4	Longyangxia to Lanzhou main stream sector	0.19	0.19	0.19	0.47
5	Huangshui River	0.19	0.18	0.18	0.68
6	Above Datong River Xiangtang	0.09	0.09	0.07	0.52
7	Lanzhou to Xiaheyan	0.31	0.31	0.29	0.17
8	Qingshui River to Kushui River	0.42	0.38	0.40	0.08
9	Above Wei River Baojixia	0.66	0.65	0.67	0.68
10	Above Jing River Zhangjiashan	0.92	0.90	0.86	0.33
11	Xiaheyan to Shizuishan	0.05	0.06	0	0.03
12	Wei River Baojixia to Xianyang	0.40	0.43	0.45	0.35
13	Interior drainage area	0.15	0.14	0.13	0.44
14	Shizuishan to the northern bank of Hekou Town	0.19	0.19	0.16	0.17
15	Shizuishan to the southern bank of Hekou Town	0.20	0.18	0.14	0.32
16	Right bank above Wubao	0.53	0.52	0.48	0.17
17	Right bank below Wubao	0.61	0.61	0.58	0.41
18	Hekou Town to left bank of Longmen	1	1	1	0.15
19	Above Beiluo River Zhuangtou	0.60	0.59	0.66	0.05
20	Fen River	0.71	0.70	0.69	0
21	Wei River Xianyang to Tongguan	0.39	0.38	0.30	0.28
22	Longmen to Sanmenxia main stream sector	0.32	0.31	0.30	0.07
23	Sanmenxia to Xiaolangdi sector	0.13	0.12	0.16	0.09
24	Qindan River	0.42	0.42	0.40	0.07
25	Yiluo River	0.24	0.25	0.24	0.15
26	Xiaolangdi to Huayuankou main stream sector	0.08	0.08	0.12	0.02
27	Jindi River and Natural Wenyan Canal	0.23	0.24	0.25	0.05
28	Dawen River	0.29	0.28	0.33	0.26
29	Main stream sector below Huayuankou	0	0	0.02	0.19

In 2010, the Loess Plateau area of the middle reaches of the Yellow River showed a severe spatial conflict of PLES in the Yellow River Basin (Figure 6). In this area, the conflict levels in the region of Hekou Town to the left bank of Longmen and the region above Jing River Zhangjiashan were level 4, indicating that the region was seriously out of control; the spatial conflict of PLES level in the Fen River Basin area was level 3, which was within basically out-of-control status; most other areas in the Loess Plateau were considered level 2, meaning that the spatial conflict of PLES was basically controllable. In comparison, the spatial conflicts of PLES in most of the other basin regions such as the lower reaches of the Yellow River, the Ningxia plain, the Inner Mongolia irrigation area, the inland flow area, the Hehuang area, and the source region of the Yellow River were relatively mild and in a stably controllable state of level 1.

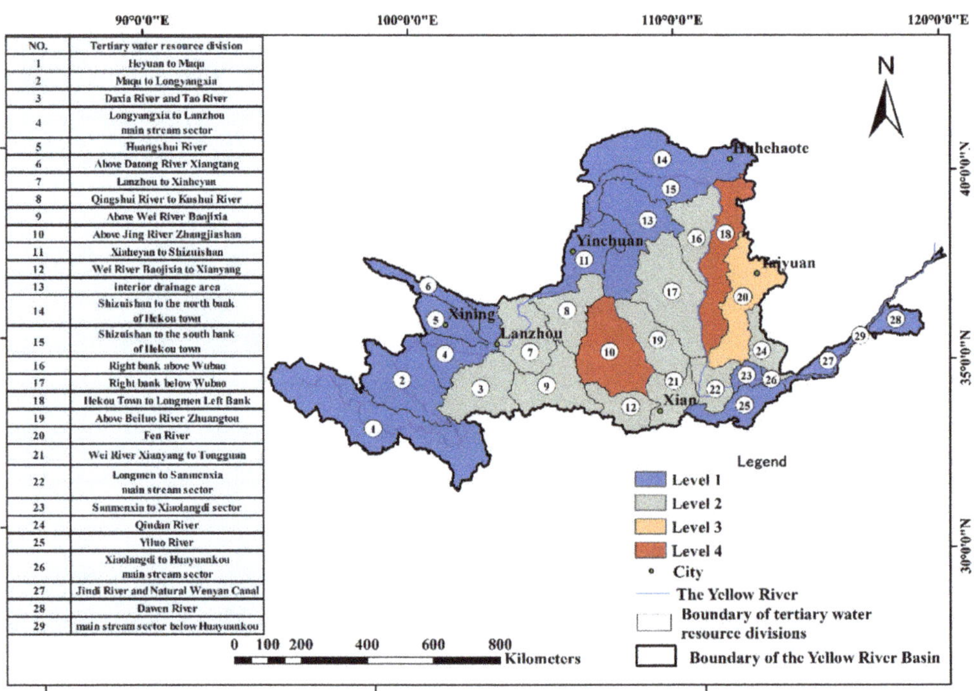

NO.	Tertiary water resource division
1	Heyuan to Maqu
2	Maqu to Longyangxia
3	Daxia River and Tao River
4	Longyangxia to Lanzhou main stream sector
5	Huangshui River
6	Above Datong River Xiangtang
7	Lanzhou to Xiaheyun
8	Qingshui River to Kushui River
9	Above Wei River Baojixia
10	Above Jing River Zhangjiashan
11	Xiaheyun to Shizuishan
12	Wei River Baojixia to Xianyang
13	Interior drainage area
14	Shizuishan to the north bank of Hekou town
15	Shizuishan to the south bank of Hekou town
16	Right bank above Wubao
17	Right bank below Wubao
18	Hekou Town to Longmen Left Bank
19	Above Beiluo River Zhuangtou
20	Fen River
21	Wei River Xianyang to Tongguan
22	Longmen to Sanmenxia main stream sector
23	Sanmenxia to Xiaolangdi sector
24	Qindan River
25	Yiluo River
26	Xiaolangdi to Huayuankou main stream sector
27	Jindi River and Natural Wenyan Canal
28	Dawen River
29	main stream sector below Huayuankou

Figure 6. Spatial conflict of PLES in the Yellow River Basin in 2010.

The spatial distribution of spatial conflict of PLES in the Yellow River Basin in 2015 (Figure 7) was basically the same as that in 2010, and the out-of-control regions, the ones measured as level 3 and level 4, which were located in the middle reaches of the Yellow River Loess Plateau area, still had not been improved, while the spatial conflict of PLES in the region from Heyuan to Maqu area had deteriorated from the stably controllable level 1 to the basically controllable level 2.

In 2018, the spatial conflict of the PLES situation in the Yellow River Basin (Figure 8) improved—only the region from Hekou Town to the left bank of Longmen remained at level 4, indicating a seriously out-of-control status. The region above Jing River Zhangjiashan changed from level 4 to level 3 status, the region Fen River improved from level 3 to level 2 status, and the region from Heyuan to Maqu was at that point at level 1 instead of level 2. Only spatial conflict of PLES of the region Dawen River deteriorated from level 1 to level 2, and the other areas were relatively stable.

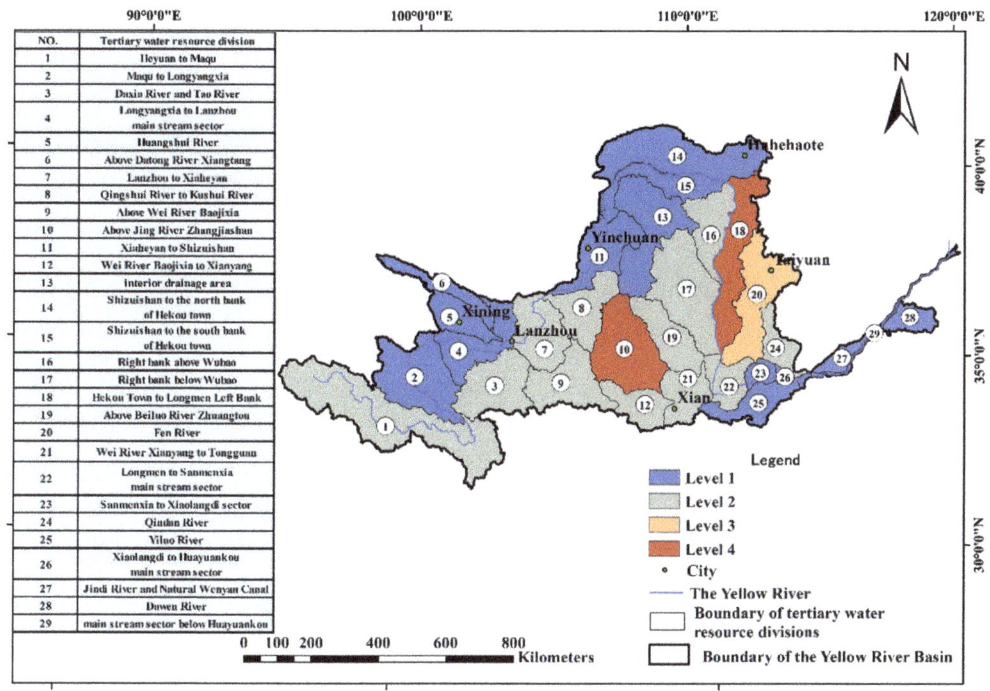

Figure 7. Spatial conflict of PLES in the Yellow River Basin in 2015.

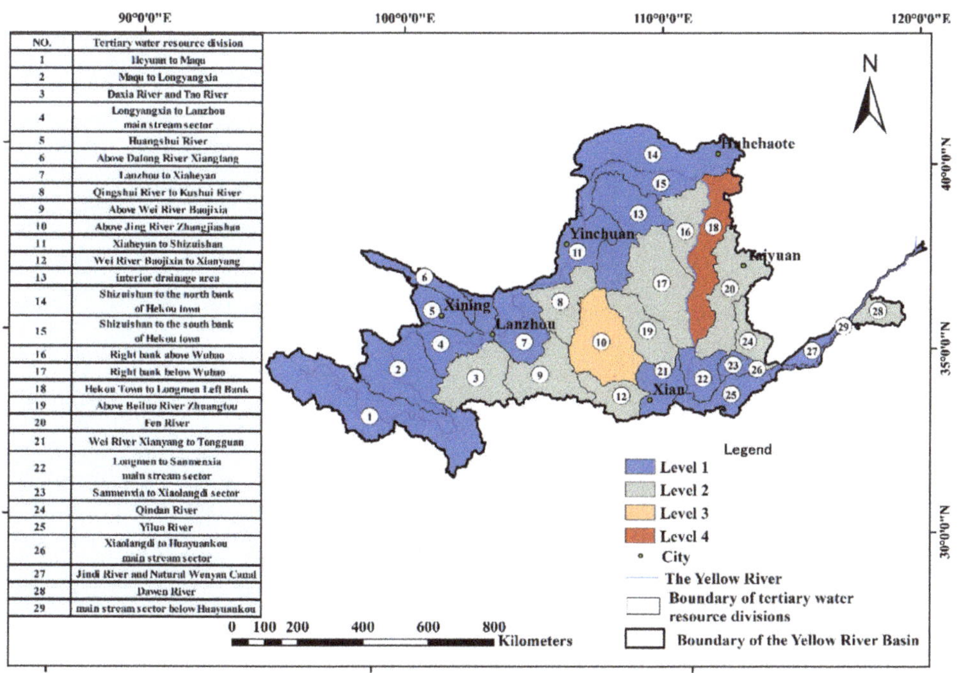

Figure 8. Spatial conflict of PLES in the Yellow River Basin in 2018.

As is shown in the simulation results of spatial conflict of PLES in the Yellow River Basin in 2025 (Figure 9), the spatial conflict of PLES in the Yellow River Basin will continue to improve, and the situation in the Loess Plateau and most of the lower reaches will be in a level 1 or level 2 controllable state. However, the spatial conflict of PLES in the upper reaches of the Yellow River is deteriorating: the conflict level of the Daxia River and Tahoe River region has risen to the seriously out-of-control state of level 4, and the source region of the Yellow River shows different degrees of deterioration, from level 1 to the basically controllable state of level 2.

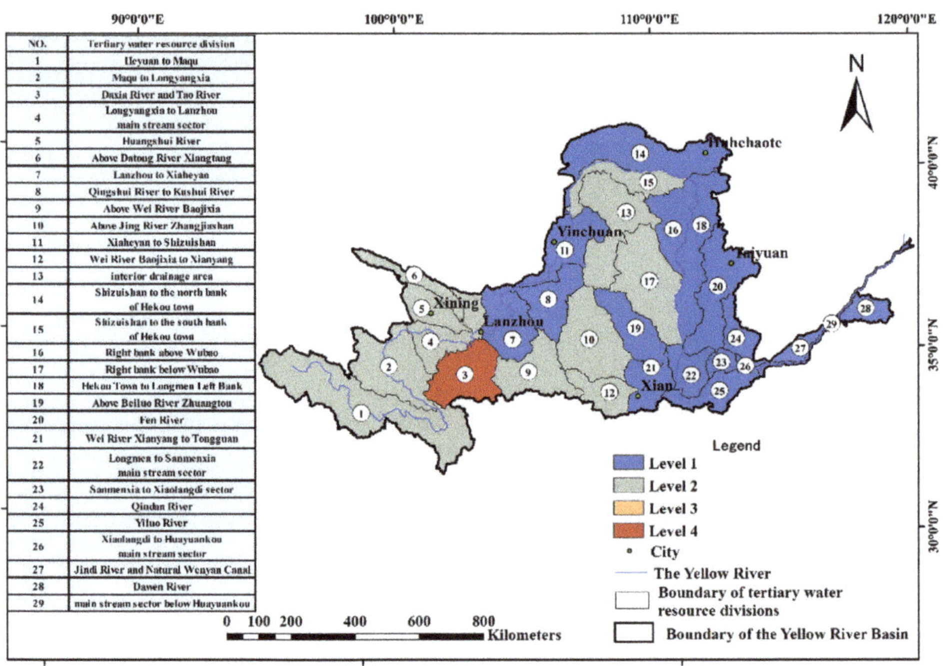

Figure 9. Predicted spatial conflict of PLES in the Yellow River Basin in 2025.

As can be seen from Figure 10, from 2010 to 2025, the spatial conflict of PLES in the Yellow River Basin is in the basically controllable level 1. Among the 29 tertiary water resource divisions, except for the regions of Daxia River and Tao River, above Jing River Zhangjiashan, Hekou Town to the left bank of Longmen, and Fen River, the *SCCI* values are in a stable and controllable state. From the development trend of spatial conflict of PLES in the Yellow River Basin, from 2010 to 2025, 15 of the 29 tertiary water resource divisions in the Yellow River Basin show a significantly decreasing trend of *SCCI*, with a percentage of more than 51%. There are eight regions where spatial conflict of PLES fluctuates (decreasing and then increasing or increasing and then decreasing), accounting for about 28%. At the same time, there are six regions in the basin where spatial conflict of PLES reveals a gradually strengthening trend, accounting for 21%.

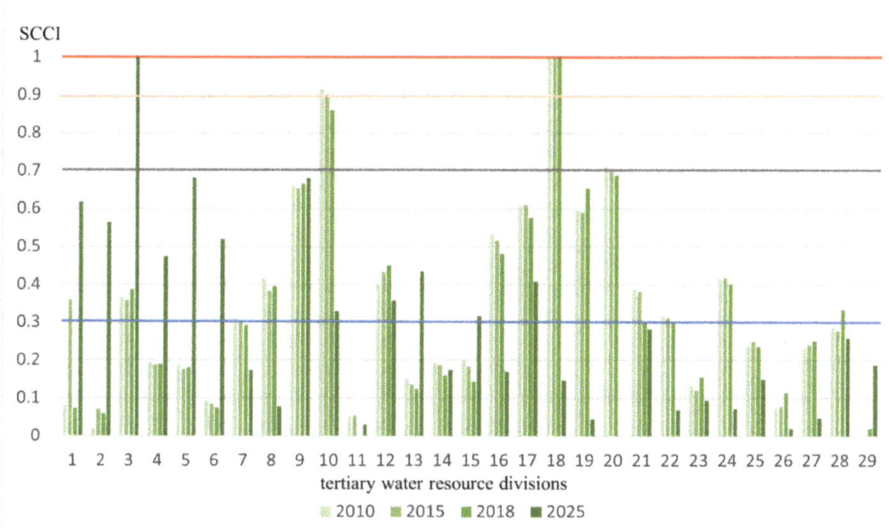

Figure 10. PLES spatial conflict changes in 29 tertiary water resource divisions from 2010 to 2025.

4. Discussion

The analysis of the spatio-temporal changes in PLES in the Yellow River Basin from 2010 to 2025 indicates that the urbanization process in the basin has further intensified—the LPS area has further increased, and metropolitan areas with a certain scale have gradually formed around provincial capital cities with population and economic siphoning effects, such as Jinan, Zhengzhou, Xi'an, Taiyuan, Hohhot, Yinchuan, and Lanzhou. In addition, the development of energy and mining industries in the Yellow River Basin continues to be the pillar industries supporting industrial and economic development, and therefore, their areas basically remain stable. The Yellow River Basin still has a pivotal and important role in securing China's energy and mineral resources. The area of PES is still unchanged, compared with 2010. PES is distributed more concentrated in the Loess Plateau of the middle reaches and the upper reaches of the Yellow River, which proves that the construction of concentrated contiguous high-standard farmland for modern agricultural development has achieved its initial results. The EPS in the Yellow River Basin is basically stable, but the EPS connectivity and agglomeration degree increased significantly. The unobstructed degree of biological habitat was strongly guaranteed, so an imminent increase in biodiversity of the basin could be expected. Saline and desertified land management in the upper reaches of the Yellow River Basin and parts of the Loess Plateau continues to be effective, as their total areas continued to decrease, resulting in a slight decrease in ES in the basin. At the same time, owing to the positive impact of the implementation of China's nature reserve policy, the disturbance of ES in the upper Yellow River Basin will be significantly reduced by people's work and living activities, and the ES area will increase significantly. As a result, large areas with national representative natural ecosystem values will be effectively protected.

Judging from the seriousness of spatial conflict of PLES, the encroachment of LPS on other PLES is the most prominent in the Yellow River Basin. With the urbanization of the basin and the strengthening of energy and mineral resources development, the encroachment of LPS on other types of PLES is increasing, which is mainly reflected in the encroachment of PES, EPS, and ES around the periphery of urban development zones and townships, and the phenomenon of "pie spreading" caused by the excessive and disorderly development of towns. The development of energy and chemical bases encroaches on regional EPS and ES, giving rise to regional pasture degradation, soil erosion, land desertification, and soil pollution. This is seen mostly in the regions from Hekou Town to the left bank of Longmen, Fen River, and Shizuishan to the south bank of Hekou Town. In addition, LPS encroachment on EPS and ES in the upper reaches of the Yellow River Basin mainly manifested as the encroachment of urban development, overgrazing, water conservancy, transportation facilities construction, etc. on the plateau grassland meadow, wetlands, and other natural water space, resulting in degradation of grassland meadow and peat swamp wetland and other ecological impacts, concentrated in the area of Daxia and Tao River and other areas.

As revealed from the area of spatial conflict of PLES, the spatial conflict of PLES in the Yellow River Basin mainly focuses on the conflict between PES and EPS, specifically the conversion of land-use types between grassland and cropland. In space, the PES and EPS change areas are highly spatially coordinated. On the one hand, the Ningmeng irrigation area, the Fenwei basin, and the lower Yellow River plain are the main agricultural production regions in China, responsible for the mission of ensuring national food security. The continuous expansion of arable land has inevitably caused encroachment on EPS, especially in some ecologically fragile areas of the Loess Plateau where water and soil resources do not align with each other. Excessive agricultural cultivation has caused the destruction of surface vegetation, increased soil erosion, and deterioration of ecosystem services. In recent years, owing to the continuous promotion of the national project of returning farmland to forest and high standard farmland construction, the original arable land with unsuitable water resources carrying capacity or mismatched soil and water conditions has been gradually withdrawn. Thus, the level of agricultural modernization has continuously increased, alleviating the problem of PES encroachment on EPS to some extent. On the other hand, in recent years, with the overlapping impacts of industrial transformation, population migration, agricultural price fluctuations, etc., the Loess Plateau and other areas that used to be agriculturally dominated regions have lost a large number of their rural population. The phenomenon of abandonment of arable land is very common, resulting in the loss of a large amount of suitable arable land, which formed the passive encroachment of EPS on PES, especially in the region above Jing River Zhangjiashan.

From the perspective of the research scale, there have been few studies on spatial conflicts of PLES in the Yellow River Basin, which mainly focus on administrative units such as urban agglomerations, provinces, and cities [65–67]. However, this study adopted tertiary water resource divisions as the basic research unit to analyze the distribution characteristics of spatial conflicts of PLES, thus enriching the scales of research related to the Yellow River Basin. At the same time, the CA–Markov model was used to simulate the spatial pattern of conflicts, which is highly practical for policymakers to formulate corresponding land-use optimization plans. In terms of research methods, GIS and RS technology are the main means to monitor land-use changes by using raster and vector data [68,69]. In this study, land-use change in the Yellow River Basin was analyzed using classified raster data. Compared with other studies [70–72], this method, based on landscape ecology, obtained good credibility, since it focused on revealing the spatio-temporal evolution of PLES conflicts from the perspective of spatial morphological changes by using relatively few volumes of data.

Water resources are the core resource elements for socio-economic development and ecological protection since the ecosystem service functions such as water connotation, soil conservation, sand fixation, and flood regulation are closely related to water resources. Therefore, this study took the 29 tertiary water resource divisions in the Yellow River Basin as the basic research unit. Moreover, in this research, analyses of PLES patterns and internal mechanisms were carried out, considering the distribution of administrative regions, topography, national economic development, watershed size, and maintaining the unity, combination, and integrity of administrative regions and basin zoning. To analyze the crux of ecological protection and high-quality economic and social development in the Yellow River Basin effectively, the study approach was to examine the coupling relationship between man and land system, based on the internal mechanism of spatial conflict of PLES. Meanwhile, other factors, including the laws of physical geography and socio-economic development, the carrying capacity of resources and environment, the rigid constraint of ecological protection on water resources and high-quality development of the Basin, and the administrative requirements of regional ecological protection and socio-economic construction, were also considered.

Due to climatic challenges, the changes in land use in the Yellow River Basin had more profound impacts on the surface water cycle now. In recent years, the climate in the Loess Plateau area of the middle reaches of the Yellow River Basin has become warmer and drier. Furthermore, a decrease in atmospheric precipitation recharge and an increase in terrestrial evapotranspiration have led to a decrease in available land serving as water resources. Meanwhile, the Chinese government has implemented a large-scale greening action in the Loess Plateau region [73], increasing the conversion from PES to EPS. Large-scale afforestation improves the regional ecological environment, but also changes the underlying surface structure of the region, affects the local water cycle process, and decreases the gradual surface runoff. According to relevant studies [74], due to the impact of climate change and human activities, the carrying capacity of water resources in the Loess Plateau has been on the verge of overload. At the same time, land-use change in the Yellow River Basin will also affect the local climate by changing the carbon cycle.

There is a large amount of unused land (ES) in the upper reaches of the Yellow River. In the future, following the principle of not affecting the ecological environment, these types of unused land can be fully utilized for the development of renewable energy, including wind, solar, and biomass. Firstly, it can reduce the occupation of limited construction and cultivated land resources in socio-economic development, and lower the conflict between ES, PES, and LPS, and secondly, it will reduce the dependence on fossil fuels to promote carbon neutrality, thus mitigating the adverse effects of climate change.

The spatial conflict of PLES resonates with the game process of ecological protection and socio-economic development in the process of territorial space development and utilization. The macroscopic natural geographical background conditions lay the geographical foundation for the construction of the spatial development and protection pattern of the Yellow River Basin, while socio-economic development, urbanization and industrialization processes, exploitation of mineral resources, and other human activities are the key driving force behind the spatial pattern of land space, accelerating the process of change in spatial patterns of the Yellow River Basin. The ecological conditions in Yellow River Basin are fragile, manifested in its serious shortage of water resources, ecologically sensitive areas and fragile areas, and massive pressure on ecological protection under climate change conditions. Meanwhile, in the process of rapid socio-economic development, disordered and uncontrolled urbanization, industrialization, and exploitation activities of mineral resources have caused a disproportionate spatial pattern of PLES and deterioration in the quality of the ecological environment. Therefore, it is of great practical significance to understand the spatio-temporal pattern of PLES in the Yellow River Basin and identify the characteristics of spatial conflict of PLES scientifically. In this way, the PLES layout of the Yellow River Basin, the efficiency of PLES comprehensive utilization, and the PLES service function will be collectively improved, which eventually will assist in achieving the strategic goals of ecological protection and high-quality development in the Yellow River Basin.

5. Conclusions

Ecological protection and high-quality development of the Yellow River Basin is an essential national strategy in China. The spatial pattern of the Yellow River Basin is evolving from a dominant space of production to one with a coordinated pattern of development of production–living–ecological space. In this study, we simulated the pattern of PLES in the Yellow River Basin from 2010 to 2025. Based on the scientific understanding of the spatial pattern of the Yellow River Basin, we adopted the landscape ecology method to identify the spatial conflict of PLES in the Yellow River Basin and quantitatively analyze the severity of spatial conflict of PLES and its spatio-temporal pattern; thus, the following conclusions were obtained:

(1) As revealed by the spatio-temporal pattern of PLES in the Yellow River Basin in the past 15 years (2010–2025), the distribution of PLES in the Yellow River Basin had obvious spatially divergent characteristics. EPS has the highest percentage of area among all PLES types and is concentrated in most areas in the upper and middle reaches of the Yellow River, showing an inverted U-shaped changing trend in 2010–2025, while the total area remains stable in terms of fluctuation. PES is distributed in the Ningmeng Irrigation Area, Fenwei Plain, part of the Loess Plateau, and most of the lower reaches of the Yellow River, presenting a U-shaped trend from 2010 to 2025, with the total area remaining stable in terms of fluctuation. Due to the urbanization and industrialization of the Yellow River Basin, the LPS area increases yearly, mainly seen around the metropolitan areas of provincial capitals and secondary cities in nine provinces in the Yellow River Basin. In contrast, due to the industrial construction and ecological restoration projects, the ES area, which can be found in the upper and middle parts of the Yellow River Basin, indicates a decreasing trend year by year. In terms of PLES type conversion relationship, the conversion between EPS and PES is the most frequent, and the conversion area accounts for the highest percentage.

(2) During 2010–2025, the spatial conflict of PLES in the Yellow River Basin is mainly reflected in the encroachment of LPS on other PLES, seen mostly in the areas from Hekou Town to the left bank of Longmen, Fen River, Shizuishan to the southern bank of Hekou Town, and Daxia and Tao River. In addition, from the spatial conflict of the PLES area, the conflict between PES and EPS accounts for the largest area, which is concentrated in certain regions of the Loess Plateau and the region above Jing River Zhangjiashan. In terms of the degree of spatial conflict of PLES, from 2010 to 2025, the average *SCCI* of the Yellow River Basin lies within 0.7, meaning a basically controllable degree. From the development trend of spatial conflict of PLES, 15 of the 29 study regions have major decreasing trends of *SCCI*, 8 regions are in a state of fluctuation, and 6 regions show gradually increasing trends, accounting for 51%, 28%, and 21%, respectively.

(3) From the analysis of the attribution of spatial conflict of PLES, it was revealed that natural ecological conditions are the important foundation of PLES patterns, while human activities are the driving force guiding the evolution of PLES patterns, which accelerates the process of change in spatial patterns. In recent years, the implementation of major ecological protection actions in the Yellow River Basin, especially the large-scale project of returning farmland to forest and grass, the construction of nature reserve systems, and major ecological restoration projects, have played important roles in alleviating the spatial conflict of PLES. Thus, the area and severity of spatial conflict of PLES have been decreasing year by year. However, overexploitation of resources (including agricultural irrigation areas and energy bases) and disorderly construction due to urbanization remain the main causes of spatial conflict of PLES in the basin. Thus, a scientific, efficient, and reasonable pattern of land space development and utilization is key to optimizing PLES in the Yellow River Basin.

(4) This research applied the CA–Markov model and landscape ecology method to evaluate and analyze the evolution of spatial conflict of PLES and accurately identified the spatio-temporal pattern of spatial conflict of PLES in the Yellow River Basin. The study provides important theoretical references and decision-making principles for later analysis of PLES formation mechanisms and internal evolution mechanisms; research on the driving mechanism, formulation of measures and countermeasures to optimize the spatial development and protection pattern of the land; targeting natural resource management, spatial planning and use control of the land; and promoting ecological and environmental protection and high-quality economic and social development in the basin.

(5) This research emphasizes the relationship revealed in the human–land coupled system in the basin, which affects the layout of PLES. Based on the important role of water resources carrying capacity for ecological protection and high-quality socio-economic

development, a new perspective for basin PLES research was proposed, taking into account tertiary water resource divisions as the basic study unit. It is also worth noting that the *SCCI* of the basic study unit is a standardized relative value, so the different study scales will have a direct impact on the study results. At the same time, the differences in PLES classification and the use of different simulation model parameters can lead to some bias in the study results. This research analyzed the spatio-temporal pattern of PLES conflicts in the Yellow River Basin only from the perspective of spatial morphology. Spatial suitability was not considered; thus, it is impossible to dissect the main influencing factors of the conflicts. In addition, due to the limitation of space, in this paper, we did not discuss the optimal adjustment strategy to deal with the spatial conflict in PLES. These research directions and contents remain to be discussed in depth by subsequent scholars.

Author Contributions: F.X. and J.S. contributed to all aspects of this work. R.W., J.Z. and Y.Y. conducted data analysis and cartographic expression. F.X. and J.S. wrote the main manuscript text. N.W. and Z.W. gave some useful suggestions for this work. All authors have read and agreed to the published version of the manuscript.

Funding: This work was supported by the National Natural Science Foundation of China (Grant No. 41907402).

Institutional Review Board Statement: Not applicable.

Informed Consent Statement: Not applicable.

Data Availability Statement: The data presented in this study are available on request from the author.

Conflicts of Interest: The authors declare no conflict of interest.

References

1. Campbell, D.J.; Gichohi, H.; Mwangi, A.; Chege, L. Land use conflict in Kajiado District, Kenya. *Land Use Policy* **2020**, 17, 337–348. [CrossRef]
2. Aero, E.; Anderson, G.; Augustinus, C.; Aero, E.; Anderson, G.; Augustinus, C. *Land and Conflict: Lessons from the Field on Conflict Sensitive Land Governance and Peacebuilding*; United Nations Human Settlements Programme (UN-Habitat): Nairobi, Kenya, 2018. [CrossRef]
3. Berry, D.; Plaut, T. Retaining agricultural activities under urban pressures: A review of land use conflicts and policies. *Policy Sci.* **1978**, 9, 153–178. [CrossRef]
4. Zou, L.; Liu, Y.; Wang, J.; Yang, Y.; Wang, Y. Land use conflict identification and sustainable development scenario simulation on China's southeast coast. *J. Clean. Prod.* **2019**, 238, 117899. [CrossRef]
5. Ishiyama, N. Environmental justice and American Indian tribal sovereignty: Case study of a land–use conflict in Skull Valley, Utah. *Antipode* **2003**, 35, 119–139. [CrossRef]
6. Coyle, D.J. This land is your land, this land is my land: Cultural conflict in environmental and land-use regulation. In *Politics, Policy, and Culture*; Routledge: Milton Park, UK, 2019; pp. 33–50. [CrossRef]
7. Ayhan, Ç.K.; Taşlı, T.C.; Özkök, F.; Tatl, H. Land use suitability analysis of rural tourism activities: Yenice, Turkey. *Tour. Manag.* **2020**, 76, 103949. [CrossRef]
8. Willemen, L.; Hein, L.; van Mensvoort, M.E.; Verburg, P.H. Space for people, plants, and livestock? Quantifying interactions among multiple landscape functions in a Dutch rural region. *Ecol. Indic.* **2010**, 10, 62–73. [CrossRef]
9. Guo, B.; Wang, X.; Pei, L.; Su, Y.; Zhang, D.; Wang, Y. Identifying the spatiotemporal dynamic of PM2.5 concentrations at multiple scales using geographically and temporally weighted regression model across China during 2015–2018. *Sci. Total Environ.* **2021**, 751, 141765. [CrossRef]
10. Andrew, J.S. Potential application of mediation to land use conflicts in small-scale mining. *J. Clean. Prod.* **2003**, 11, 117–130. [CrossRef]
11. Wehrmann, B. *Land Conflicts: A Practical Guide to Dealing with Land Disputes*; Deutsche Gesellschaft für Technische Zusammenarbeit: Eschborn, Germany, 2008; pp. 12–13. [CrossRef]
12. Pavón, D.; Ventura, M.; Ribas, A. Land use change and socio-environmental conflict in the Alt Empordà county (Catalonia, Spain). *J. Arid Environ.* **2003**, 54, 543–552. [CrossRef]
13. De Groot, R. Function-analysis and valuation as a tool to assess land use conflicts in planning for sustainable, multi-functional landscapes. *Landsc. Urban Plan.* **2006**, 75, 175–186. [CrossRef]
14. Ioja, C.I.; Nita, M.R.; Vanau, G.O.; Onose, D.A.; Gavrilidis, A.A. Using multicriteria analysis for the identification of spatial land-use conflicts in the Bucharest Metropolitan Area. *Ecol. Ind.* **2014**, 42, 112–121. [CrossRef]

15. Kim, I.; Arnhold, S. Mapping environmental land use conflict potentials and ecosystem services in agricultural watersheds. *Sci. Total Environ.* **2018**, *630*, 827–838. [CrossRef] [PubMed]

16. Adam, Y.O.; Pretzsch, J.; Darr, D. Land use conflicts in central Sudan: Perception and local coping mechanisms. *Land Use Policy* **2015**, *42*, 1–6. [CrossRef]

17. Delgado-Matas, C.; Mola-Yudego, B.; Gritten, D.; Kiala-Kalusinga, D.; Pukkala, T. Land use evolution and management under recurrent conflict conditions: Umbundu agroforestry system in the Angolan Highlands. *Land Use Policy* **2015**, *42*, 460–470. [CrossRef]

18. Petrescu-Mag, R.M.; Petrescu, D.C.; Azadi, H.; Ioan, V. Agricultural land use conflict management—vulnerabilities, law restrictions and negotiation frames. A wake-up call. *Land Use Policy* **2018**, *76*, 600–610. [CrossRef]

19. Wu, Z.Y. Space optimization of "Production-Living-Ecological lives" and ecological environment protection of Beijing, Tianjin and Hebei. *City* **2014**, *12*, 26–29. [CrossRef]

20. Lin, G.; Jiang, D.; Fu, J.Y. Discussion on scientific foundation and approach for the overall optimization of "Production-Living-Ecological" space. *J. Nat. Resour.* **2021**, *36*, 1085–1101.

21. He, Y.H.; Tang, C.L.; Zhou, G.H.; He, S.; Qiu, Y.H.; Shi, L.; Zhang, H.Z. The Analysis of Spatial Conflict Measurement in Fast Urbanization Region from the Perspective of Geography—A Case Study of Changsha-Zhuzhou-Xiangtan Urban Agglomeration. *J. Nat. Resour.* **2014**, *29*, 1660–1674. [CrossRef]

22. Carpenter, S.L.; Kennedy, W.J.D. Environmental conflict management: New ways to solve problems. *Mt. Res. Dev.* **1981**, *1*, 65–70. [CrossRef]

23. Cabrera-Barona, P.; Murphy, T.; Kienberger, S.; Blaschke, T. A multi-criteria spatial deprivation index to support health inequality analyses. *Int. J. Health Geogr.* **2015**, *14*, 11. [CrossRef]

24. Ouyang, W.; Wang, B.; Tian, L.; Niu, X. Spatial deprivation of urban public services in migrant enclaves under the context of a rapidly urbanizing China: An evaluation based on suburban Shanghai. *Cities* **2017**, *60*, 436–445. [CrossRef]

25. Fang, C.L.; Liu, H.Y. The Spatial Privation and the Corresponding Controlling Paths in China's Urbanization Process. *Acta Geogr. Sin.* **2007**, *62*, 849–860. [CrossRef]

26. Debolini, M.; Valette, E.; Francois, M.; Chery, J.P. Mapping land use competition in the rural-urban fringe and future perspectives on land policies: A case study of Meknès (Morocco). *Land Use Policy* **2015**, *47*, 373–381. [CrossRef]

27. Moein, M.; Asgarian, A.; Sakieh, Y.; Soffianian, A. Scenario-based analysis of land-use competition in central Iran: Finding the trade-off between urban growth patterns and agricultural productivity. *Sustain. Cities Soc.* **2018**, *39*, 557–567. [CrossRef]

28. Xu, C.X.; Wang, F.Y.; Wang, K.Y.; Li, P. Exploring destination spatial competition rules: A case study of Hunan province. *Geogr. Res.* **2017**, *36*, 321–335. [CrossRef]

29. Zhang, L.; Wan, R.R.; Hu, H.B.; Dong, Y.W. Environmental function and spatial integration of ecological land—A case study of Nanjing. *Resour. Environ. Yangtze Basin* **2011**, *20*, 1222–1227.

30. Wagner, P.D.; Bhallamudi, S.M.; Narasimhan, B.; Kantakumar, L.N.; Sudheer, K.P.; Kumar, S.; Schneider, K.; Fiener, P. Dynamic integration of land use changes in a hydrologic assessment of a rapidly developing Indian catchment. *Sci. Total Environ.* **2016**, *539*, 153–164. [CrossRef]

31. Tu, S.S.; Long, H.L.; Zhang, Y.N.; Ge, D.Z.; Qu, Y. Rural restructuring at village level under rapid urbanization in metropolitan suburbs of China and its implications for innovations in land use policy. *Habitat Int.* **2018**, *77*, 143–152. [CrossRef]

32. Richardson, H.W. Optimality in city size, systems of cities and urban policy: A skeptic's view. *Urban Stud.* **1972**, *1*, 29–48. [CrossRef]

33. Mubarak, F.A. Urban growth boundary policy and residential suburbanization: Riyadh, Saudi Arabia. *Habitat Int.* **2004**, *28*, 567–591. [CrossRef]

34. Zhang, J.X. Urban and region governance research and application in China. *Urban Probl.* **2000**, *6*, 40–44.

35. Gu, C.L.; Shen, J.F.; Yao, X. *Urban Governance: Concept, Theory, Method and Empirical Research*; Southeast University Press: Nanjing, China, 2003.

36. Peng, J.J.; Zhou, G.H.; Tang, C.L.; He, Y.H. The Analysis of Spatial Conflict Measurement in Fast Urbanization Region Based on Ecological Security—A Case Study of Changsha-Zhuzhou-Xiangtan Urban Agglomeration. *J. Nat. Resour.* **2012**, *27*, 1507–1519. [CrossRef]

37. Fan, J. Social demands and new propositions of economic geography discipline development based on the eleventh national five-year plan. *Econ. Geogr.* **2006**, *26*, 545–550. [CrossRef]

38. Ruan, S.T.; Wu, K.N. Research of the Land Use Conflict and Mitigation Mechanism during the Urbanization in China. *China Popul. Resour. Environ.* **2013**, *23*, 388–392.

39. Jaiswal, D.; De Souza, A.P.; Larsen, S. Brazilian sugarcane ethanol as an expandable green alternative to crude oil use. *Nat. Clim. Chang.* **2017**, *7*, 788–792. [CrossRef]

40. Xie, Y.; Chen, Y.; Han, F.; Fang, J.Y. Research on Influence and Time-space Effect of New-type Urbanization on Urban Eco-environmental Quality. *Manag. Rev.* **2018**, *30*, 230–241.

41. Qi, X.; Wang, Y.L. An Empirical Study of the Spatial Spillover Effects of the Development of Urbanization Economy from the Perspective of Cities, Markets and Urbanization. *J. Financ. Econ.* **2013**, *39*, 84–92.

42. Zhou, K.; Wang, Q.; Fan, J. Impact of economic agglomeration on regional water pollutant emissions and its spillover effects. *J. Nat. Resour.* **2019**, *34*, 1483–1495. [CrossRef]

43. Wang, Z.B.; Liang, L.W.; Fang, C.L.; Zhuang, R.L. Study of the evolution and factors influencing ecological security of the Beijing-Tianjin-Hebei Urban Agglomeration. *Acta Ecol. Sin.* **2018**, *38*, 4132–4144. [CrossRef]
44. Liao, L.H.; Dai, W.Y.; Chen, J.; Huang, W.L.; Jiang, F.Q.; Hu, Q.F. Spatial conflict between ecological-production-living spac-es on Pingtan Island during rapid urbanization. *Resour. Sci.* **2017**, *39*, 1823–1833. [CrossRef]
45. Song, Z.J.; Li, Z.; Yang, J. Study on the calculation of land use conflict intensity in mineral grain complex area—Taking the polluted areas of Dexing Copper Mine and Yongping Copper Mine in Jiangxi Province as an example. *Chin. J. Agric. Resour. Reg. Plan.* **2018**, *39*, 78–85. [CrossRef]
46. Ma, X.G.; Wang, A.M.; Yan, X.P. A Study on land use conflict in the process of urban spatial reconstruction—Taking Guangzhou city as an example. *Hum. Geogr.* **2010**, *25*, 72–77.
47. Kalabamu, F.T. Land tenure reforms and persistence of land conflicts in Sub-Saharan Africa-the case of Botswana. *Land Use Policy* **2019**, *81*, 337–345. [CrossRef]
48. Yang, Y.F.; Zhu, L.Q. The Theory and Diagnostic Methods of Land Use Conflicts. *Resour. Sci.* **2012**, *34*, 1134–1141.
49. Qin, K. The Analysis of Land Use Spatial Conflicts Based on Ecological Security—A Case Study of the Urban Agglomeration Around. Master's Thesis, Wuhan University, Wuhan, China, 2017.
50. Yao, J.; Chen, J.L.; Yao, S.M. Research on Coordination of Spatial Plans Based on New Regionalism: A Case Study of the Coastal Region in Jiangsu Province. *China Soft Sci.* **2011**, *7*, 102–110. [CrossRef]
51. Wang, H.Y.; Qin, F.; Zhang, X.C. The Scenario Analysis on Urban Ecological Land Spatial Conflict and Ecological Security Hidden Danger in Guangzhou. *J. Nat. Resour.* **2015**, *30*, 1304–1318. [CrossRef]
52. Min, J.; Wang, Y.; Bai, R.Y.; Zhang, J.; Kong, X.Y. The Research on Identification of Land Use Potential Conflict Based on Multiobjective Suitability Evaluation on Mountainous City: A Case Study of Qijiang District of Chongqing Municipality. *J. Chongqing Norm. Univ.* **2018**, *35*, 82–89. [CrossRef]
53. Liu, Q.Q.; Zhao, H.F.; Wu, K.N.; Yu, X.; Zhang, Q.J. Identifying Potential Land use Conflict Based on Competitiveness of Different Land Use Types in Beijing, China. *Resour. Sci.* **2014**, *36*, 1579–1589.
54. Wang, A.M.; Ma, X.G.; Yan, X.P. Land Use Conflicts and Their Governance Mechanics on Actors Network Theory: A Case of Fruit Tree Protection Zone of Haizhu District, Guangzhou City. *Sci. Geogr. Sin.* **2010**, *30*, 80–85.
55. Feng, Y.; Bi, R.T.; Wang, J.; Lv, C.J.; Zhang, Q.Q. Spatial Conflict of Land Use Caused by Mining Exploitation and Optimal Allocation Scheme of Land Resources in River Basin. *China Land Sci.* **2016**, *30*, 32–40. [CrossRef]
56. Huang, J.C.; Lin, H.X.; Qi, X.X. A literature review on optimization of spatial development pattern based on Ecological-Production-Living space. *Prog. Geogr.* **2017**, *36*, 378–391. [CrossRef]
57. Zhao, X.; Tang, F.; Zhang, P.T.; Hu, B.Y.; Xu, L. Dynamic simulation and characteristic analysis of county production-living-ecological spatial conflicts based on CLUE-S model. *Acta Ecol. Sin.* **2019**, *39*, 5897–5908.
58. Cohen-Shacham, E.; Walters, G.; Janzen, C. Nature-based solutions to address global societal challenges. *IUCN Gland Switz.* **2016**, *97*, 2016–2036. [CrossRef]
59. Liu, J.L.; Liu, Y.S.; Li, Y.R. Classification evaluation and spatial-temporal analysis of "production-living-ecological" spaces in China. *Acta Geogr. Sin.* **2017**, *72*, 1290–1304. [CrossRef]
60. Zhang, H.; Xu, E.; Zhu, H. Ecological-living-productive land classification system in China. *J. Resour. Ecol.* **2017**, *8*, 121–128. [CrossRef]
61. Liao, G.; He, P.; Gao, X.; Deng, L.; Zhang, H.; Feng, N.; Deng, O. The Production–Living–Ecological Land Classification System and Its Characteristics in the Hilly Area of Sichuan Province, Southwest China Based on Identification of the Main Functions. *Sustainability* **2019**, *11*, 1600. [CrossRef]
62. Zhou, G.H.; Peng, J.J. The Evolution Characteristics and Influence Effect of Spatial Conflict. A Case Study of Changsha-Zhuzhou-Miangtan Urban Agglomeration. *Prog. Geogr.* **2012**, *31*, 717–723.
63. Zhou, D.; Xu, J.C.; Wang, L. Process of Land Use Conflict Research in China during the Past Fifteen Years. *China Land Sci.* **2015**, *29*, 21–29. [CrossRef]
64. Sang, L.; Zhang, C.; Yang, J.; Zhu, D.; Yu, W. Simulation of land use spatial pattern of towns and villages based on CA–Markov model. *Math. Comput. Model.* **2011**, *54*, 938–943. [CrossRef]
65. Lu, C.P.; Ji, W.; Liu, Z.L.; Mao, J.H.; Li, J.Z.; Li, J.Z.; Xue, B. The spatial and temporal pattern of functional space in Gansu section of the Yellow River Basin. *Sci. Geogr. Sin.* **2022**, *44*, 1–10.
66. Wang, X.F.; Zhang, X.; Wang, Y.; Yan, Y. The spatial evolution and driving force analysis of the Loess Plateau. *J. Anhui Agric. Univ.* **2022**, *49*, 112–121. [CrossRef]
67. Li, J.S.; Sun, W.; Yu, J.H. The Evolution and regional difference of Sansheng Space in the Yellow River Basin- based on the comparison of resource-based and non-resource-based cities. *Resour. Sci.* **2020**, *42*, 2285–2299.
68. Wang, T.; Kazak, J.; Han, Q. A framework for path-dependent industrial land transition analysis using vector data. *Eur. Plan. Stud.* **2019**, *27*, 1391–1412. [CrossRef]
69. Kumar, P.; Dobriyal, M.; Kale, A. Temporal dynamics change of land use/land cover in Jhansi district of Uttar Pradesh over past 20 years using Landsat TM, ETM+ and OLI sensors. *Remote Sens. Appl. Soc. Environ.* **2021**, *23*, 100579. [CrossRef]
70. Wei, W.; Yin, L.; Xie, B.; Bo, L.M. The evolution characteristics and mechanism of "three zones" in the Yellow River Basin under the background of territorial spatial planning. *Econ. Geogr.* **2022**, *42*, 44–55+86. [CrossRef]

71. Chang, T.Y.; Zhang, Z.W.; Qiao, X.N.; Zhang, Y.F. Land use transformation and its ecological environment effect from 2000 to 2020 in the Yellow River Basin. *Bull. Soil Water Conserv.* **2021**, *41*, 268–275. [CrossRef]

72. Zhang, Y.Z.; Chen, Y.; Wang, J.; Ye, J.P.; Zhang, B.B. Functional coordination measure of "three students" in the Yellow River Basin. *J. Agric. Eng.* **2021**, *37*, 251–261+321.

73. Chen, H.; Fleskens, L.; Baartman, J.; Wang, F.; Moolenaar, S.; Ritsema, C. Impacts of land use change and climatic effects on streamflow in the chinese loess plateau: A meta-analysis. *Sci. Total Environ.* **2020**, *703*, 134989. [CrossRef]

74. Feng, X.; Fu, B.; Piao, S. Revegetation in China's Loess Plateau is approaching sustainable water resource limits. *Nat. Clim. Chang.* **2016**, *6*, 1019–1022. [CrossRef]

Article

Land-Use Classifying and Identification of the Production-Living-Ecological Space of Island Villages—A Case Study of Islands in the Western Sea Area of Guangdong Province

Rui Bai [1,2], Ying Shi [1,3,†] and Ying Pan [1,3,*,†]

[1]   School of Architecture, South China University of Technology, Guangzhou 510641, China;
       201910101105@mail.scut.edu.cn (R.B.); aryshi@scut.edu.cn (Y.S.)
[2]   Municipal Key Laboratory of Landscape Architecture, Guangzhou 510641, China
[3]   The Key Laboratory for Subtropical Architecture Science of the Ministry of Education Research Center,
       Guangzhou 510641, China
*   Correspondence: panying@scut.edu.cn; Tel.: +86-13380056897
†   These authors contribute equally to this work.

Abstract: Accurately identifying the rural production-living-ecological space (PLES) of different islands can help reveal their distinct natural resources and land-use situations, which is significant for the sorted management, subarea utilization, and protection of islands. At present, studies on the PLES of island villages are deficient. For instance, the existing land-use classification system is incomplete; the PLES is poorly identified; and the dominant function of multiple land-use types based on different island geomorphology types is insufficiently investigated. Therefore, a case study was conducted on the island villages of the western sea area of Guangdong Province, based on remote sensing, spatial analysis, and land classification, with field research and the relevant data. In this study, before establishing the PLES system, the islands were classified, including six bedrock islands, 10 sedimentary islands, and one volcanic island. When the PLES system of the island villages was classified, the ecological and utilized areas of the intertidal zone and neritic region should be combined with the island–continent part, and the distinct industrial types should be emphasized, before forming 22 secondary types of PLES. Furthermore, it is found that each island generally has its own dominant space and land-use type. Ecological space (ES) dominates the bedrock islands, and production space (PS) is prominent for sedimentary islands and volcanic islands. Forestland, aquaculture pond, and dryland are the prominent land-use types for bedrock islands, sedimentary islands, and volcanic islands, respectively. The rural residential lands are the main component of living space (LS) in all islands, and the most urban residential lands are distributed on the bedrock islands. The main driving factors for the formation and distribution of island rural PLES are the altitudinal gradient and geomorphic characteristics. The research shows that the main problems of PLES are that the intertidal zones are threatened by aquaculture ponds at various levels, and the development of LS in these islands is generally backward.

Keywords: production-living-ecological space; identification; island exploitation; perspective of geomorphology

Citation: Bai, R.; Shi, Y.; Pan, Y. Land-Use Classifying and Identification of the Production-Living-Ecological Space of Island Villages—A Case Study of Islands in the Western Sea Area of Guangdong Province. *Land* **2022**, *11*, 705. https://doi.org/10.3390/land11050705

Academic Editor: Dong Jiang

Received: 6 April 2022
Accepted: 6 May 2022
Published: 8 May 2022

Publisher's Note: MDPI stays neutral with regard to jurisdictional claims in published maps and institutional affiliations.

1. Introduction

The dense population and rapid economic development in coastal zones in China cause frequent changes in land-use structures of offshore islands and intensify the conflicts among the production space, living space, and ecological space [1,2]. In order to achieve sustainable development of geographical space, in 2012, the Chinese government proposed the principle of the national territorial space development: optimizing the allocation of natural and socioeconomic resources for high-efficiency and intensive production, comfortable living, and eco-friendly territorial space [3–6]. With the discussion and popularization of the concept of production-living-ecological space (PLES) [7,8], the identification of PLES [9,10], the evolution of PLES [11,12], and PLES functionality [13,14] have become research hotspots in the fields of landscape planning and urban planning.

Current studies on the identification of PLES mainly focus on the national land space [15], provincial spaces [16,17], and urban spaces [18,19]. However, PLES identification was rarely conducted in rural spaces, especially in island villages. With the increasing anthropogenic activities, the ecology of the islands has been experiencing critical deterioration. Moreover, the island ecosystem, characterized by a small area, low ecological diversity, and low self-regulating and restoring capacity, is fragile and challenging to repair [20–23]. Therefore, island land planning and management should be carried out from an integrative research perspective. PLES covers the ecological, production, and living spaces, and the classification of PLES is a comprehensive land spatial zoning [24]. PLES identification on island villages could help to find out the conflict and contradiction among the production space (PS), living space (LS), and ecological space (ES), with achieving co-ordinated developments.

Generally, the identification and division of rural PLES can be achieved by two mainstream methods: (1) the evaluation index system (EIS); (2) land-use classification (LUC) [25–27]. The method of EIS is usually used to divide the PLES by evaluating the function and suitability of each administrative unit [28,29]. However, it cannot be used to identify the PLES within the village because of the difficulty in obtaining the socioeconomic statistics at the village level. The method of LUC can identify rural PLES by classifying land-use types with the same dominant function and can be used to identify the multispatial characteristics of rural PLES within the village [30]. For example, Duan et al. [24] identified the multispatial rural PLES in Ertai Town of Zhangjiakou from the perspective of the villagers' behavior.

The existing LUC-related studies on PLES mainly used Chinese "Land-use status classification" (GB/T21010-2017) for spatial identification [31]. Nevertheless, the surface features of island areas cannot be highlighted by using this taxonomy. In this case, scholars who participated in the Special Project for Comprehensive Investigation and Evaluation of China's Offshore Ocean (908 Special Project) formulated an LUC system of islands [2], which can emphasize the surface features of the island–continent part. However, each island contains the island–continent part, island intertidal zone, and neritic region [32]. Thus, it is essential to combine the ecology and the utilized spaces of the intertidal zone and neritic region while classifying the land-use of islands. Furthermore, rural spaces in different islands have distinct resource combinations and utilization characteristics. Consequently, in order to find out the advantages and limitations of resources and utilization in various islands, island classification is necessary.

With the increasing exploitation, the island villages in the western sea area of Guangdong Province are commonly confronted with the following issues: (1) development and utilization behaviors such as reclamation and arbitrary quarrying which changed the topography and geomorphology of some islands; (2) severe destruction of the ecological environment; and (3) extensive utilization of natural resources. With a narrow economic base, restricting the protection and management of the islands. Previous studies in this area mainly focused on qualitative planning strategies, the islands' intertidal zones and neritic regions were rarely considered in the LUC-related studies, and research from a geomorphological perspective is scarce. Hence, the present study aims to fill these knowledge gaps using a remote sensing-based approach. The specific objectives of this study are to: (1) establish a PLES system for the island villages; (2) map the spatial distribution of secondary types of PLES; (3) explain the formation mechanism from the geographical perspective; and (4) figure out the conflict among the PLES and propose space-optimization strategies.

2. Materials and Methods

2.1. Study Area

The study area covers 17 populated offshore islands (within approximately 30 nautical miles) in the western sea area of Guangdong Province (Figure 1). The largest island in the study area is Donghai Island, with an area of 286 km^2, and the smallest island is Gonggang Island, with a mere area of 1.26 km^2, and all the islands are located at a distance of about 0.4 km~18 km from the coastline. The study area belongs to the subtropical climate. Compared with the coastwise mainland, the islands have an oceanic climate, with an annual temperature of 1~1.5 °C lower than that of the adjacent mainland; the number of days with at least Level-8 gale on annual average is 3~8 times that of the adjacent mainland; the annual amount of evaporation is 20~200 mm higher than that of the adjacent mainland [33].

Figure 1. Distribution map of island in the western sea area of Guangdong Province.

2.2. Data Processing and Classification Method

This study uses satellite images and auxiliary data, and employs a hybrid classification approach for image classification. The auxiliary data includes those from field investigation, yearbooks, local chronicles, Google images, and the land use/land cover maps of the Resource and Environment Science and Data Center (https://www.resdc.cn (accessed on 13 December 2021)). The auxiliary data were mainly used to identify training samples for image classification, testing samples for accuracy evaluation, and visually modified classified images. A hybrid classification method, which was developed for image classification, was found to be an effective approach to enhancing the accuracy of image classification [34]. This method includes automatic classification (random forest) and visual modification. The classification process is shown in Figure 2. See the following steps for details.

2.2.1. Data Processing and Classification Method

This study adopts the physical boundary of islands as the division, including the ecological and utilized areas of the islands' intertidal zones and neritic regions. The Geospatial Data Cloud (http://www.gscloud.cn/ (accessed on 15 December 2021)) server was used to download Landsat 8 Operational Land Imager (OLI), and Thermal Infrared Sensor (TIRS) database was used for the images of the 2014s and 2015s. All images used in this study are from the dry season period (November–January), during which zero cloud cover allowed for high image quality. The raw images have a spatial resolution of 30 m.

Preprocessing is an essential step to correct atmospheric effects and minimize geometric and radiometric errors before image classification. This study used the ENVI software to undertake the radiometric calibration and atmospheric correction of all the images. A fusion or pan-sharpened multispectral (MS) image provides an improved image of high spatial resolution and can also improve the classification results [35,36]. Therefore, the panchromatic (PAN) band with a resolution of 15 m and MS images were adopted to be fused by the Gram–Schmidt (GS) method in this study. Subsequently, four pan-sharpened MS images with a spatial resolution of 15 m of the studied area were generated.

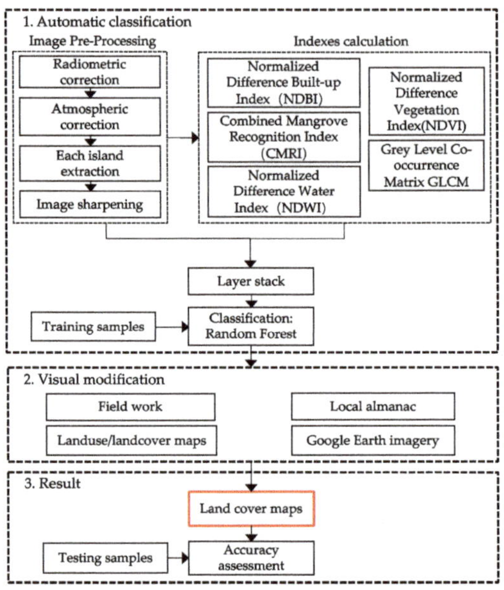

Figure 2. Flow chart of island land-use classification in the study area.

2.2.2. Island Classification

Referring to the 908 Special Project and "Law of the People's Republic of China on the Protection of Islands", according to their results of the classification of islands by geological composition, the islands were classified into three types: bedrock islands, volcanic islands, and sedimentary islands (Table 1). Furthermore, while classified, two or more islands connected via aquaculture pond(s) were deemed as one island.

Table 1. Island classification and image characteristics (Data resources: *Google Earth*).

Primary Classification	Secondary Classification	Image Characteristics	Descriptions
Island	Bedrock island		The island consists of bedrocks.
	Volcanic island (High Island)		The island consists of volcanic ejecta (lava and volcanic ash).
	Sedimentary island		The island consists of incompact substances (mud and sand).

2.2.3. Island PLES System Construction

The identification of a rural place or village varies in different countries. The term in this paper is identified as a region that is outside the urban built-up area and is a system of the territorial complex with natural, social, and economic functions [37,38].

Based on the island land-use classification formulated by the 908 Special Project (Table 2), the primary industries were emphasized and the ecological and utilized area of the intertidal zones and neritic regions were supplemented. Meanwhile, to allow for the resolution of remote sensing (RS) images, some microscale land-use types that were difficult to acquire from the images were merged and adjusted. The islands' PLES-classification-system fit for the studied area was formulated afterward (Table 3).

Table 2. Island land-use classification of study area.

First-Level Land-Use Type	Second-Level Land-Use Type
Arable land	Paddy field
	Dryland
Garden	Orchard
	Tea garden
Forestland	Forestland
	Mangrove forest
Industrial and mining storage land	Industrial land
	Salt pan
Residential land	-
Transportation land	Highway land
	Rural road
Water area and water conservancy facilities land	Harbor land
	Reservoir
	Pond
	Ditch
Other lands	Aquaculture pond
	Mudflat aquaculture
	Sandy land
	Bare land

The production space (PS) refers to the space dominated by the production functions. The PS includes paddy field, dryland, orchard land, and aquaculture pond, which supports agriculture and marine production. The reservoirs and ponds of the islands in the studied area mainly serve agricultural production and hence are classified as PS.

The living space (LS) refers to the space guaranteeing the survival of humankind and carrying human culture. Many time-honored villages on the islands in the studied area have historical, cultural, scientific, artistic, social, and economic value [39]. Thus, in this classification, the rural residential land was separated from the urban residential land by considering the further assessment and preservation of the traditional villages, and was classified into LS alongside land for the harbor and wharf.

The ecological space (ES) refers to the natural space which can maintain regional ecological security [40]. The mangrove forest is a critical ecological space in the study area, listed in the list of Ramsar Convention wetlands of international importance. Besides, wetlands, sandy beaches, and mudflats in the intertidal zones were highlighted in the PLES classification system. Due to the weak capacity of the islands to withstand natural disasters, this classification also stresses the shelterbelts, which defend against natural disasters, protect production, and maintain ecological balance, so as to highlight the particular ecosystem of the subtropical islands.

Table 3. PLES classification and image features (Data resources: *Google Earth*).

PLES	Secondary Types of PLES	Interpretation Signal	Description
Production space	Paddy field		Cultivated land for aquatic crops such as rice. Including the areas that rotated by aquatic and xerophytic crop.
	Dryland		Cultivated land for xerophilous crops with no irrigation facilities, and the cultivation mainly relies on natural precipitation.
	Tea garden		Site for tea production.
	Salt pan		Salt production sites mainly use the evaporation method, including mixed land-use by fish-farming and salt production.
	Aquaculture pond		The site for aquaculture includes its ancillary facilities, which are located above the shoreline or in the intertidal zones.
	Industrial land		Site for industrial production.
	Reservoir		Site for water storage by artificial intercept (area $> 1 \times 10^4$ m^2), including its ancillary facilities.
	Pond		Site ponding area or bottomland (area $< 1 \times 10^4$ m^2), the aquaculture pond is not included.
	Ditch		Site for the channels (width ≥ 1 m) used for drainage and irrigation, including ditches, embankments, and surrounding shelterbelt.

Table 3. *Cont.*

PLES	Secondary Types of PLES	Interpretation Signal	Description
Ecological space	Forestland		Site for the forest with canopy density $\geq$ 0.2.
	Stream		A stream is formed naturally or excavated artificially.
	Shelterbelt		Site for shelterbelt that is distributed on the coastal zone.
	Mangrove forest land		Site for the semimangrove or mangrove forest.
	Wetland		Site for the herb growth in the inland swamp.
	Beach		The dry shoal consists of gravel or sand.
	Rock foreshore		The dry shoal consists of rock.
	Mudflat		The dry shoal consists of sand or mudflat.

Table 3. *Cont.*

PLES	Secondary Types of PLES	Interpretation Signal	Description
Living space	Rural residential land		Villages' house.
	Urban residential land		Urban residential community.
	Harbor land		Site for berthing ships and storage of goods, including the ancillary buildings.
	Highway land		Site for national roads, provincial roads, county roads, and township roads.
	Bare land		Areas with no dominant vegetation cover on at least 90% of the area or areas covered by lichens/moss.

2.2.4. Imagery Feature Extraction

The construction of the categorical dataset requires the spectral signature, index feature calculation, gray-level co-occurrence matrix extraction, and feature fusion [41]. The spectral signature is the foremost feature in remote-sensing image classification [42], while the index feature can effectively enhance image classification accuracy. Index features include Normalized Difference Vegetation Index (NDVI) [43], Normalized Difference Water Index (NDWI) [44], and Normalized Difference Built-up Index (NDBI) [45,46]. Besides, this study adopts the Combined Mangrove Recognition Index (CMRI), which can enhance the discrimination between mangrove forests and nonmangrove vegetations [47] (Table 4).

The existing studies suggest that the gray-level co-occurrence matrix (GLCM) can effectively enhance the classification accuracy of various land-use types and diminish the classification errors due to similar spectral signatures [48]. Combined with the test, the study sets the window size of statistical pixels as 9×9 while extracting textures and selects the value of grayscale quantization level as 32 to calculate six textural features of images using the GLCM (Table 4).

After the calculations above, this step adopted the integrating the multifeatures method [49], including spectral features, texture features (GLCM), and index features. Each feature obtained in this study corresponds to a layer, and the method of layer-overlay is used for feature fusion (Figure 3). The integrating multifeatures method can significantly enrich the information content of remote sensing data, and one of the most commonly used methods to quantify the importance of features is decision trees.

Table 4. The characteristic attributes involved in classification.

Index Features	Normalized Difference Vegetation Index (NDVI)	$NDVI = \frac{NIR-Red}{NIR+Red}$
	Normalized Difference Water Index (NDWI)	$NDWI = \frac{Green-NIR}{Green+NIR}$
	Normalized Difference Built-up Index (NDBI)	$NDBI = \frac{MIR-NIR}{MIR+NIR}$
	Combined Mangrove Recognition Index (CMRI)	$CMRI = NDVI - NDWI$
Gray-Level Co-Occurrence Matrix (GLCM)	GLCM Homogeneity	$\sum_{i=1}^{k}\sum_{j=1}^{k}\frac{P(i,j)}{1+(i+j)^2}$
	GLCM Contrast	$\sum_{n=0}^{k-1}n^2\sum_{|i-j|=1}P(i,j)$
	GLCM Dissimilarity	$\sum_{i=1}^{k}\sum_{j=1}^{k}|i-j|P(i,j)$
	GLCM Entropy	$-\sum_{i=1}^{k}\sum_{j=1}^{k}P(i,j)\log[P(i,j)]$
	GLCM Mean	$\sum_{i=1}^{k}\sum_{j=1}^{k}\frac{\sum_{i=1}^{k}\sum_{j=1}^{k}P(i,j)}{n^2}$
	GLCM Std. Dev.	$\sum_{i=1}^{k}\sum_{j=1}^{k}[P(i,j)-\mu_{n*n}]^2$

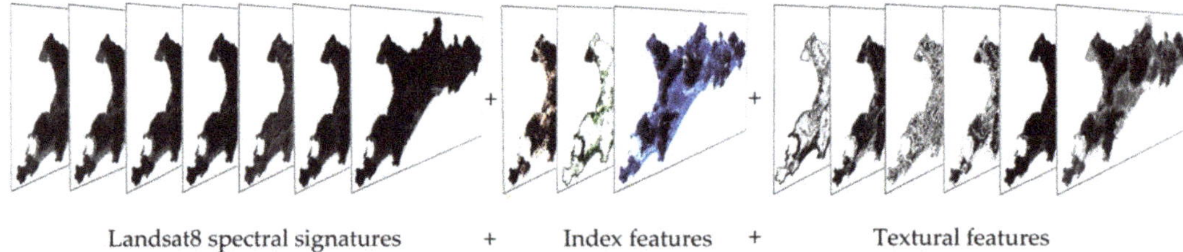

Figure 3. Multifeature fusion diagram (take Hailing Island as an example).

2.2.5. Random Forest Classification

Random forest (RF) is an integrated learning technology that can generate a large number of decision trees for the training, calculation, and classification of samples. The bagging method is adopted in RF to generate independent, identically distributed training sample sets for each decision tree, and the final classification result of the RF depends on the voting of all decision trees.

The RF algorithm is amplified as follows: (1) obtain N (N is a random positive number) training sample sets from a large number of original samples, by drawing with replacement N times; (2) select m (m is a random positive number) classification features randomly from the total features in each sample set; (3) divide the nodes of the decision trees by complete segmentation methodology, and then build a great number of decision trees. After completing the classification for each decision tree, the classes of new samples are determined by a majority vote according to the classification results of the decision trees [50].

The main idea of using the random forest to measure the importance of features is to evaluate the contribution of each feature in each decision tree, and to calculate the average values. Subsequently, the contribution value of features can be compared.

2.2.6. Visual Modification

This study involves various classification types, with the existence of the phenomenon of "different objects with the same spectrum". This leads to unavoidable ineffectiveness when distinguishing surface features. Therefore, in combination with the auxiliary data, the classification maps generated by RF were visually modified to enhance the classification accuracies.

3. Results

3.1. Classification Accuracy Evaluation

The accuracy assessment gave overall accuracies of 87.06~97.38% for the images of all the islands. Meanwhile, the kappa coefficient was 0.84~0.96. Images of larger islands coincide with higher accuracy, whereas images of smaller islands coincide with lower accuracy. The user and producer accuracies of the various thematic classes were over 81.3%, except for ponds (70.3%), reservoirs (72.7%), and bare lands (71.2%).

3.2. Types and Quantity of PLES Classification of the Islands in the Western Sea Area of Guangdong Province

Through the result of PLES classification, the number of PLES types on a single island range between 6 and 19. The PS types divided from the Landsat images include paddy field, dryland, salt pan, aquaculture pond, industrial land, tea garden, reservoir, pond, and ditch. The ES types include forestland, shelterbelt, mangrove forest land, wetland, beach, rock foreshore, mudflat, and stream. The LS types include rural residential land, urban residential land, harbor land, bare land, and highway land. Some secondary types of PLES are distributed only on certain islands. For example, in the PS, the industrial land and salt pan only appear on the sedimentary islands, a vast majority of irrigation ditches are distributed on the bedrock islands, and tea gardens are distributed only in a small area on the bedrock islands. Most mangrove forests and mudflats in the ES are distributed on the sedimentary islands. Only a few streams are distributed on the bedrock islands, while rock foreshores are mainly distributed in the relatively large area on the volcanic islands (Figures 4–6).

3.3. PLES Distribution Characteristics of the Three Types of Islands in the Western Sea Area of Guangdong Province

Overall, the areas of LS are less than both PS and ES in all types of islands. Rural residential lands are the main component of LS and are distributed dispersedly, while the urban residential land is mainly distributed on the bedrock islands. The aquaculture pond is the dominant land-use type in PS of the intertidal zone in the study area, with various proportion levels.

The bedrock islands are dominated by ES, with an area of 264.24 km^2, or 70.35% of the area of bedrock islands. The ES is mainly composed of forestland, which takes 48.0%~89.4% over the island area, and is concentrated distributed on the island–continent part. The ES of the intertidal zone is made up of rock foreshore, with a small area. The PS occupies a total area of 54.95 km^2, representing 14.63% of the bedrock island. The arable land of the island–continent part in bedrock islands takes up a large proportion, from 1.2% to 16.1%, except for Fengtou Island. The LS area of bedrock islands is 22.52 km^2, accounting for only 6.0% of the bedrock island's total area, and the quantity of the harbor land is more than other islands, as shown in Figure 7 and Table 5.

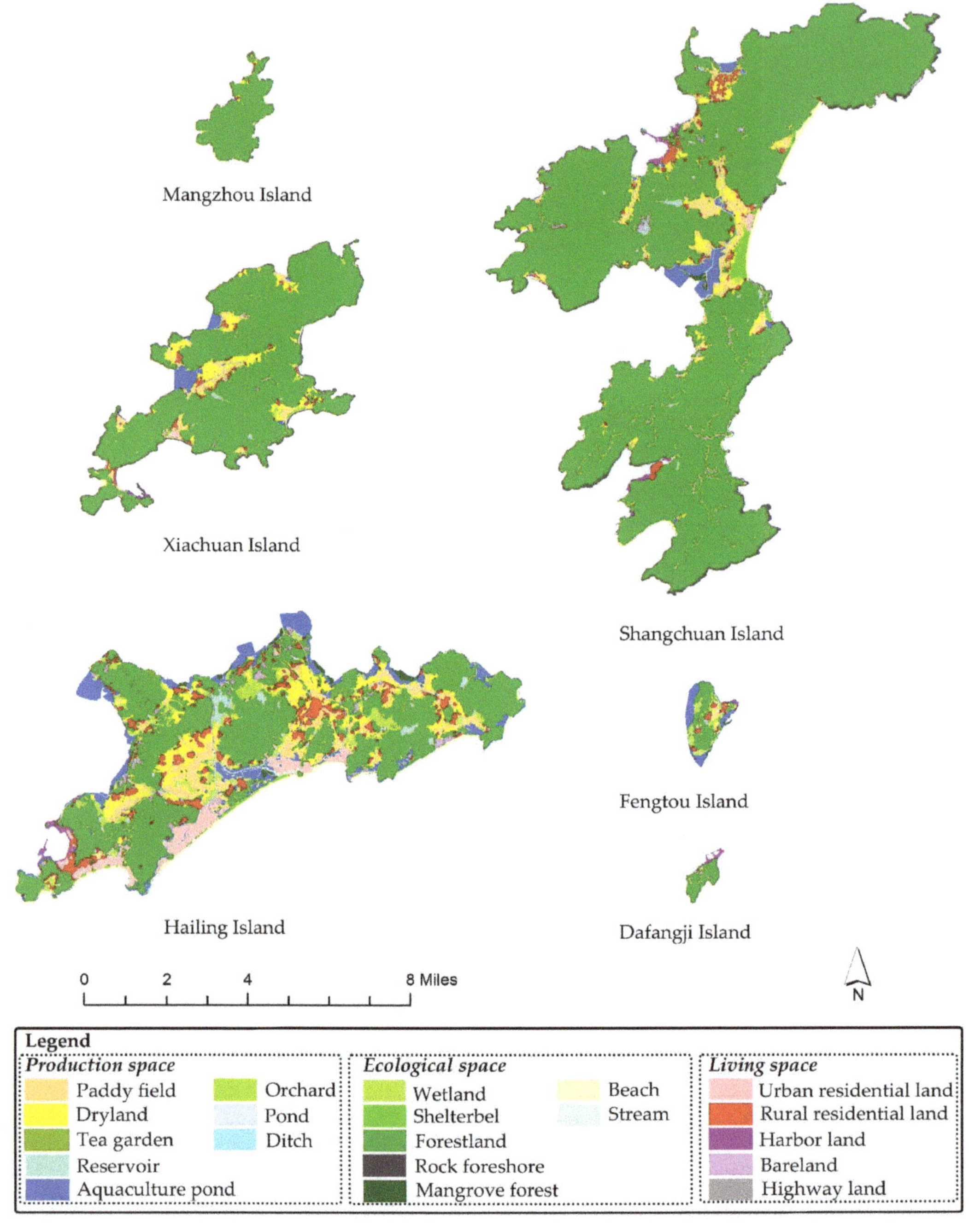

Figure 4. PLES classification of bedrock islands.

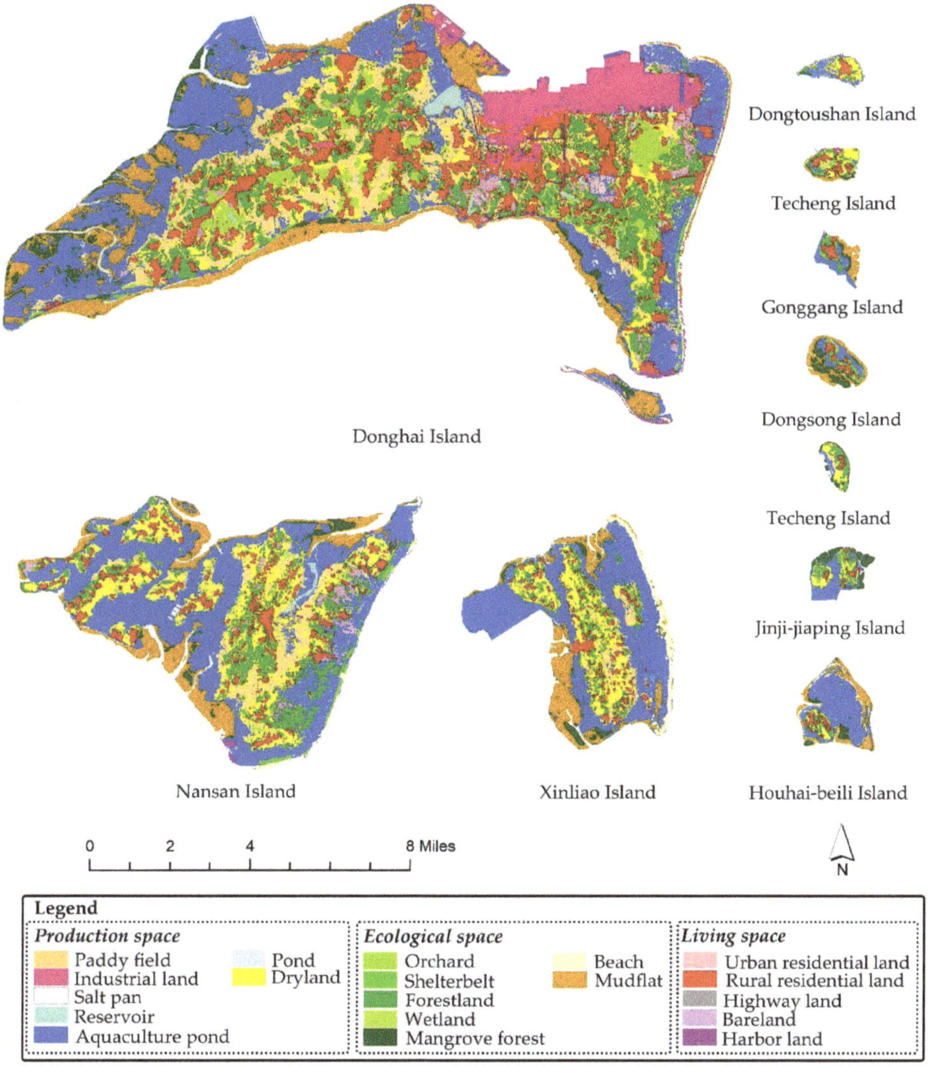

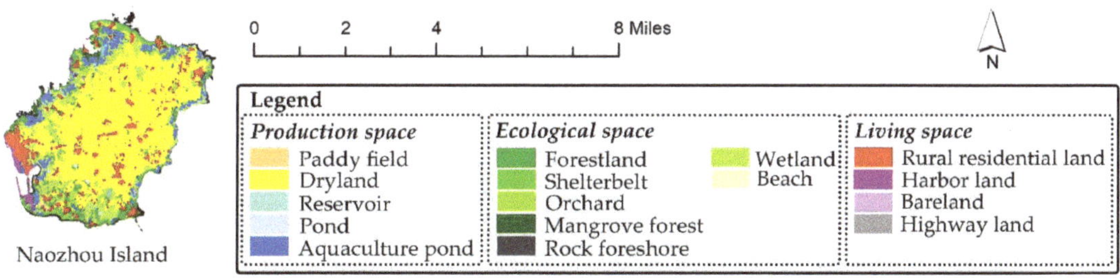

Figure 5. PLES classification of sedimentary island.

Figure 6. PLES classification of volcanic island.

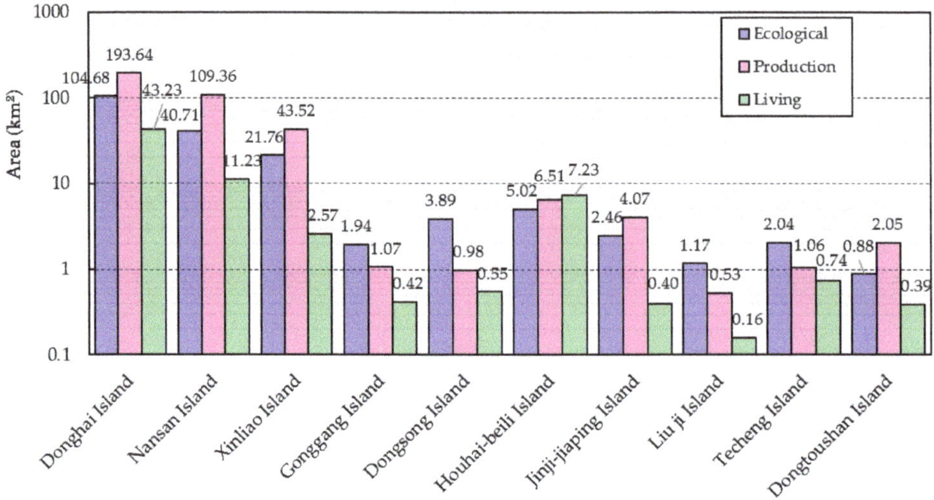

Figure 7. PLES distribution characteristics of sedimentary islands.

Table 5. The PLES pattern and connection between the island sizes and the PLES diversity.

Type of the Islands	Name of the Islands	The Area of the Island (km²)	The Number of PLES	The Dominant Space of PLES	Dominant Plaque Type of PLES	Percentage of Dominant Patch Types in PLES (of Each Island)	Area of Dominant Patch Types in PLES (km²)
Volcanic island	Naozhou Island	56.40	16	Production space	Dry farming	53.20%	27.89
Bedrock island	Hailing Island	108.90	19	Ecological space	Forestland	47.00%	51.64
	Shangchuan Island	157.00	18	Ecological space	Forestland	75.30%	103.36
	Xiachuan Island	98.32	13	Ecological space	Forestland	76.25%	63.39
	Mangzhou Island	6.82	9	Ecological space	Forestland	89.40%	5.38
	Fengtou Island	3.67	9	Ecological space	Forestland	41.37%	1.57
	Dafangji Island	1.23	5	Ecological space	Forestland	82.25%	1.06
Sedimentary island	Donghai Island	286.60	18	Production space	Aquafarm	20.27%	70.68
	Techeng Island	3.13	12	Production space	Dry farming	19.97%	0.77
	Nansan Island	123.40	15	Production space	Aquafarm	28.13%	41.26
	Dongtoushan Island	2.91	8	Production space	Aquafarm	27.36%	0.91
	Gonggang Island	1.26	7	Production space	Aquafarm	30.00%	1.02
	Dongsong Island	2.80	8	Ecological space	Mudflat	33.50%	1.81
	Houhai-beili Island	2.76	7	Production space	Aquafarm	50.69%	6.10
	Jinji-jiaping Island	4.54	6	Production space	Aquafarm	33.89%	2.09
	Liuji Island	1.94	6	Ecological space	Forestland	34.57%	0.82
	Xinliao Island	40.70	12	Production space	Aquafarm	33.17%	22.91

The sedimentary islands are dominated by PS, with an area of 362.79 km^2, or 77.26% of the sedimentary islands. In PS, the aquaculture pond is the dominant land-use type and is mainly distributed in the intertidal zone, and the area of aquaculture ponds on some islands even takes up 50%, while the paddy field occupies a large proportion of the island–continent part. The proportion of the intertidal zone of ES occupies an area of 184.55 km^2, or 39.3% of the island's total area, and is formed by mudflats and mangrove forests. In comparison, the ES of the island–continent part is composed of forest and shelterbelt. LS occupies an area of 66.92 km^2 (or 14.25%) of the sedimentary islands, as illustrated in Figure 8 and Table 5.

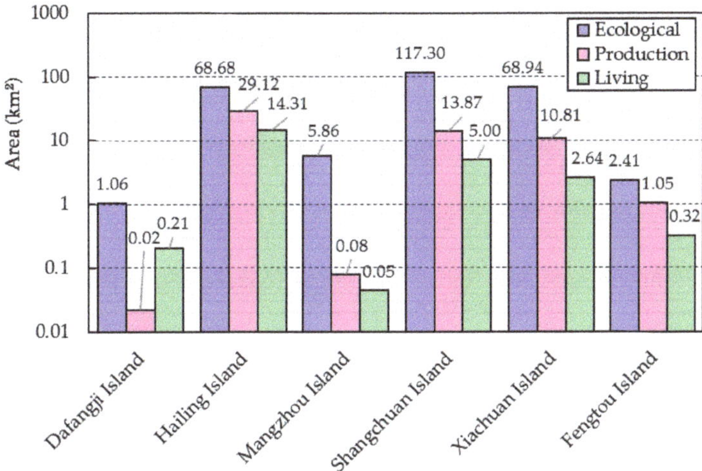

Figure 8. PLES distribution characteristics of bedrock islands.

The number of the volcanic island samples is less than that of other island types. The overall distribution of their PLES is similar to the sedimentary islands. PS takes an area of 32.87 km^2 or 58.69% of the island. The dominant land-use type in PS is day land (53.20%), and the intertidal zone is mainly occupied by aquaculture ponds. The ES on the island–continent part occupies 25.35% of the volcanic island and comprises wetland (13.80%), while rock foreshore takes a large proportion of the intertidal zone. The LS of volcanic islands occupies an area of 9.41% (see Figure 9 and Table 5).

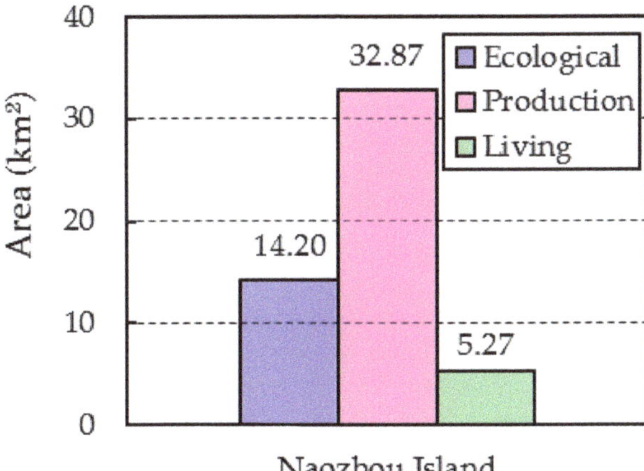

Figure 9. PLES distribution characteristics of volcanic islands.

4. Discussion

4.1. Satellite-Image-Classification-Based Method and Classification Accuracy

In this study, the accuracy of most images can reach a threshold of above 90%. Although some of the user and producer accuracies are lower than 80%, most have an accuracy between 81.3% and 99.5%, and such an accuracy threshold is deemed satisfactory for the studied area with complex and diverse land features categories [51].

4.2. The Formation and Distribution Mechanism of PLES from the Geomorphic Perspective

The PLES diversity in the studied area is closely related to island sizes. Usually, islands with a larger area have relatively complete geomorphic combinations with more developable geomorphic types [52]. This kind of island has more diversified PLES. In contrast, the landform of small islands is unitary. Some small bedrock islands only have hills and sandy beaches. Thus, the exploitation extent is limited, with less PLES (Table 5). Furthermore, the geomorphology of islands also influences the development level, dominant space, and distribution characteristics. According to the concept and method of Geomorphons [53], the geomorphic elements were extracted from DEM data by ArcGIS, and the PLES of different islands was analyzed in this study.

4.2.1. Bedrock Islands

The diversity of PLES in the bedrock islands dominated by hills is lower than that of the islands occupied by platforms. For example, the area of Hailing Island is similar to Xiachuan Island, but the PLES diversity of the former is higher because the platform occupies nearly 50% of the total area.

The hill and platform difference led to the ES and PS/LS dualistic pattern of the bedrock islands. Hills generally account for over 60% of the bedrock islands, and forest land in ES is mainly distributed in hilly areas, with an altitude of 16–517 m and steep slopes. Since hills are not developed on a large scale, a few primary and some secondary forests are preserved, with a large area of the artificial forest planted and the nature reserve established [54], leading to the idea that ES maintenance is better than that of other island types.

Generally, PS and LS are distributed on the platform at an altitude of 0–20 m. The platform area of the bedrock island is usually small; thus, the exploitation of PS and LS is limited. In PS, arable land irrigation mainly relies on reservoirs, resulting in more reservoirs reconstructed from bedrock islands than other islands under a water-scarce situation. In PS of the intertidal zone, since the shoal area is small, the aquaculture ponds cannot be built on a large scale. In LS, the natural environment of the bedrock islands is more suitable for harbor exploitation since they have a jagged coastline, deep near-shore water, and fewer barrier shoals. Thus, the number of harbor land in bedrock islands is more than that of other island types.

4.2.2. Sedimentary Islands

The altitude of the sedimentary island is 0–93 m, with over 95% of areas lower than 16 m, and the flat terrain caused a high level of exploitation indirectly, including small islands with an area of 1 km^2 to 3 km^2. Thus, the proportion of PS and LS in sedimentary islands is large.

The marine depositional plain and mudflat are the main geomorphic types of the sedimentary island, leading to the different dominant land-use types of PS and ES on the island–continent and intertidal zones. Sedimentary islands are formed by the deposition of sediment flow [55], during which more sandy lands are formed, with the low-lying wetlands evolving from tidal creeks on the island–continent. These landscapes are primarily turned into drylands and paddy fields. These islands, with a low vegetation coverage, are susceptible to desertification and storm surge invasion, so many casuarinas are cultivated for wind-breaking and sand-fixation in this area. Furthermore, the intertidal zone of the sedimentary island developed a large area of mudflat, offering the enabling environment for the mangrove forest growth and the aquaculture pond reclamation. Thus, the area and the density of aquaculture ponds far outweigh the other islands under disorderly exploitation.

4.2.3. Volcanic Island

Naozhou Island features a volcanic landform, which has the altitude ranging from 0 to 81 m and decreasing outward from the crater center with a gentle slope gradient. Thus, it is intensively exploited. Since loping fields on volcanic islands with a topographic relief gradient between 10–40 m lack surface water, many drought-tolerant crops are planted in this area, such as sweet potatoes, peanuts, and mung beans. Combined with the field research, local literature, and historical satellite maps, it is found that some lagoon and ancient lagoon landforms were developed on the coast and intertidal zone. The formation of harbor lands mainly relies on the lagoon landforms. While the wetlands mostly evolved from ancient lagoons and most reservoirs, paddy fields were exploited on this geomorphic type.

4.3. Problem Analysis of PLES and Exploitation Suggestions
4.3.1. Problem Analysis of PLES

The specialties of the small islands (e.g., low environmental carrying capacity and limited natural resources) restrict the production and living activities of the island. With the different degrees of anthropogenic activities and exploitation in various types of islands, the ecological environment has been affected at different levels: (1) The overall threat to ES in bedrock islands is less than that of other island types. Among them, ES of the island–continent part is less contaminated by agricultural and domestic sewage. The primary pollution risk is concentrated in the intertidal zone, mainly from the harbor lands and aquaculture ponds, especially in Hailing Islands. (2) The contradictions between PS, LS, and ES in the sedimentary island are more prominent. The problems are as follows: most sedimentary islands have no garbage disposal facilities except for the islands with large areas (e.g., Nansan Island and Donghai Island); much domestic and agricultural sewage threatens the ES of the island–continent part after intense exploitation, especially on the natural wetlands; and the mangrove forests and the mudflats in the intertidal zone are threatened by the high-density aquaculture ponds in disordered distribution. (3) According to the field investigation, the water shortage and overuse by tourism and agriculture are the main ecological issues for the volcanic island.

The production activities of the islanders were restricted by the frequent wind disasters and water scarcity of these islands. In order to increase their income, the islanders on the sedimentary and volcanic islands adopted large-scale and high-density aquaculture systems and drought-tolerant economic crop patterns. However, the income of islanders who participated in traditional agriculture is still unstable. This leads to the obvious population loss, farmland abandonment, and homestead hollowing of the agricultural areas. Although the islanders can increase their income by engaging in tourism on the bedrock islands, the islanders who participated in traditional agriculture face the same problems as the sedimentary islands and volcanic islands.

Overall, low exploitation can help to keep the PLES balance of small islands because it can sustain the available natural resources for production and living activities and have fewer threats to the ES. The PLES of bedrock islands maintains the balance through the low exploitation rate and comparatively abundant natural resources. However, the drastic human–land conflict causes a PLES imbalance on the sedimentary and volcanic islands.

4.3.2. Exploitation Suggestions

It is suggested that ecological carrying capacity and self-purification should be emphasized. Additionally, the core and buffer protection zones of the ecology and cultural relics should be established when further developing and planning these islands. To balance the PLES of these islands and meet the needs of the islanders, intensive agriculture, and fisheries, eco-friendly and novel industries with higher technologies can be introduced on these islands. Specifically, the development modes such as ecological tourism are more applicable to the bedrock islands. Ecological restoration should be considered before further exploiting the sedimentary islands. During the volcanic island exploitation, the agricultural planting structure and tourism mode should be fully considered for the sustainable use of water resources. Moreover, attention should be paid to the public infrastructure, such as the schools, transportation facilities, and garbage disposal facilities, to improve the livability of LS in these islands.

5. Conclusions

In the present study, Landsat images were used to identify and analyze the rural PLES of different island types in the western sea area of Guangdong Province, and the PLES distribution characteristics and formation mechanism were explored. The following conclusions can be drawn.

(1) It is found that the ecological and utilized area of the intertidal zone and neritic region should be combined with the island–continent part, and the distinct industrial type should be considered during the establishment of the rural island PLES classification system.

(2) The bedrock islands are dominated by ES, which is composed of forestlands. While the PS is the dominant space of sedimentary islands, and the aquaculture pond is the main land-use type. The dominant type of volcanic island is dryland. The rural residential lands are the main component of LS in all kinds of islands.

(3) Altitudinal gradient and geomorphic characteristics are the main driving factors for the formation and distribution of island rural PLES. In addition, the exploitation level, preference of dominant industries, and PLES contradiction of different island types are clarified from a geomorphology perspective.

(4) A comprehensive and targeted method of land-use classification and PLES establishment of the island is proposed in this study. Thus, this method is more applicable to the island areas, such as the Small Island Developing States, the small islands of other developing countries, or regions with similar natural resources. Moreover, it can be applied on the mesolevel and microlevel of space, as it is difficult to distinguish the boundaries between the cities and villages on the macrolevel. Additionally, this method could divide nonoverlapping boundaries and generate continuous geographical spatial maps with different functions. Therefore, the method is easier to accomplish in regional management and implement in the specific departments when facing practical applications.

This work only qualitatively analyzed the balance of the rural PLES of islands and how the rural PLES meet the needs of the islanders because the village-level social and economic data lacked statistics from the local statistical bureau. Such data is challenging to obtain from fieldwork. Nevertheless, a quantitative analysis of the social and economic data will be conducted in the future.

Author Contributions: Conceptualization, R.B. and Y.P.; methodology, R.B.; software, R.B.; validation, R.B.; formal analysis, R.B. and Y.P.; investigation, R.B. and Y.P.; resources, Y.S.; data curation, R.B., Y.S. and Y.P.; writing—original draft preparation, R.B.; writing—review and editing, R.B., Y.S. and Y.P.; visualization, R.B.; supervision, Y.S. and Y.P.; project administration, Y.S. and Y.P.; funding acquisition, Y.S. and Y.P. All authors have read and agreed to the published version of the manuscript.

Funding: This research was funded by National Natural Science Foundation of China, grant number 51978275.

Institutional Review Board Statement: Not applicable.

Informed Consent Statement: Not applicable.

Data Availability Statement: Not applicable.

Acknowledgments: The supports from Bureaus of Natural Resources in Yangjiang City, Zhanjiang City, and Taishan City, Agricultural Departments and local government offices of the islands in the study area, Local Chronicles Museum are acknowledged. The technological advises from Cheng Jiang from Western Sydney University are well acknowledged.

Conflicts of Interest: The authors declare no conflict of interest.

References

1. Liu, J.; Wen, J.; Huang, Y.; Shi, M.; Meng, Q.; Ding, J.; Xu, H. Human settlement and regional development in the context of climate change: A spatial analysis of low elevation coastal zones in China. *Mitig. Adapt. Strateg. Glob. Chang.* **2015**, *20*, 527–546. [CrossRef]
2. Jiang, X.W. *Remote Sensing Image Processing and Interpretation of Island and Coastal Zone in Offshore Ocean of China*, 1st ed.; Maritime Press: Beijing, China, 2016; pp. 6–7, 31–38.
3. The State Council's Opinions on Establishing and Supervising Implementation of the Spatial Planning System. Available online: http://www.gov.cn/zhengce/2019-05/23/content_5394187.htm (accessed on 19 March 2022).
4. Zou, L.; Liu, Y.; Yang, J.; Yang, S.; Wang, Y.; Hu, X. Quantitative identification and spatial analysis of land use Ecological-Production-Living functions in rural areas on China's southeast coast. *Habitat Int.* **2020**, *100*, 102182. [CrossRef]
5. Deng, Y.; Yang, R. Influence mechanism of Production-Living-Ecological space changes in the urbanization process of Guangdong Province, China. *Land* **2021**, *10*, 1357. [CrossRef]
6. Yang, Y.; Bao, W.; Li, Y.; Wang, Y.; Chen, Z. Land use transition and its eco-environmental effects in the Beijing-Tianjin-Hebei urban agglomeration: A Production-Living-Ecological perspective. *Land* **2020**, *9*, 285. [CrossRef]
7. Lin, G.; Jiang, D.; Fu, J.; Zhao, Y. A review on the overall optimization of Production-Living-Ecological space: Theoretical basis and conceptual framework. *Land* **2022**, *11*, 345. [CrossRef]
8. Fang, C.; Yang, J.; Fang, J.; Huang, X.; Zhou, Y. Optimization transmission theory and technical pathways that describe multiscale urban agglomeration spaces. *Chin. Geogr. Sci.* **2018**, *28*, 543–554. [CrossRef]
9. Zhang, H.; Xu, E.; Zhu, H. An Ecological-Living-Industrial land classification system and its spatial distribution in China. *Resour. Sci.* **2015**, *37*, 1332–1338.
10. Shi, Z.; Deng, W.; Zhang, S. Spatio-temporal pattern changes of land space in Hengduan mountains during 1990–2015. *J. Geogr. Sci.* **2018**, *28*, 529–542. [CrossRef]
11. Li, G.; Fang, C. Quantitative function identification and analysis of urban Ecological-Production-Living spaces. *Acta Geogr. Sin.* **2016**, *71*, 49–65.
12. Xi, J.; Wang, S.; Zhang, R. Restructuring and optimizing Production-Living-Ecology space in rural settlements: A case study of Gougezhuang village at Yesanpo tourism attraction in Hebei Province. *J. Nat. Resour.* **2016**, *31*, 425–435.
13. Wan, J.; Su, Y.; Zan, H.; Zhao, Y.; Zhang, L.; Zhang, S.; Deng, W. Land Functions, Rural Space Governance, and Farmers' Environmental Perceptions: A Case Study from the Huanjiang Karst Mountain Area, China. *Land* **2020**, *9*, 134. [CrossRef]
14. Yang, Y.; Bao, W.; Liu, Y. Coupling coordination analysis of rural Production-Living-Ecological space in the Beijing-Tianjin-Hebei region. *Ecol. Indic.* **2020**, *117*, 106512. [CrossRef]
15. Liu, J.L.; Liu, Y.S.; Li, Y.R. Classification evaluation and spatial-temporal analysis of "Production-Living-Ecological" spaces in China. *Acta Geogr. Sin.* **2017**, *72*, 1290–1304.
16. Cai, E.; Jing, Y.; Liu, Y.; Yin, C.; Gao, Y.; Wei, J. Spatial-temporal patterns and driving forces of Ecological-Living-Production land in Hubei province, central China. *Sustainability* **2018**, *10*, 66. [CrossRef]
17. Chen, L.; Zhou, S.L.; Zhou, B.B.; Lv, L.G.; Chang, T. Characteristics and driving forces of regional land use transition based on the leading function classification: A case study of Jiangsu Province. *Econ. Geogr.* **2015**, *35*, 155–162.
18. Liu, C.; Xu, Y.; Wang, Y.; Cheng, L.; Lu, X.; Yang, Q. Analyzing the value and evolution of land use functions from "Demand–Function–Value" perspective: A framework and case study from Zhangjiakou City, China. *Land* **2021**, *11*, 53. [CrossRef]
19. Feng, C.; Zhang, H.; Xiao, L.; Guo, Y. Land use change and its driving factors in the rural–urban fringe of Beijing: A Production–Living–Ecological perspective. *Land* **2022**, *11*, 314. [CrossRef]
20. Li, X.M.; Zhang, J.; Cao, J.F.; Ma, Y. Ecological risk assessment of exploitation and utilization in Chuanshan archipelago, Guangdong Province, China. *Acta Ecol Sin.* **2015**, *35*, 2265–2276.
21. Pan, Y.; Bai, R.; Shi, Y. Research on the landscape characteristics of traditional settlement under the influence of topographic factor—A case of Weizhou Island in Beibu Gulf. *Chin. Landsc. Archit.* **2021**, *37*, 33–38.
22. Island Protection Plan of Guangdong Province (2011–2020). Available online: http://nr.gd.gov.cn/zwgknew/tzgg/tz/content/post_3006827.html (accessed on 15 March 2022).
23. Ma, Y.; Zhang, J.; Li, X.M.; Hao, Y.L.; He, Z.H. Feasibility study of remote sensing technology applied to island protection and utilization planning. *Ocean. Dev. Manag.* **2009**, *26*, 92–95.
24. Duan, Y.; Wang, H.; Huang, A.; Xu, Y.; Lu, L.; Ji, Z. Identification and spatial-temporal evolution of rural "production-living-ecological" space from the perspective of villagers' behavior–A case study of Ertai Town, Zhangjiakou City. *Land Use Policy* **2021**, *106*, 105457. [CrossRef]

25. Fang, Y.G.; Shi, K.J.; Niu, C.C. A comparison of the means and ends of rural construction land consolidation: Case studies of villagers' attitudes and behaviours in Changchun City, Jilin Province, China. *J. Rural Stud.* **2016**, *47*, 459–473. [CrossRef]

26. Ma, W.; He, X.; Jiang, G.; Li, Y.; Zhang, R. Land use internal structure classification of rural settlements based on land use function. *Trans. Chin. Soc. Agric. Eng.* **2018**, *34*, 269–277.

27. Ma, X.; Li, X.; Hu, R.; Khuong, M.H. Delineation of "production-living-ecological" space for urban fringe based on rural multifunction evaluation. *Prog. Geogr.* **2019**, *38*, 1382–1392. [CrossRef]

28. Zhen, L.; Chao, S.Y.; Wei, Y.J.; Xie, G.D.; Li, F.; Yang, L. Land use functions: Conceptual framework and application for China. *Resour. Sci.* **2009**, *31*, 544–551.

29. Du, G.M.; Sun, X.B.; Wang, J.Y. Spatiotemporal patterns of multi-functions of land use in northeast China. *Prog. Geogr.* **2016**, *35*, 232–244.

30. Xia, M.; Feng, X.H.; Xia, J.L.; Zhou, W. Delineation of Production-Living-Ecological space in Lishui District of Nanjing based on land multi-functions and suitability. *Trans. Chin. Soc. Agric. Eng.* **2021**, *375*, 242–250.

31. Wu, J.; Zhang, D.; Wang, H.; Li, X. What is the future for Production-Living-Ecological spaces in the Greater Bay Area? A multi-scenario perspective based on DEE. *Ecol. Indic.* **2021**, *131*, 108171. [CrossRef]

32. Liu, C.; Cui, W.L.; Zhu, Z.T.; Ye, F.; Yu, X.J. Study on the technical methods of the delineation of island ecological red lines. *Acta Ecol. Sin.* **2018**, *38*, 8564–8573.

33. Compilation Committee of Chorography of Guangdong Province. *Guangdong Province Chorography: Ocean and Islands Chorography*, 1st ed.; Guangdong People's Press: Guangzhou, China, 2000; p. 98.

34. Dang, A.T.; Kumar, L.; Reid, M.; Nguyen, H. Remote sensing approach for monitoring coastal wetland in the Mekong Delta, Vietnam: Change trends and their driving forces. *Remote Sens.* **2021**, *13*, 3359. [CrossRef]

35. Kumar, L.; Sinha, P.; Taylor, S. Improving image classification in a complex wetland ecosystem through image fusion techniques. *J. Appl. Remote Sens.* **2014**, *8*, 083616. [CrossRef]

36. Ehlers, M.; Klonus, S.; Johan Astrand, P.; Rosso, P. Multi-sensor image fusion for pansharpening in remote sensing. *Int. J. Image Data Fusion* **2010**, *1*, 25–45. [CrossRef]

37. Law of the People's Republic of China on the Promotion of Rural Revitalization. Available online: http://www.gov.cn/xinwen/2021-04/30/content_5604050.htm (accessed on 28 April 2022).

38. Shi, H.; Zhao, M.; Simth, D.A.; Chi, B. Behind the Land Use Mix: Measuring the Functional Compatibility in Urban and Sub-Urban Areas of China. *Land* **2022**, *11*, 2. [CrossRef]

39. Ministry of Housing and Urban-Rural Development, Finance, Culture, Protection of Cultural Relics: Guidelines on Strengthening the Protection of Traditional Settlements of China. Available online: http://www.gov.cn/zhengce/2016-05/22/content_5075656.htm (accessed on 16 March 2022).

40. Bohua, L.; Can, Z.; Yindi, D.; Peilin, L.; Chi, C. Change of human settlement environment and driving mechanism in traditional villages based on living-production-ecological space: A case study of Lanxi Village, Jiangyong County, Hunan Province. *Prog. Geogr.* **2018**, *37*, 677–687.

41. Li, J.M.; Wang, M.C.; Wang, F.Y.; Chen, X.Y.; Ding, W. Urban land use classification of multi-features random forest. *Sci. Surv. Mapp.* **2021**. in press (In Chinese)

42. Mu, Y.N.; Ding, L.X.; Li, N.; Lu, L.Y.; Wu, M. Classification of coastal wetland vegetation in Hangzhou Bay with an object-oriented, random forest model. *J. Zhejiang A F Univ.* **2018**, *35*, 1088–1097.

43. Tucker, C.J. Red and photographic infrared linear combinations for monitoring vegetation. *Remote Sens. Environ.* **1979**, *8*, 127–150. [CrossRef]

44. Gautam, V.K.; Gaurav, P.K.; Murugan, P.; Annadurai, M.J.A.P. Assessment of surface water dynamicsin bangalore using WRI, NDWI, MNDWI, supervised classification and KT transformation. *Aquat. Procedia* **2015**, *4*, 739–746. [CrossRef]

45. Zha, Y.; Gao, J.; Ni, S. Use of normalized difference built-up index in automatically mapping urban areas from TM imagery. *Int. J. Remote Sens.* **2003**, *24*, 583–594. [CrossRef]

46. Measuring Vegetation (NDVI and EVI). Available online: https://earthobservatory.nasa.gov/features/MeasuringVegetation (accessed on 19 March 2022).

47. Gupta, K.; Mukhopadhyay, A.; Giri, S.; Chanda, A.; Majumdar, S.D.; Samanta, S.; Hazra, S. An index for discrimination of mangroves from non-mangroves using LANDSAT 8 OLI imagery. *MethodsX* **2018**, *5*, 1129–1139. [CrossRef]

48. Xie, J.; Wang, Z.M.; Mao, D.H.; Ren, C.Y.; Han, J.X. Remote sensing classification of wetlands using object-oriented method and multi-season HJ-1 images—A case study in the Sanjiang plain north of the Wandashan mountain. *Wetl. Sci.* **2012**, *10*, 429–438.

49. Guo, X.; Zhang, C.; Luo, W. Urban impervious surface extraction based on multi-features and random forest. *IEEE Access* **2020**, *8*, 226609–226623. [CrossRef]

50. Breiman, L. Random forests. *Mach. Learn.* **2001**, *45*, 5–32. [CrossRef]

51. Tran, H.; Tran, T.; Kervyn, M. Dynamics of land cover/land use changes in the Mekong Delta, 1973-2011: A remote sensing analysis of the Tran Van Thoi District, Ca Mau Province, Vietnam. *Remote Sens.* **2015**, *7*, 2899–2925. [CrossRef]

52. Compilation Committee of Island Annals of China. *Chinese Island Chronicle (Guangdong Volumes)*, 1st ed.; Maritime Press: Beijing, China, 2013; pp. 9–10.

53. Jasiewicz, J.; Stepinski, T.F. Geomorphons-a pattern recognition approach to classification and mapping of landforms. *Geomorphology* **2013**, *182*, 147–156. [CrossRef]

54. Li, Y.; Zhou, H.C. Characteristics and exploitation policy for island resources in Guangdong. *Coast. Eng.* **2010**, *29*, 75–82.
55. Ratter, B.M. *Geography of Small Islands*, 2nd ed.; Springer: Cham, Switzerland, 2018; p. 42.

Article

Optimization of Production–Living–Ecological Space in National Key Poverty-Stricken City of Southwest China

Di Wang [1,*]**, Jingying Fu** [2,3] **and Dong Jiang** [2,3,4]

[1] School of Management, Tianjin University of Commerce, Tianjin 300131, China
[2] Institute of Geographic Sciences and Natural Resources Research, Chinese Academy of Sciences, Beijing 100101, China; fujy@igsnrr.ac.cn (J.F.); jiangd@igsnrr.ac.cn (D.J.)
[3] College of Resources and Environment, University of Chinese Academy of Sciences, Beijing 100049, China
[4] Key Laboratory of Carrying Capacity Assessment for Resource and Environment, Ministry of Land & Resources, Beijing 100101, China
* Correspondence: wangdi@tjcu.edu.cn

Abstract: Trade-offs and conflicts among different sectors of production, living, and ecology have become important issues in regional sustainable development planning due to both the versatility and limitation of land resources, especially in poverty-stricken mountainous areas. This study builds an optimization model to assist policymakers in simulating land demand and allocation in the future. The model takes socioeconomic and demographic development into consideration and couples local planning policy with land use data from the perspective of system integration. The model was employed for a case study of Zhaotong city to optimize production–living–ecological (PLE) space. The results show that the model provides a feasible method to explore the sustainable development pattern of territorial space, especially in distressed regions.

Keywords: PLE space; trade-offs and conflicts; sustainable development; system dynamic model; FLUS

Citation: Wang, D.; Fu, J.; Jiang, D. Optimization of Production–Living–Ecological Space in National Key Poverty-Stricken City of Southwest China. *Land* **2022**, *11*, 411. https://doi.org/10.3390/land11030411

Academic Editor: Carlos Parra-López

Received: 28 February 2022
Accepted: 9 March 2022
Published: 11 March 2022

Publisher's Note: MDPI stays neutral with regard to jurisdictional claims in published maps and institutional affiliations.

1. Introduction

Interactions between human and the environment have attracted increasing attention [1]. In general, cross-sectoral issues involve a variety of social and natural knowledge [2]. Gaps between natural and social sciences imply that understanding the mechanisms underlying human–environment systems from a systematic perspective is crucial to sustainable development [3]. To a certain extent, several studies have reflected the fact that the problems are multi-scale and complex while the solutions are diverse [4,5]. Effective domestic policymaking hinges on diverse stakeholders, which plays a critical role in accelerating the localization of the SDGs (Sustainable Development Goals, proposed by the United Nations in 2015) [6]. National development plans for the 2030 Agenda in many countries try to position human, social, environmental, economic, and institutional objectives at the same level [7–9]. However, a scientific challenge that obviously exists in sustainable issues is trade-offs and compensation. The achievement of one SDG is often at the cost of sacrificing or assisting another [10–13].

By 2020, China achieved the goal of eliminating poverty, which was accompanied by economic growth and rapid urbanization and contributed to SDG 1 [14,15]. Unfortunately, inconsistent and rough terrestrial land development patterns in production space and living space have placed pressure on ecological protection and resource security for ecological spaces [16–19]. To control the intensity of terrestrial land development and adjust the spatial structure, the concept of PLE (production–living–ecology) space was proposed in the report of the 18th National Congress of the Communist Party of China. The objective is to construct intensive and efficient production space, livable and appropriate living space, and protected and beautiful ecological space with beautiful mountains and clear water based on the principle of balancing economic, social, and ecological benefits. It aims to delimit the boundaries of multi-function terrestrial land development and establish a system for future sustainable land development [20]. To a certain extent, PLE space can be considered a combination of SDG indicators from a land function point of view (Figure 1). It is highly relevant to the 17 SDGs that involve environmental integrity, social equity, and economic prosperity, which comprise the triple bottom line approach of human wellbeing [21,22].

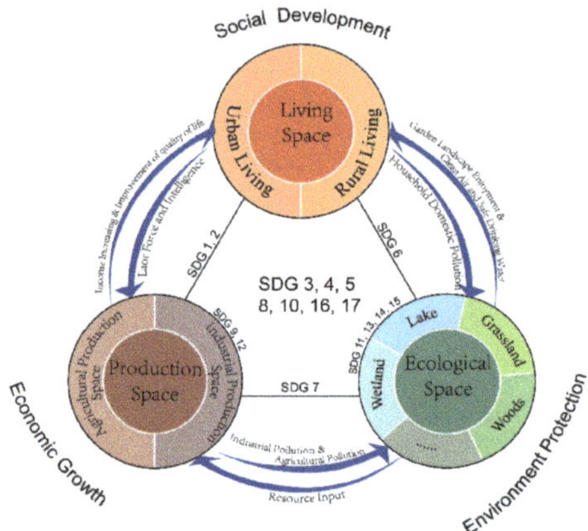

Figure 1. The relationship of PLE space and SDGs.

International academic research has discussed food security [23], tourism [24], agriculture [25], and urban regeneration [26] under the background of urban sustainability but rarely focuses on PLE space. After all, the concept has been proposed by the Chinese government in the context of China. Most studies are conducted on a domestic scope. Huang et al. claimed that the assessment of regional spatial carrying capacity and suitability is an important guideline for PLE space optimization [20]. Wu et al. indicated that the carrying capacity determines the upper limit of the PLE space in quantity, while suitability evaluation is related to the spatial layout structure [27]. Some studies have provided qualitative and empirical optimization suggestions based on the results of the carrying capacity and suitability evaluation [28–30]. In general, the suggestions point out that rules and regulations should be further enhanced and technology standards need to be improved in the future. However, it is uncertain whether these suggestions are feasible and effective. It is necessary to develop quantified and visualized tools to simulate future scenarios, especially for policymakers who face the difficulty of applying social and ecological approaches to decision-making [31].

Land resources are the most constrained factor for PLE space optimization. One parcel of land may be used as production space or as living space. In addition, land policy has become an indispensable means of macroeconomic regulation under China's national policies because land resources have both natural and economic properties. Therefore, some studies approach PLE space optimization as a mathematical problem of multi-objective optimization, guiding maximum economic benefits, social benefits, and ecological benefits [32]. However, two major gaps remain in the literature. Firstly, multi-objective optimization algorithms still face some challenges in flexibility and convergence, especially in preference adaptation for various formulations [33,34]. Secondly, difficulty and complexity increase when referencing spatial data, although addressing quantitative issues has significant advantages.

The system dynamics model seeks opportunities and ways to optimize the structure of the system from a holistic perspective based on the feedback characteristics of the internal components [35–37]. It fills the first gap presented above. In contrast to multi-objective optimization equations, the system dynamics model simulates the real world by establishing relationships between social and economic factors. This method can introduce more factors and equations and perform dynamic simulations. SD models have been used in resource management, such as future urbanization and water scarcity [16], energy consumption [38], and water resource management systems [39]. However, the ability of the SD model to be applied in spatial allocation is very weak.

As for the other gap, the FLUS (Future Land Use Simulation) model (https://geosimulation.cn/FLUS.html, accessed on 15 February 2022) is used to simulate human activities and natural influences on land-use change and future scenarios. The model introduces an artificial neural network algorithm (ANN)-based probability calculation of suitability for various land use types based on traditional meta-automata. It proposes an adaptive inertial competition mechanism based on roulette selection (a stochastic selection method) [40], which can effectively deal with the uncertainty and complexity of multiple land-use types when they are transformed under the joint influence of natural effects and human activities, meaning the FLUS model has high simulation accuracy and can obtain similar results to the real land-use distribution [41]. To date, the model has been successfully applied in many cases, such as the simulation of future urban sprawl boundaries [42,43] and the simulation of flooding risks in rapid urbanization [44]. Furthermore, the input of future demand for land in the FLUS model can be determined by SD models, which means that the two can be well coupled. Sustainable development issues require a systematic approach to integrate various socioeconomic and environmental components that interact across regional levels, space, and time [45]. Some studies have integrated the idea of system science into the optimization of resource allocation [46–48] but rarely have focused on PLE space optimization. This paper aims to build an optimization model based on the SD and FLUS models, in which PLE space can be planned quantitatively and spatially under future scenarios.

Trade-offs and conflicts of the PLE function are more obvious and intense in the poor areas of southwest China. Yunnan Province has the highest number of poverty-stricken counties and includes Zhaotong city, which is located in Wumeng Mountain and was one of the 14 concentrated contiguous poverty-stricken areas in China until 2020 (Figure 2). The GDP per capita is far below the national average level. People's living standards and social development are limited by natural geographical conditions and ecosystem protection [49,50]. It is necessary to balance development and conservation when making future policies [51]. This paper has two main tasks, one is to develop a model applicable to the problem, and the other is to use the model to provide solutions for the future development of Zhaotong city, and attempt to answer the following questions: (1) In this impoverished region, what is the main trend in PLE space changes over the past few years? (2) Which scenario can meet the planning target of 2030? (3) How can the development pattern be optimized?

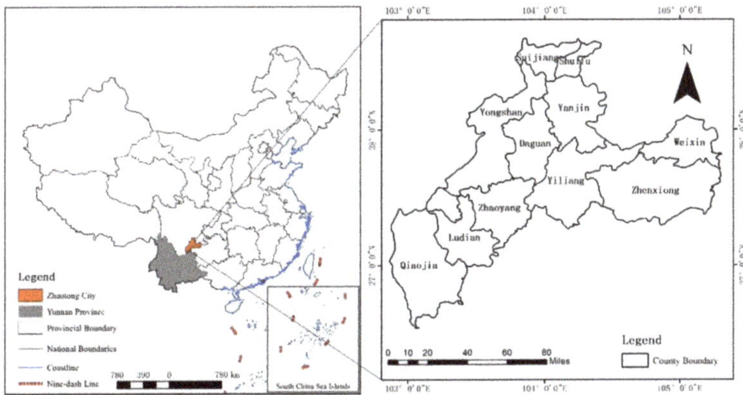

Figure 2. Location and slope map of Zhaotong city.

2. Methods and Data

This paper aims to build an optimization model based on system dynamics and FLUS in which land resources can be planned quantitatively and spatially in terms of PLE space. A system dynamics model is employed to simulate the structure of the PLE space system and predict quantitative demand in 2030. FLUS is applied to allocate the land parcels at the spatial scale. The workflow is shown in Figure 3. The work consists of 4 main parts:

(1) Classification of PLE space. PLE space is divided into production, living, and ecological space based on land-use data. Then, space is further divided into subclasses according to the function assessment, which serves as the basis for the subsequent PLE space optimization.

(2) Establishment of the SD Model. We simulate the demand quantity of each PLE space and piece of land in 2030 by constructing a dynamic system model of PLE space, including the population, economy, and land system.

(3) FLUS Model Settings. We take the quantity of demand obtained by (2) as the input data of the FLUS model and carry out spatial allocation of various PLE classes.

(4) Constrained scenario design and analysis. Based on verifying the accuracy of the models in (2) and (3), a variety of future optimization scenarios are designed for simulation, and the simulation results are evaluated from multiple perspectives.

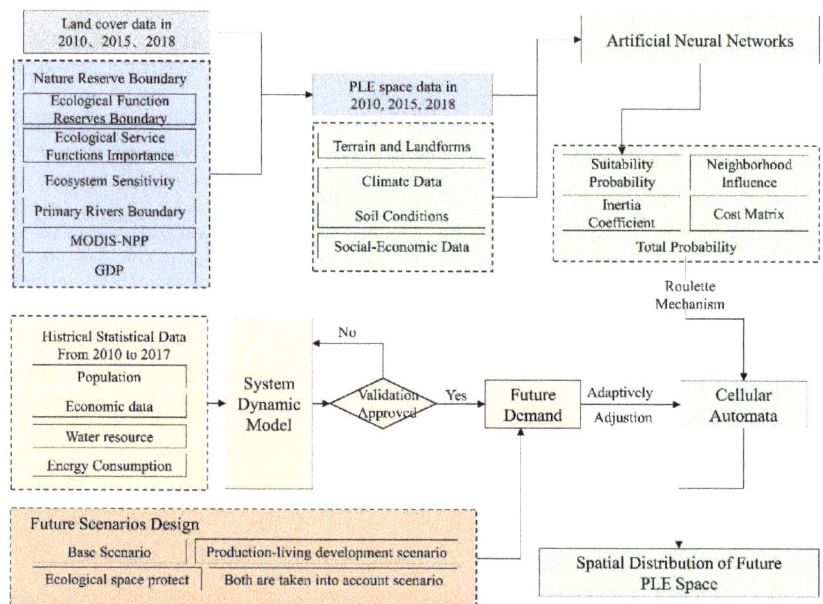

Figure 3. Workflow for the optimization.

2.1. Classification of PLE Space

The classification of PLE space is the basis for the optimization of space layout. For PLE space optimization, it is necessary to clarify the priority. This study proposes a new classification system. According to the production–living–ecological function, the first level of classification was carried out, and then the second category was determined based on the land-use data from 2010, 2015, and 2018.

2.1.1. Detailed Data Sources

Based on the dataset of the multiperiod land cover dataset of China (MLC), the classification system of the "PLE space" is established. Detailed data sources are shown in Table 1. The MLC dataset was obtained by manual visual interpretation using Landsat remote sensing image data from the US. The land-use types include the six primary types of cropland, forestland, grassland, water, residential land, and unused land, and 25 secondary types. All raster datasets were resampled to 30 m by software Arcgis 10.2. As shown in Table 1, some datasets that are shared online publicly are limited due to temporal resolution.

Table 1. Data sources for the "PLE Spaces" classification.

Datasets	Type	Spatial Resolution	Year	Data Source
Multi-period land cover dataset of China	Raster	30 m	2010 2015 2018	Resource and Environment Science and Data Center http://www.resdc.cn/ (accessed on 15 May 2020)
Nature reserve boundary data of China	Shapefile	——	2018	
Dataset of primary rivers spatial distribution in China	Shapefile	——	2000	Resource and Environment Science and Data Center http://www.resdc.cn/ (accessed on 15 May 2020)
Ecological Function Reserves of China	Shapefile	——	2010	
Importance of Ecological Service Functions of China	Raster	1 km	2010	Ecosystem Assessment and Ecological Security Database http://www.ecosystem.csdb.cn/ (accessed on 15
Ecosystem Sensitivity of China	Raster	1 km	2010	May 2020)
GDP (Gross domestic product)	Raster	1 km	2010 2015 2018	Resource and Environment Science and Data Center http://www.resdc.cn/ (accessed on 15 May2020)
NPP (Net primary productivity)	Raster	500 m	2010 2015 2018	LAADS DAAC https://ladsweb.nascom.nasa.gov/search (accessed on 15 May 2020)

2.1.2. Classification System

As shown in Table 2, according to the importance of ecological functions, ecological space is further subdivided into four secondary categories: restricted ecological space (ER), priority ecological space (EP), general ecological space (EG), and accommodated ecological space (EA). Restricted ecological space refers to areas that perform important ecological functions and require restricted protection, including national nature reserve areas with extremely important ecological services, first-class rivers (including waters in the land cover data collection), and extremely sensitive ecological areas.

Table 2. Relationship between the classification system of PLE spaces and MLC classification.

Class	Subclass	MLC Classification and Codes
Ecological Space	Restricted ecological space (ER)	National nature reserves area, extremely important ecological services areas, first-class rivers area, ecologically extremely sensitive areas, and waters area (4)
	Priority ecological space (EP)	High-coverage grass (31), Mid-coverage grass (32), Wooded land (21)
	General ecological space (EG)	Shrubland (22), Sparse woodland (23), Low-coverage grass (33)
	Accommodated ecological space (EA)	Marshland (64), Bare rocky gravel land (66)
Production Space	Priority agricultural production space (PP)	Both arable land (1) and other woodland (24) with above-average GDP or NPP
	General agricultural production space (PG)	Both arable land (1) and other woodland (24) with below-average GDP or NPP
	Industrial production space (PI)	Other construction land (53)
Living Space	Urban living space (LU)	Urban building land (51)
	Rural living space (LR)	Rural settlements (52)

According to different production functions, the production space is divided into industrial production space (PI) and agricultural production space (PA). Based on GDP and NPP, the agricultural production space is further subdivided into priority agricultural production space (PP) and general production space (PG). The economic value (GDP) and net primary productivity (NPP) are evaluated, and those that are higher than the average GDP or NPP are regarded as priority agricultural production spaces, and the others are regarded as general agricultural production spaces. General agricultural production space and general ecological space give priority to land conversion in the subsequent optimization process.

2.2. Establishment of the SD Model

The structure optimization model of PLE space is realized by constructing the system dynamic model. First, it is necessary to clarify the system boundary and system structure. The system boundary affects the model's complexity, and the boundary cannot be too large or too small. The system structure refers to the variables involved in the model and the relationship between them. The purpose of this research is to understand the interaction between the main elements in PLE space. Therefore, the system boundary was confined to Zhaotong from 2010 to 2017, and the model was constructed by selecting factors related to population, social fixed asset investment, regional GDP, land resources, and more. The PLE space system is divided into a population–economy subsystem and a land-use subsystem. According to data recorded by the Statistical Yearbook of Zhaotong and Statistical Bulletin of Zhaotong, 33 variables were finally selected. The system structure is shown in Figure 4. Specific variable names and units are detailed in Table S1 of the supplementary file.

The relationship between system variables and mathematical equations is the core of the model construction, and it influences the accuracy of the model. In this paper, the comprehensive empirical method, regression analysis method, logical inference method, weighted average method, and other methods are used to determine the equation parameters. Moreover, the Lookup function and conditional function provided by Vensim software were fully utilized, and 33 groups of equations were finally determined (see Equations (S1)–(S33) in the methods and data of the supplementary file for details).

The simulated values were compared with the actual values of some variables from 2010 to 2017, and the results are shown in Table S2. The absolute values of the errors are in the range of 0–5%, which indicates that the constructed model can respond to the interrelationships among the elements within a reasonable range and can be used for future simulation.

2.3. FLUS Model

2.3.1. Parameters

The FLUS model is employed to realize the spatial layout of PLE land use. The detailed theory and process can be found in Liu's article [41]. Parameters mainly used in the model are described as follows:

(1) Suitability probability

The suitability probability of different land types on each parcel can be calculated based on the powerful, intelligent prediction function of the FLUS model's neural network by inputting the historical data of driving factors related to land-use evolution. The amount of future land demand is determined by other methods. In this study, SD is coupled with FLUS to simulate the spatial layout of the PLE space of Zhaotong city in 2030.

The selection of driving factors is crucial to the ANN-based suitability probability. This paper references previous literature studies [52–54]. Sixteen driving factors of different aspects of land use were taken into account, such as topography (including elevation, slope, and slope direction), natural meteorology (including soil conditions, precipitation, and annual mean air temperature), and social economy (including population, GDP, distance from the city center, commuting time). Data sources are shown in Table 3. All raster datasets were resampled to 30 m by the software Arcgis 10.2.

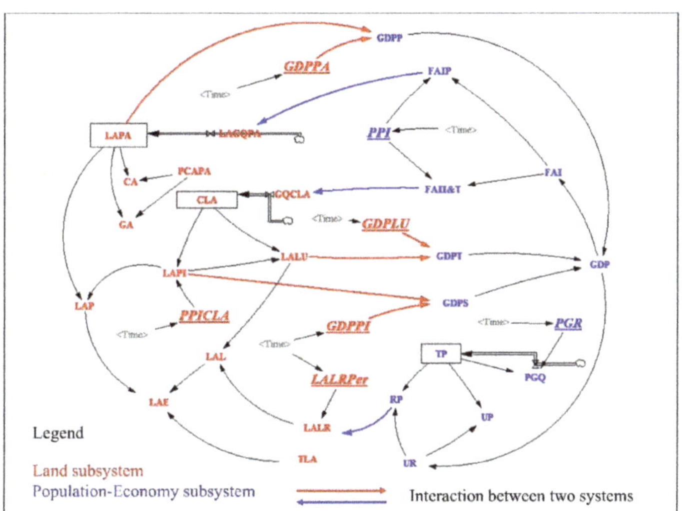

Figure 4. Interrelationship diagram of the PLE system dynamics model. Variables in underlined italics are control variables.

Table 3. Driving factors and data sources.

Categories	Data	Spatial Resolution	Data Source
Terrain and Landforms	Elevation, slope, and direction	90 m	Resource and Environment Science and Data Center http://www.resdc.cn/ (accessed on 2 April 2020)
Soil conditions	Nutrient availability Nutrient retention capacity Rooting conditions Oxygen availability to roots Excess salts Toxicity Workability (constraining field management)	10 km	Harmonized World Soil Database v 1.2 http://www.fao.org/soils-portal/data-hub/soil-maps-and-databases/harmonized-world-soil-database-v12/en/ (accessed on 2 April 2020)
Precipitation and Temperature	Annual average precipitation Annual average temperature	1 km	Resource and Environment Science and Data Center http://www.resdc.cn/ (accessed on 2 April 2020)
Socio-economic	Population	1 km	The Gridded Population of the World, Version 4 (GPWv4) https://doi.org/10.7927/H4F47 M65(accessed on 2 Apri 2020)
	GDP per capita	1 km	Dryad: Gridded global datasets for Gross Domestic Product and Human Development Index over 1990–2015 https://doi.org/10.5061/dryad.dk1j0 (accessed on 2 April 2020)
	Distance to city center	100 m	Euclidean distance calculated using city coordinate data
	Travel Time	1 km	European Commission Joint Research Centre Global Environment Monitoring Unit http://forobs.jrc.ec.europa.eu/products/gam/(2008) (accessed on 2 April 2020)

(2) Conversion cost matrix

The cost matrix reflects the conversion rules between different land types, in which 0 means that one type of land is not allowed to be converted to another, 1 means that it can be converted. The conversion cost matrix for different land types in this paper varies during model validation and scenario simulation.

(3)   Neighborhood factor values

Referring to the method of Ou et al. [55] and expert empirical knowledge, the neighborhood weights of the nine land types in this paper were determined, as shown in Table 4. The weight values range from 0 to 1. A value closer to 1 represents a stronger expansion capacity.

Table 4. Neighborhood factor values for different PLE subclasses.

ER	EP	EG	EA	PP	PG	PI	LU	LR
0.01	0.01	0.01	0.01	0.19	0.67	0.81	1.00	1.00

(4)   Other parameters

Another parameter is the time t in the formula, representing the number of iterations. More iterations mean more time and lower efficiency. To fully carry out spatial simulation and take simulation efficiency into account at the same time, the number of iterations is set as 600 and the acceleration factor as 0.5.

2.3.2. Verification

The feasibility and effectiveness of the FLUS model should be verified at first. In this paper, land use data from 2010 and 2015 are used as the base period data to simulate those of 2015 and 2018, respectively.

Firstly, the suitability probabilities of different land types in 2010 and 2015 were calculated using artificial neural networks (ANNs). Accuracy can be measured by the indicators RMSE (root mean square error), ROC (receiver operating characteristic), and AUC (area under curve) [41]. Smaller RMSE values and larger AUC values indicate higher model accuracy. Results showed that the RMSE was 0.2464 in 2010 and 0.2471 in 2015. AUC values corresponding to the ROC curves are shown in Table 5. ROC curves are shown in Figures S1 and S2.

Table 5. AUC values based on ANN suitability probability calculation results.

AUC	ER	EP	EG	EA	PP	PG	PI	LU	LR
2010	0.85	0.63	0.61	0.88	066	0.73	0.90	0.97	0.78
2015	0.84	0.63	0.62	0.91	0.68	0.72	0.84	0.98	0.77

Secondly, conversion cost matrices used for verification are obtained from the transition matrix between the years 2010~2015 and 2015~2018. If there is a conversion happened between two land classes during the above period, it is assigned as 1; otherwise, it is assigned as 0. They are shown in Tables S3 and S4, respectively. Other parameters are the same as those described in the Parameters section.

Results showed that compared with the actual data, the kappa coefficient (an indicator used to measure accuracy) of the 2015 simulation is 0.91, with an overall accuracy of 93.35%. The confusion matrix (a specific table layout that allows visualization of the performance of an algorithm in the field of machine learning) between the actual data and simulation is shown in Tables S5 and S6. Similarly, the kappa coefficient of the 2018 simulation is 0.85, with an overall accuracy of 88.99%. Both kappa coefficients are greater than or equal to 0.85, indicating that the FLUS model has high simulation accuracy and can be used for future simulations.

2.4. Constrained Scenarios

Base scenarios and optimization scenarios are designed to fully simulate the future PLE space layout of Zhaotong city in 2030. The optimization scenarios include Scenario A, in which production and living development are given priority; Scenario B, in which ecological protection is given priority; and Scenario C, in which both are considered. In addition, three levels (high level, medium level, and low level) are designed in every optimization scenario. The details are shown in Table 6.

Table 6. Scenario design and solution code.

Scenarios Name	Abbreviations		
Base Scenario	BS		
Optimization Scenarios	High level	Medium level	Low level
Giving priority to production-living development scenario (A)	A1	A2	A3
Giving priority to ecological protection scenario (B)	B1	B2	B3
Both are considered in scenario (C)	C1	C2	C3

There are two main aspects in which the above scenarios differ. On the one hand, the control variables' values are different in the system dynamics model for quantitative optimization. On the other hand, the transformation rules are different when using the FLUS model for spatial simulation.

Five different approaches are chosen as optimization projects in the PLE space system dynamics model, including land-use efficiency promotion, industrial structure adjustment, agricultural production space protection, intensive development of construction land, and controlling population growth. Relevant control variables in the model are listed in Table 7. The variables associated with the land-use efficiency promotion are mainly related to the GDP output per land (including GDPPA, GDPPI, and GDPLU). In the base scenario, i.e., at the previous rate of development, the values are 0.05, 11.58, and 9.10, respectively, by 2030. Based on the expert experience, on this basis, the optimization is carried out assuming that at high, medium, and low levels; GDPPA increases by 100%, 60%, and 30%, respectively; GDPPI increases by 60%, 30%, and 10%, respectively, and GDPLU increases by 120%, 80%, and 50% respectively. The values obtained are presented in Table 7. The values of the other variables for the different scenarios are also listed in Table 7.

Table 7. Base scenario and parameter settings at different levels.

Approaches	Variables [1]	Base Scenario	High Level	Medium Level	Low Level
Land-use efficiency promotion	GDPPA	0.05	0.05	0.04	0.03
	GDPPI	11.58	18.04	14.66	12.40
	GDPLU	9.10	11.58	9.47	7.89
Industrial structure adjustment	PPICLA	47	32	36	41
Agricultural production space protection	PPI	2.35	0.85	1.00	1.50
Controlling population growth	PGR	0.85	1.15	1.00	0.85
Intensive development of construction land	LALRPer	0.54	0.90	0.75	0.60

[1] Definition of variables can be found in Table S1.

The cost matrix of the base scenario is simple; that is, the urban living land use cannot be converted into another land use, and the conversion between other different land types is unrestricted. See Table S7 for the detailed cost matrix. No masking of the restricted area is performed.

The development scenario of giving production-living development priority (Scenario A) ensures that the improvement of production space and living space is fully considered. The expansion of production-living space comes at the expense of occupying ecological space. In the stage of spatial simulation, urban living space cannot be converted into others and is the same as priority agricultural production space and general agricultural production space (see Tables S8 and S9 for the detailed cost matrix). At the high and medium levels, the restricted ecological space vector scope of 2018 is used as a mask area; that is, the land parcels within the mask scope are no longer involved in the subsequent land-use conversion process. However, there is no mask at low levels.

The ecological space is fully protected in the context of Scenario B. For spatial simulation, urban living space cannot be converted into other land use and is the same as priority ecological space and general ecological space (see Tables S10 and S11 for the detailed cost matrix). Other settings are like Scenario A.

A balanced development scenario (Scenario C) means that the priority ecological space and priority agricultural production space are fully protected. During the spatial simulation, urban living space cannot be converted into another land use, and the same requirements are made for the priority ecological space and priority agricultural production space (see Tables S12 and S13 for the detailed cost matrix). Other settings are similar to Scenario A.

Only the industrial production space area, urban living space area, and rural living space area can be directly obtained from the simulation results of the system dynamic model. Other PLE space needs to be calculated according to the proportion of subclasses to the upper-class land in 2018, especially in the base scenario and Scenario B, in which ecological space protection is given priority. In Scenario A, the areas of priority agricultural production space and general agricultural production space were the same as those in 2018, and others were obtained by the above method. In Scenario C, the priority ecological space and priority agricultural production space were not less than those in 2018, and others were obtained by the above method.

3. Results

3.1. Spatiotemporal Pattern of PLE Space: Economic Development Is Related to the Decline of Ecological Space and General Agricultural Production Space

The spatial distribution of PLE spaces is shown in Figure 5. Restricted ecological space (ER) is mainly distributed in the western part of Zhaotong city, but it is more concentrated in the southwestern part. The priority ecological space (EP) is distributed in the south and northwest of the city. The general ecological space (EG) is more scattered. The area of accommodated ecological space (EA) is small. Priority agricultural production space (PP) is distributed in the city's central part and is more concentrated and contiguous, while general agricultural production space (PG) is in the northeastern part. In addition, industrial production space (PI), urban living space (LU), and rural living space (LR) are more scattered.

The land areas of the PLE spaces in 2010, 2015, and 2018 are shown in Table S14. Detailed subclass areas are shown in Table S15. Table S20 shows that the ecological space extends over an area of more than 16,200 km^2, accounting for more than 72% of the total, followed by the production space at approximately 6000 km^2, accounting for approximately 27%, and then by the living space, the smallest area at only approximately 1%. Among ecological spaces, the areas of priority ecological space and general ecological space accounted for more than 90%. Living spaces are mainly rural living space, accounting for approximately 70%.

In terms of changing trends, the ecological space area decreased from 2010 to 2018 but not significantly. Figure 6 shows that between 2010 and 2018, Zhaotong's GDP continued to increase, while at the same time, the space for priority ecological space continued to decrease. Conversely, production space decreased significantly from 6031 km^2 in 2010 to 6000 km^2 in 2018, which was mainly due to the decrease in general agricultural production space. There was a clear trend of growth in living space, from 194.67 km^2 in 2010 to 232.97 km^2 in 2018, with an increase of approximately 20%. This increase is mainly due to the expansion of urban living space, which was only 40.78 km^2 in 2010, reaching 72.76 km^2 in 2018, with an increase of approximately 78%.

The transfer matrix between 2010 and 2015 is shown in Table S16. The most significant growth between 2010 and 2015 was in industrial production space, which increased from 15.74 to 32.75 km^2. The expansion of industrial production space mainly encroached on priority agricultural production space (8.69 km^2) and priority ecological space (4.96 km^2), accounting for 51% and 29% of the expansion space, respectively. The main distribution of industrial space is in the central region. The most significant growth between 2015 and 2018 was in urban living space, which increased from 42.97 to 72.76 km^2 (Table S17). Urban living space expansion mainly encroached on priority agricultural production space (16.74 km^2) and industrial production space (12 km^2), accounting for 56% and 41% of the total expansion space, respectively. This change in use was concentrated in the central region.

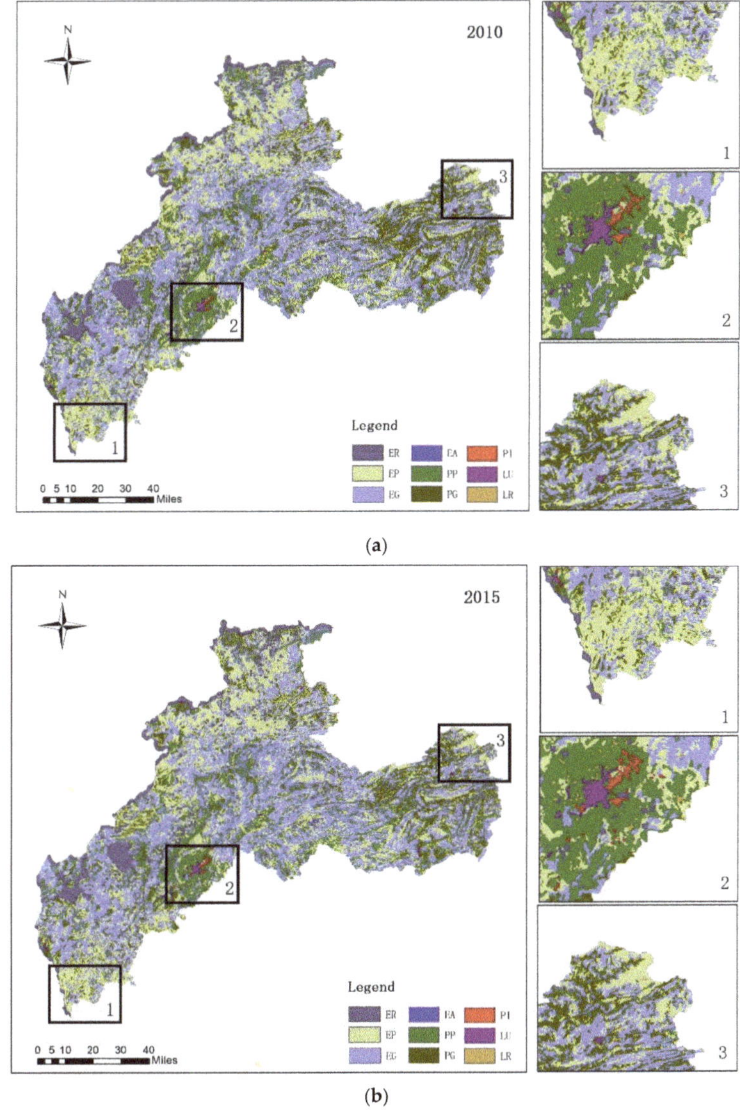

(a)

(b)

Figure 5. *Cont.*

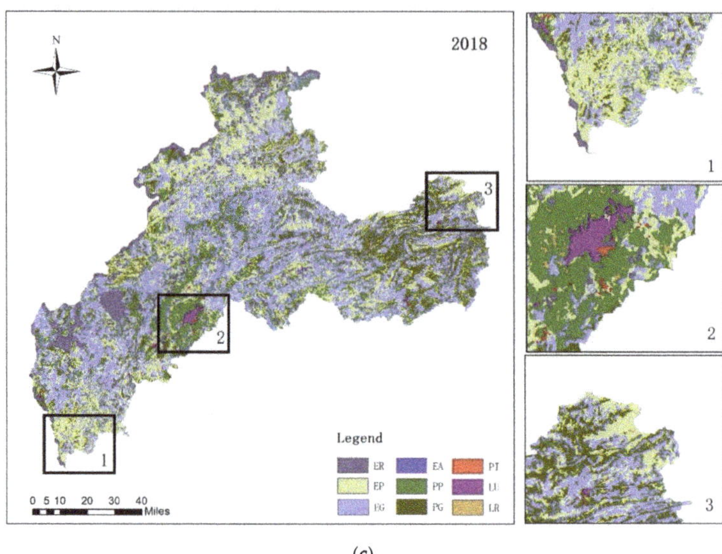

(c)

Figure 5. (**a–c**) represent the spatial distribution of PLE space in 2010, 2015, and 2018, respectively.ER is restricted ecological space. EP is priority ecological space. EG is general ecological space. EA is accommodated ecological space. PP is priority agricultural production space. PG is general agricultural production space. PI is industrial production space. LU is urban living space. LR is rural living space.

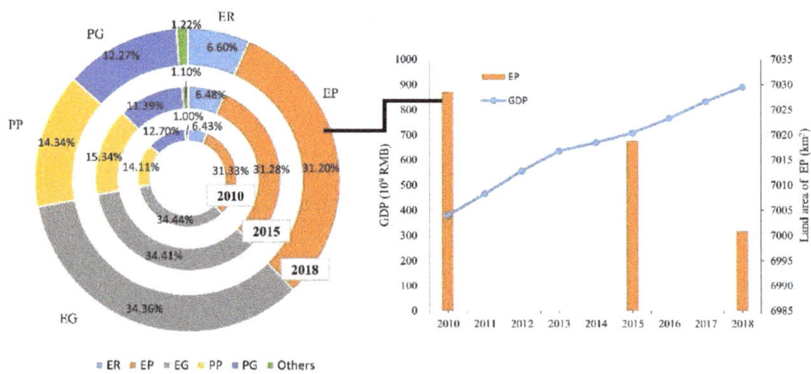

Figure 6. Changing trends of PLE space and GDP from 2010 to 2018.

3.2. Assessment of the Optimization Scenarios from the View of Total Amount: A High Level of Development Is More Conducive for Achieving the Planned Target of 2030

The planning goal for economic and social development in the "Urban Master Planning of Zhaotong City (2011–2030)" was used as a basis for reference. The economic development target is to reach a regional GDP of 410 billion RMB and a per capita GDP of approximately 67,000 RMB by 2030. For the social development target, the total population is 6.15 million, and the urbanization rate is 55%.

Based on the constructed SD model of PLE space, the simulated values of the main variables under different scenarios in 2030 were obtained, as shown in Table S18. This result indicates that under the base scenario, the regional GDP is 340.508 billion yuan by 2030, with a total population of 6,245,700 people, the GDP per capita of 54,500 yuan, and an urbanization rate of 68.33%. This means that under the base scenario, the regional GDP and GDP per capita cannot meet the planning target, although the population and urbanization rate can.

As shown in Table S18 and Figure 7, A1, B1, and C1 can meet the planning requirements. The remaining six scenarios show the opposite. Under Scenario A1, the regional GDP is 461.444 billion yuan by 2030, the total population is approximately 6.424 million, the GDP per capita is 71.8 thousand yuan, and the urbanization rate is 71.27%. Since the difference between Scenario B and Scenario C is mainly reflected in the constraints on agricultural space during spatial allocation, the quality is the same. Under Scenarios B and C, the regional GDP is 460.220 billion yuan, the total population is 6,424,400 people, the GDP per capita is 71,600 yuan, and the urbanization rate is 71.24%.

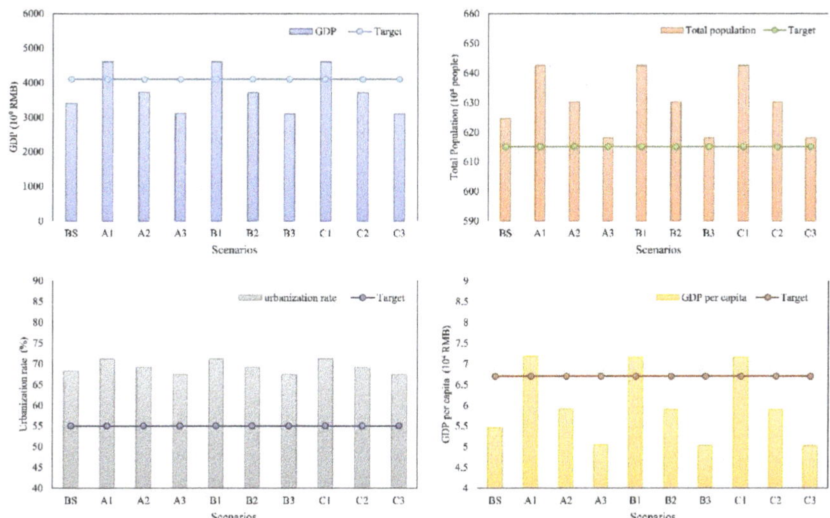

Figure 7. Development target in 2030 and simulation results of different scenarios.

3.3. Assessment of the Optimization Scenarios from the View of Spatial Configuration: High Levels of Development Will Lead to Further Expansion of Industrial Production Space and Living Space

The land area of PLE space in 2030 under different scenarios and the changes in terms of percentage compared with 2018 are listed in Table 8. Under the base scenario, urban living space (LU) is approximately 1.2 times larger than that in 2018, while rural living space (LR) decreases by approximately one-third (33.32%). The production land-use area shrinks from 6000.09 to 5760.48 km^2, a decrease of approximately 4%, and priority agricultural production space and general agricultural production space shrink by 5.91%, while industrial production space grows by 3.8 times. This implies that the expansion of urban living space and industrial production space is very evident in the base scenario, while priority agricultural production space and industrial production space are the most occupied.

Under Scenario A1, because of the protection of agricultural production space, the expansion of living space and industrial production space leads to different degrees of reduction in ecological space. Ecological space decreases by 1.49%, industrial production space and urban living space increase by 241.56% and 194.98%, and rural living space increases by 3.7%. Under Scenario B1, there is a slight decrease in ecological space (approximately 0.01%). The expansion of living space and industrial production space is at the expense of agricultural production space, resulting in a 3.66% decrease. Compared with Scenario B1, the priority agricultural production space is protected with constraints in Scenario C1, leading to the general agricultural production space being further reduced by 7.93%.

Table 8. Land area of the PLE spaces in 2030 and changes compared with 2018 under different scenarios.

	ER	EP	EG	EA	PP	PG	PI	LU	LR
2018	1481.17	7000.82	7710.14	11.74	3218.38	2752.14	29.57	72.76	160.21
BS	1499.88	7089.26	7807.54	11.89	3028.17	2589.49	142.82	161.06	106.83
Change%	1.26	1.26	1.26	1.28	−5.91	−5.91	382.99	121.36	−33.32
A1	1481.17	6896.57	7595.33	11.57	3218.38	2752.14	101.00	214.63	166.14
Change%	0.00	−1.49	−1.49	−1.49	0.00	0.00	241.56	194.98	3.70
A2	1481.17	6910.55	7610.73	11.59	3218.38	2752.14	110.44	196.33	145.60
Change%	0.00	−1.29	−1.29	−1.29	0.00	0.00	273.49	169.84	−9.12
A3	1466.79	6932.87	7635.31	11.63	3218.38	2752.14	122.65	176.49	120.67
Change%	−0.97	−0.97	−0.97	−0.97	0.00	0.00	314.77	142.57	−24.68
B1	1481.17	7000.36	7709.63	11.74	3100.69	2651.50	100.98	214.58	166.29
Change%	0.00	−0.01	−0.01	−0.01	−3.66	−3.66	241.49	194.91	3.80
B2	1481.17	7012.31	7722.80	11.76	3103.00	2653.48	110.41	196.29	145.72
Change%	0.00	0.16	0.16	0.16	−3.58	−3.58	273.39	169.77	−9.05
B3	1489.38	7039.63	7752.89	11.81	3085.17	2638.23	122.61	176.44	120.77
Change%	0.55	0.55	0.55	0.55	−4.14	−4.14	314.65	142.50	−24.62
C1	1481.17	7000.82	7709.17	11.74	3218.38	2533.80	100.98	214.58	166.29
Change%	0.00	0.00	−0.01	−0.01	0.00	−7.93	241.49	194.91	3.80
C2	1481.17	7012.31	7722.80	11.76	3218.38	2538.10	110.41	196.29	145.72
Change%	0.00	0.16	0.16	0.16	0.00	−7.78	273.39	169.77	−9.05
C3	1489.38	7039.63	7752.89	11.81	3218.38	2505.02	122.61	176.44	120.77
Change%	0.55	0.55	0.55	0.55	0.00	−8.98	314.65	142.50	−24.62

3.4. Optimization Pattern of the Coordinated Development: Production–Living Space Should Be Given Priority

The spatial optimization efficiency is evaluated by whether the spatial allocation can reach the future total amount required. When the FLUS model is used for the spatial layout, the number of requirements may not be reached for the same 600 iterations due to the different constraint rules. The percentage difference between the actual number of configurations and the demands is used as the spatial optimization efficiency indicator.

The future demand area of different land types needs to be proportionally converted to the number of land parcels in the FLUS model for spatial simulation. The gap between the actual allocation of parcels and the demand parcels is also presented in Table S19. In the base scenario, the expansion of rural living space far exceeds the planned demand by approximately 38%. In contrast, the number of allocated parcels of industrial production space differs from the number of demands by −22.63%, which means that it does not meet the target. In Scenario A1, 25.37% of the urban living space does not meet the future demand in the spatial configuration. In Scenario B1, neither urban living space nor industrial production space meets future needs; for example, the gap in urban living space is −22.14%. This is more obvious in Scenario C1, where 55.23% of the urban living space does not meet the demand.

Since the differences in several scenarios are mainly reflected in urban living space and industrial production space, the central and eastern regions are compared in separate enlargements (see Figure 8). Compared with 2018, the expansion of urban living space in the base scenario is mainly in the central region, centering on the original urban living space and expanding further outward. The expansion of industrial production space is also distributed in the central part but in a scattered manner around the central city, with the exception of two concentrated locations in the eastern region (see Figure 8).

Unlike the base scenario, the urban living space expands to the southwest with a small-scale agglomeration in Scenario A, and the same phenomenon occurs in the eastern part. In contrast, the expansion of industrial production space tends to extend to the south compared to the base scenario. The expansion of urban living space is very similar at different levels under Scenario B. Only Scenario B1 shows a clustering distribution in the northern part of the original urban living space and a relatively smaller degree of clustering in the eastern part. However, the industrial production space is not obvious in other scenarios except for a more obvious expansion at B1, especially in the northeastern region. In Scenario C, the expansion of urban living space and industrial production space is somewhat limited and only slightly expands compared to 2018, again mainly in the central and eastern regions.

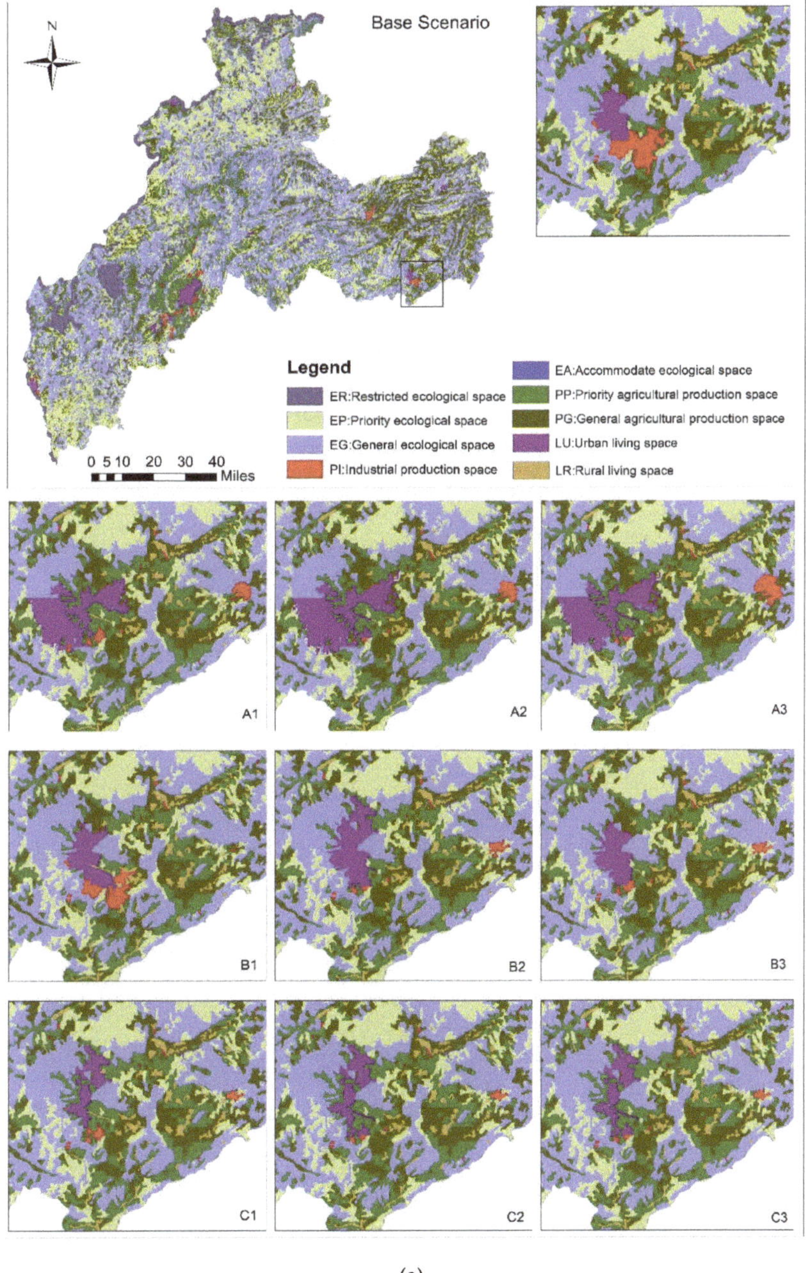

(**a**)

Figure 8. *Cont.*

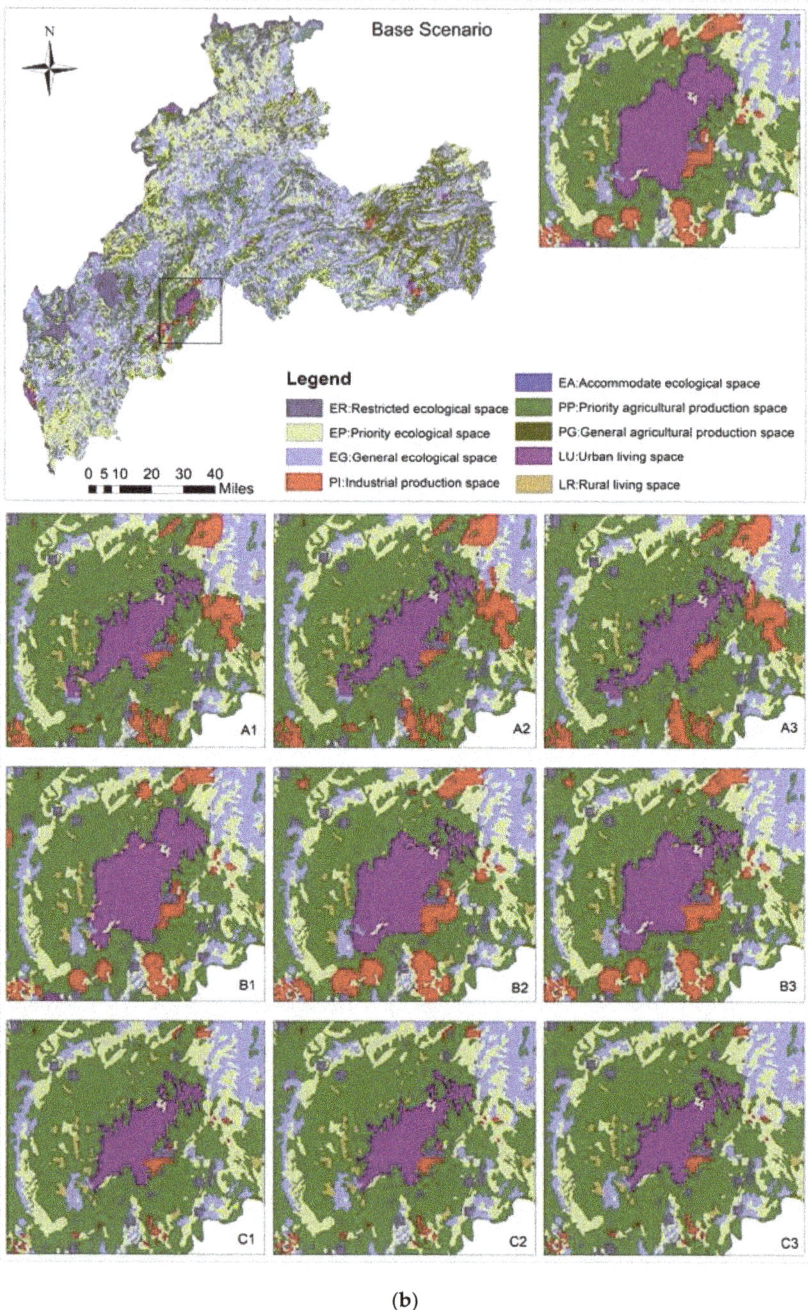

(**b**)

Figure 8. Spatial distribution of PLE space in different scenarios: (**a**) represent eastern region; (**b**) represent central region. And A1–C3 represent differen scenarios.

4. Conclusions and Implications

4.1. Findings

Results show that the model has high accuracy and can be combined with local policy to simulate the PLE space under future scenarios:

(1) Combined with the actual situation of Zhaotong city, the accuracy of the SD model is more than 95%. Absolute values of the errors are in the range of 0–5%, which indicates that the constructed model can respond to the interrelationships among the social and natural elements within a reasonable range and can be used for future simulation.

(2) The suitability probabilities of different land types were calculated using artificial neural networks (ANNs). Accuracy in terms of RMSE is both about 0.24. Additionally, AUC values corresponding to the ROC curves are all between 0.6~1. To a certain extent, it means that selective factors related to land use suitability are effective and the ANN method performs well.

(3) In the stage of spatial simulation by FLUS, confusion matrices describe the wrong allocation between simulated and actual spatial distribution. The related kappa coefficients are both above 0.85.

As a typical area with a backward economy and limited geographical conditions, Zhaotong city has unique characteristics of PLE space. Talking about the future development direction and whether the planning objectives can be achieved, the model constructed in this paper gives these answers:

(1) The ratio of ecological space, production space, and living space is approximately 7.2:2.7:0.1. Almost all the production space is agricultural production space, accounting for approximately 99%. Living space is dominated by rural living space, accounting for approximately 70%. Between 2010 and 2018, Zhaotong's GDP continued to increase, while at the same time, the space for priority ecological space continued to decrease. Zhaotong city has seen a significant expansion of living space and industry space in recent decades, which has greatly encroached on agricultural production space, especially in the central region, which is the core of the city.

(2) The planning requirements for 2030 can only be met under Scenarios A1, B1, and C1. This means that Zhaotong needs further improvements in land-use efficiency, industrial structure adjustments, and intensive development. However, considering suitability and space allocation rules, the demand for urban living space and industrial production space and the spatial allocation of land types could not be fully realized in the simulation time (600 iterations). Except for A3, where the configuration of all nine land types could meet the requirements, the demand for urban living space or industrial production space in other scenarios is far from satisfied.

(3) Production and living space should be given priority in future development. A higher level of development is still needed in terms of policies and measures. When planning future PLE space, the layout of urban living space and industrial production space should be given priority, especially in the central and south-central regions as well as the eastern region.

4.2. Contributions

The study of the PLE space is important for the planning of regional sustainable development. Complicatedly, it involves geographical, ecological, and economic knowledge and experiments, requiring integrated research. This paper built an integrated optimization model by which PLE space can be simulated quantitatively and spatially, which enriches the current study of PLE. Additionally, it can be a powerful tool.

The SD model constructed takes local demographic and economic factors into account from a system science perspective. The mathematical relationships between the main factors are quantified by establishing equation and simulation parameters as a way to simulate the mechanism of urban operation. The FLUS model, based on the suitability probability using an ANN, is carried out in a roulette wheel way by considering the cost matrix, neighborhood influence, and other factors. For example, some constraints, as well as planning objectives, are embedded into the model in a parametric way, and the land allocation operations in the actual planning process are restored by different land class conversion rules.

In contrast to the multi-objective optimization model, which is often used to find the optimal solution [32], the coupled model in this paper can simulate the urban operation mechanism at a more microscopic scale and perform dynamic derivation in both time and space dimensions simultaneously with more flexible parameter adjustment. Moreover, this paper sets a variety of development scenarios in the form of parameters for the relevant indicators in local policies and assists in decision-making based on the simulation results. For Zhaotong city, economic development and environmental protection are advocated to be given equal importance. For other cities that also face urban development planning problems, the case study of this paper can provide inspiration.

4.3. Limitations and Future Research

Like all research, there are some limitations to this study. For example, there is no clearly defined relationship between the urbanization rate and GDP, and the urbanization rate is influenced by many factors [38,56]. To simplify the SD model, the equation parameters between the two can only be determined using regression analysis and referring to the development trend of other cities. In addition, adequate surveys involving stakeholders and local policy can also improve the rationality of the scenario simulation. However, this requires more comprehensive and in-depth preliminary research, which is also an aspect that can be considered for further studies.

In addition, only taking Zhaotong, a city located in southwest China, as a case study is far from enough. For planning problems in other cities with different geographical and natural endowments, the applicability of the model needs more work to be carried out to verify the model. Additionally, urban sustainability is a common issue around the world, but the solutions to different problems need to be contextualized in a variety of ways. Regardless of the approach taken, the rights of stakeholders should be considered when making decisions [57,58].

Supplementary Materials: The following supporting information can be downloaded at: https://www.mdpi.com/article/10.3390/land11030411/s1, Table S1: Names, units and data sources of the variables included in the PLE space system dynamics model; Table S2: Comparison of simulated and real values of some key variables; Table S3: Cost Matrix for the 2015 simulation using 2010 PLE spatial distribution data; Table S4: Cost Matrix for the 2015 simulation using 2010 PLE spatial distribution data; Table S5: Confusion matrix between the actual and simulated values of PLE in 2015; Table S6: Confusion matrix between the actual and simulated values of PLE in 2018; Table S7: Cost Matrix for Base Scenario; Table S8: Cost Matrix for Scenario A1 and A2; Table S9: Cost Matrix for Scenario A3; Table S10: Cost Matrix for Scenario B1 and B2; Table S11: Cost Matrix for Scenario B3; Table S12: Cost Matrix for Scenario C1 and C2; Table S13: Cost Matrix for Scenario C3; Table S14: Ratio of land area for production, living and ecological space in 2010, 2015, and 2018; Table S15: Land area and percentage of 9 subclasses in PLE space in 2010, 2015, and 2018; Table S16: Land-use transition matrix from 2010 to 2015; Table S17: Land-use transition matrix from 2015 to 2018; Table S18: Projections of the economy and population under different scenarios in 2030; Table S19: Difference between future demand and allocated values for different scenarios; Figure S1: ROC curves and AUC values for 2010 suitability probability results based on ANN; Figure S2: ROC curves and AUC values for 2015 suitability probability results based on ANN.

Author Contributions: D.W.: conceptualization; methodology; validation; formal analysis; investigation; writing—original draft preparation; writing—review and editing; visualization. J.F.: conceptualization; writing—review and editing; supervision; project administration. D.J.: supervision; project administration; funding acquisition. All authors have read and agreed to the published version of the manuscript.

Funding: This research was funded by the Strategic Priority Research Program of the Chinese Academy of Sciences (XDA19040305), Youth Innovation Promotion Association (Grant No. 2018068).

Institutional Review Board Statement: Not applicable.

Informed Consent Statement: Not applicable.

Data Availability Statement: Not applicable.

Conflicts of Interest: The authors declare no conflict of interest.

References

1. Galvani, A.P.; Bauch, C.T.; Anand, M.; Singer, B.H.; Levin, S.A. Human-environment interactions in population and ecosystem health. *Proc. Natl. Acad. Sci. USA* **2016**, *113*, 14502–14506. [CrossRef]
2. Lin, C.L. Evaluating the urban sustainable development strategies and common suited paths considering various stakeholders. *Environ. Dev. Sustain.* **2022**, 1–42. [CrossRef]
3. What are Coupled Human-Environment System? GEOG: 30N: Environment and Society in a Changing World. Available online: https://www.e-education.psu.edu/geog30/node/325 (accessed on 15 February 2022).
4. Scholz, R.W.; Binder, C.R. Princles of Human-Environment Systems (HES) Research. In Proceedings of the 2nd International Congress on Environmental Modelling and Software, Osnabrück, Germany, 1–7 June 2004; Volume 116. Available online: https://scholarsarchive.byu.edu/iemssconference/2004/all/116 (accessed on 26 February 2022).
5. Turner, B.L.; Matson, P.A.; McCarthy, J.J.; Corell, R.W.; Christensen, L.; Eckley, N.; Hovelsrud-Broda, G.K.; Kasperson, J.X.; Kasperson, R.E.; Luers, A.; et al. Illustrating the coupled human-environment system for vulnerability analysis: Three case studies. *Proc. Natl. Acad. Sci. USA* **2003**, *100*, 8080–8085. [CrossRef]
6. Katramiz, T.; Okitasari, M. *Accelerating 2030 Agenda Integration: Aligning National Development Plans with the Sustainable Development Goals*; United Nations University Institute for the Advanced Study of Sustainability (UNU-IAS): Tokyo, Japan, 2021; Volume 25.
7. Australian Government, Department of Agriculture, Water and the Environment. 2030 Agenda for Sustainable Development and the Sustainable Development Goals. Available online: https://www.awe.gov.au/environment/international/2030-agenda (accessed on 15 February 2022).
8. Government of Canada. Towards Canada's 2030 Agenda National Strategy. Available online: https://www.canada.ca/en/employment-social-development/programs/agenda-2030/national-strategy.html#h2.06 (accessed on 15 February 2022).
9. The Department of Strategic Policy, Planning and Aid Coordination, Republic of Vanuatu. *Vanuatu 2030 the People Plan: National Sustainable Development Plan 2016 to 2030*; The Department of Strategic Policy, Planning and Aid Coordination, Republic of Vanuatu: Port Vila, Vanuatu, 2016. Available online: https://www.nab.vu/sites/default/files/documents/Vanuatu%20Sustainable%20Dev.%20Plan%202030-EN_0.PDF (accessed on 15 February 2022).
10. Barbier, E.B.; Burgess, J.C. Sustainable development goal indicators: Analyzing trade-offs and complementarities. *World Dev.* **2019**, *122*, 295–305. [CrossRef]
11. United Nations. *The Sustainable Development Goals Report 2018*; United Nations: New York, NY, USA, 2018.
12. Nilsson, M.; Griggs, D.; Visbeck, M. Map the interactions between Sustainable Development Goals. *Nature* **2016**, *534*, 320–322. [CrossRef]
13. Von Stechow, C.; Minx, J.C.; Riahi, K.; Jewell, J.; McCollum, D.L.; Callaghan, M.W.; Bertram, C.; Luderer, G.; Baiocchi, G. 2 degrees C and SDGs: United they stand, divided they fall? *Environ. Res. Lett.* **2016**, *11*, 15. [CrossRef]
14. Wang, Y.P.; Zhou, X.N. The year 2020, a milestone in breaking the vicious cycle of poverty and illness in China. *Infect. Dis. Poverty* **2020**, *9*, 8. [CrossRef]
15. Yang, Y.Y.; de Sherbinin, A.; Liu, Y.S. China's poverty alleviation resettlement: Progress, problems and solutions. *Habitat Int.* **2020**, *98*, 13. [CrossRef]
16. Bao, C.; He, D.M. Scenario Modeling of Urbanization Development and Water Scarcity Based on System Dynamics: A Case Study of Beijing-Tianjin-Hebei Urban Agglomeration, China. *Int. J. Environ. Res. Public Health* **2019**, *16*, 19. [CrossRef]
17. Chen, J. Rapid urbanization in China: A real challenge to soil protection and food security. *Catena* **2007**, *69*, 1–15. [CrossRef]
18. Bai, X.M.; Chen, J.; Shi, P.J. Landscape Urbanization and Economic Growth in China: Positive Feedbacks and Sustainability Dilemmas. *Environ. Sci. Technol.* **2012**, *46*, 132–139. [CrossRef] [PubMed]
19. Wang, Q. Effects of urbanisation on energy consumption in China. *Energy Policy* **2014**, *65*, 332–339. [CrossRef]
20. Huang, J.C.; Lin, H.X.; Qi, X.X. A literature review on optimization of spatial development pattern based on ecological-production-living space. *Prog. Geogr.* **2017**, *36*, 378–391.
21. Sachs, J.D. From Millennium Development Goals to Sustainable Development Goals. *Lancet* **2012**, *379*, 2206–2211. [CrossRef]
22. Bansal, P. Evolving sustainably: A longitudinal study of corporate sustainable development. *Strateg. Manag. J.* **2005**, *26*, 197–218. [CrossRef]
23. Saci, H.; Berezowska-Azzag, E. Food security and urban sustainability of alternative food models: Multicriteria analysis based on Sustainable Development Goals and Sustainable Urban Planning. *Cah. Agric.* **2021**, *30*, 35. [CrossRef]
24. Grah, B.; Dimovski, V.; Peterlin, J. Managing Sustainable Urban Tourism Development: The Case of Ljubljana. *Sustainability* **2020**, *12*, 792. [CrossRef]
25. Khosravi, S.; Lashgarara, F.; Poursaeed, A.; Najafabadi, M.O. Modeling the relationship between urban agriculture and sustainable development: A case study in Tehran city. *Arab. J. Geosci.* **2022**, *15*, 97. [CrossRef]

26. Balletto, G.; Ladu, M.; Milesi, A.; Camerin, F.; Borruso, G. Walkable City and Military Enclaves: Analysis and Decision-Making Approach to Support the Proximity Connection in Urban Regeneration. *Sustainability* **2022**, *14*, 457. [CrossRef]

27. Wu, Z.Y.; Shan, J.J. Study on the Evolution and Optimization Countermeasures of "Ecological-production-living Space" Pattern. *City* **2019**, *10*, 15–26.

28. Zhang, C.H. Study on Conflict Measure and Optimization of Coastal Zone Based on the Ecological-Production-Living Spaces—A Case Study of Zhuang He in Dalian. Ph.D. Thesis, Liaoning Normal University, Dalian, China, June 2018.

29. Wang, J. Study on spatial reconstruction and optimization of "sansheng" in mining area—Take dexing copper mine in jiangxi province as an example. Master's Thesis, East China University of Technology, Nanchang, China, June 2016.

30. Zhou, L.Q. The Study on the Optimization and Function Improvement of "Ecological-Production-Living" Space in Zhengzhou. Master's Thesis, Zhengzhou University, Zhengzhou, China, May 2018.

31. Anderson, C.B.; Seixas, C.S.; Barbosa, O.; Fennessy, M.S.; Diaz-Jose, J.; Herrera-F., B. Determining nature's contributions to achieve the sustainable development goals. *Sustain. Sci.* **2019**, *14*, 543–547. [CrossRef]

32. Ye, Y.C. Spatial Layout Optimization of Industrial-Living-Ecological Land in Yingtan City Based on Spatial Decision-Making. Ph.D. Thesis, Jiangxi Agricultural University, Nanchang, China, June 2018.

33. Gunantara, N.J.C.E. A review of multi-objective optimization: Methods and its applications. *Cogent Eng.* **2018**, *5*, 1502242. [CrossRef]

34. Wang, H.D.; Olhofer, M.; Jin, Y.C. A mini-review on preference modeling and articulation in multi-objective optimization: Current status and challenges. *Complex Intell. Syst.* **2017**, *3*, 233–245. [CrossRef]

35. Yeh, S.C.; Wang, C.A.; Yu, H.C. Simulation of soil erosion and nutrient impact using an integrated system dynamics model in a watershed in Taiwan. *Environ. Modell. Softw.* **2006**, *21*, 937–948. [CrossRef]

36. Wei, S.K.; Yang, H.; Song, J.X.; Abbaspour, K.C.; Xu, Z.X. System dynamics simulation model for assessing socio-economic impacts of different levels of environmental flow allocation in the Weihe River Basin, China. *Eur. J. Oper. Res.* **2012**, *221*, 248–262. [CrossRef]

37. Wingo, P.; Brookes, A.; Bolte, J. Modular and spatially explicit: A novel approach to system dynamics. *Environ. Modell. Softw.* **2017**, *94*, 48–62. [CrossRef]

38. Gu, C.L.; Ye, X.Y.; Cao, Q.W.; Guan, W.H.; Peng, C.; Wu, Y.T.; Zhai, W. System dynamics modelling of urbanization under energy constraints in China. *Sci. Rep.* **2020**, *10*, 16. [CrossRef]

39. Abadi, L.S.K.; Shamsai, A.; Goharnejad, H. An analysis of the sustainability of basin water resources using Vensim model. *KSCE J. Civ. Eng.* **2015**, *19*, 1941–1949. [CrossRef]

40. Lloyd, H.; Amos, M. Analysis of Independent Roulette Selection in parallel ant colony optimization. In Proceedings of the Genetic and Evolutionary Computation Conference; Association for Computing Machinery: New York, NY, USA, 2017; pp. 19–26.

41. Liu, X.P.; Liang, X.; Li, X.; Xu, X.C.; Ou, J.P.; Chen, Y.M.; Li, S.Y.; Wang, S.J.; Pei, F.S. A future land use simulation model (FLUS) for simulating multiple land use scenarios by coupling human and natural effects. *Landsc. Urban Plan.* **2017**, *168*, 94–116. [CrossRef]

42. Liang, X.; Liu, X.P.; Li, X.; Chen, Y.M.; Tian, H.; Yao, Y. Delineating multi-scenario urban growth boundaries with a CA-based FLUS model and morphological method. *Landsc. Urban Plan.* **2018**, *177*, 47–63. [CrossRef]

43. Liang, X.; Liu, X.P.; Li, D.; Zhao, H.; Chen, G.Z. Urban growth simulation by incorporating planning policies into a CA-based future land-use simulation model. *Int. J. Geogr. Inf. Sci.* **2018**, *32*, 2294–2316. [CrossRef]

44. Lin, W.B.; Sun, Y.M.; Nijhuis, S.; Wang, Z.L. Scenario-based flood risk assessment for urbanizing deltas using future land-use simulation (FLUS): Guangzhou Metropolitan Area as a case study. *Sci. Total Environ.* **2020**, *739*, 10. [CrossRef] [PubMed]

45. Mooney, H.A.; Duraiappah, A.; Larigauderie, A. Evolution of natural and social science interactions in global change research programs. *Proc. Natl. Acad. Sci. USA* **2013**, *110*, 3665–3672. [CrossRef] [PubMed]

46. Mortada, S.; Abou Najm, M.; Yassine, A.; El Fadel, M.; Alamiddine, I. Towards sustainable water-food nexus: An optimization approach. *J. Clean. Prod.* **2018**, *178*, 408–418. [CrossRef]

47. Veldhuis, A.J.; Yang, A.D. Integrated approaches to the optimisation of regional and local food-energy-water systems. *Curr. Opin. Chem. Eng.* **2017**, *18*, 38–44. [CrossRef]

48. Nie, Y.L.; Avraamidou, S.; Xiao, X.; Pistikopoulos, E.N.; Li, J.; Zeng, Y.J.; Song, F.; Yu, J.; Zhu, M. A Food-Energy-Water Nexus approach for land use optimization. *Sci. Total Environ.* **2019**, *659*, 7–19. [CrossRef] [PubMed]

49. Zhaotong Bureau of Statistics. *Zhaotong Statistical Yearbook 2018*; Zhaotong Bureau of Statistics: Zhaotong, China, 2018.

50. Forest Resources Overview of Yunnan Province in 2018. Available online: http://lcj.yn.gov.cn/html/2019/zuixindongtai_1219/55160.html (accessed on 2 April 2020).

51. Xie, H.L.; Zhang, Y.W.; Zeng, X.J.; He, Y.F. Sustainable land use and management research: A scientometric review. *Landsc. Ecol.* **2020**, *35*, 2381–2411. [CrossRef]

52. Li, M.; Wu, J.J.; Deng, X.Z. Identifying Drivers of Land Use Change in China: A Spatial Multinomial Logit Model Analysis. *Land Econ.* **2013**, *89*, 632–654. [CrossRef]

53. Ni, J.P.; Shao, J.A. The Drivers of Land Use Change in the Migration Area, Three Gorges Project, China: Advances and Prospects. *J. Earth Sci.* **2013**, *24*, 136–144. [CrossRef]

54. Xu, X.M.; Jain, A.K.; Calvin, K.V. Quantifying the biophysical and socioeconomic drivers of changes in forest and agricultural land in South and Southeast Asia. *Glob. Chang. Biol.* **2019**, *25*, 2137–2151. [CrossRef]

55. Ouyang, X.; He, Q.; Zhu, X. Simulation of Impacts of Urban Agglomeration Land Use Change on Ecosystem Services Value under Multi-Scenarios: Case Study in Changsha-Zhuzhou-Xiangtan Urban Agglomeration. *Econ. Geogr.* **2020**, *40*, 93–102.
56. Chen, M.X.; Zhang, H.; Liu, W.D.; Zhang, W.Z. The Global Pattern of Urbanization and Economic Growth: Evidence from the Last Three Decades. *PLoS ONE* **2014**, *9*, 15. [CrossRef] [PubMed]
57. Stein, S. *Capital City: Gentrification and the Real Estate State*; Verso Books: Brooklyn, NY, USA, 2019.
58. Sklair, L. *The Icon Project: Architecture, Cities, and Capitalist Globalization*; Oxford University Press: New York, NY, USA, 2017.

Article

Units of Military Fortification Complex as Phenomenon Elements of the Czech Borderlands Landscape

Jiří Kupka [1], Adéla Brázdová [2,*] and Jana Vodová [1]

[1] Department of Environmental Engineering, Faculty of Mining and Geology, VSB–Technical University of Ostrava, 17. listopadu 2172/15, 708 00 Ostrava, Czech Republic; jiri.kupka@vsb.cz (J.K.); jana.prymusova@gmail.com (J.V.)

[2] Department of Building Materials and Diagnostics of Structures, Faculty of Civil Engineering, VSB–Technical University of Ostrava, 17. listopadu 2172/15, 708 00 Ostrava, Czech Republic

* Correspondence: adela.brazdova@vsb.cz

Abstract: This paper is focused on selected units of casemates with enhanced fortification in the military fortification complex of the Czech borderlands landscape as specific forms of brownfields. They represent a functional system that interacts with surrounding nature, landscape character, and human society. Four approaches were chosen to study the function and potential of selected individual abandoned casemates with enhanced fortification, where each of them corresponds to one of the four landscape layers: genius loci, socio-economic sphere, functional relationship (between human and the landscape), and natural conditions. There is a corresponding research method for each of the landscape layers (guided interview with respondents, data analysis on abandoned casemates with enhanced fortifications as brownfields, analysis of their landscape functions, and zoological survey of interior). The main results could show that abandoned casemates with enhanced fortifications can play important roles in all landscape layers: stories and genius loci, abandoned casemates with enhanced fortification as a special type of military brownfield but also as a semi-natural ecosystem, and the same time as a habitat for invertebrates. The analyses and surveys conducted clearly demonstrate that abandoned casemates with enhanced fortification as units of military fortification complex of the Czech borderlands landscape perform several hidden important functions in the landscape for which they cannot be viewed as brownfields. This hidden functional potential is most likely best described by the concept of hidden singularity, which offers itself for integration into basic approaches to brownfields.

Keywords: brownfields; military fortification brownfields; casemates with enhanced fortification; historical and fabricated stories; semi-natural ecosystem; hidden curriculum; butterflies and moths (Lepidoptera); land snails (Gastropoda); hidden singularity

Citation: Kupka, J.; Brázdová, A.; Vodová, J. Units of Military Fortification Complex as Phenomenon Elements of the Czech Borderlands Landscape. *Land* **2022**, *11*, 79. https://doi.org/10.3390/land11010079

Academic Editors: Dong Jiang, Jinwei Dong and Gang Lin

Received: 23 November 2021
Accepted: 1 January 2022
Published: 5 January 2022

Publisher's Note: MDPI stays neutral with regard to jurisdictional claims in published maps and institutional affiliations.

1. Introduction

Brownfields in general can represent one of the key environmental problems. Although they may not always be associated with ecological burden, they always interact with the human, the landscape, and with the surrounding nature. These may be sites that are abandoned, underused, but may also be historically or architecturally significant. The regeneration of brownfields is one of the basic strategies for improving conditions not only in the urban environment (regeneration strategies vary across Europe, as does the definition of brownfields itself) [1]. It is the interactions between people, brownfields and their associated stories, landscape, and nature which this paper addresses.

After the departure of the Soviet army and following the fall of the Iron Curtain (in 1991), abandoned military buildings became a major issue in the Czech Republic, e.g., complexes of buildings such as barracks, shooting ranges, and other buildings and lands too [2].

The former line of Czechoslovak fortifications consists of Casemates with Enhanced Fortification (CEFs), heavy fortifications, and artillery forts. All of these units have since lost their significance and have become abandoned—and they can now be marked as specific types of military brownfields [3]. The same line of Czechoslovak fortifications, also consists of Abandoned Casemates with Enhanced Fortifications (A-CEFs), has become an integral part of the cultural landscape, which itself is the result of millennia of interaction between nature, man, and his activities. The cultural landscape has been influenced by the military landscape.

Similar military landscapes from different periods of history can be found in other parts of Europe (e.g., Vallo Alpino), and consequently throughout the world (e.g., Great Wall of Gorgan) [4,5]. Some relics of post-military landscapes are even included in the World Heritage List (e.g., Atlantic Wall, The Great Wall in China) [6]. Military landscape (more precisely post-military landscape) is the result of the interaction of natural and anthropogenic factors (economic, technical, political, and cultural human influence) that are bound to a specific area with a common history [7–13]. Human military activity affects not only the appearance but also the structure and function of the landscape. This is evident in the case of the fortification lines [11,14]. The post-military landscape also becomes part of the evidence of the historical development and therefore part of the cultural heritage of the area [15]. In the case of abandoned military objects (in our study A-CEFs units) the landscape acquires new character and function, possible variability of the use of these objects but an 'atmosphere' connected with these objects 'remains' in them [16,17].

The landscape can be characterized by layers, which we perceive as the result of the relationships of its individual components that change dynamically over time [9,18,19]. We can recognize four layers of landscape—genius loci, socio-economic sphere, functional relationship (between human and the landscape), and natural conditions. **Genius loci** are the spirit of the place, respectively of the landscape [20]. This layer is also the first aspect that causes the interaction of humans with the landscape (evokes emotions). We consider the **'socio-economic sphere'** as a cultural heritage, human creations and their history but also recent use of the landscape, and the spiritual perception of the landscape. The socio-cultural sphere describes the functional relationship of the landscape with humans. The **'Functional relationship'** of the landscape with humans should be defined by the socio-cultural sphere. **'Natural conditions'** include living and non-living nature, including natural processes and occurrences.

Military fortification units represent an immense fortification system of Czechoslovakia (1918–1938). These units were built in 1935–1938, just before the outbreak of the Second World War (WWII), and this line was never fully completed [21,22]. In general, it is composed of a system of strategically placed CEFs units (Figure 1), heavy fortifications, and artillery forts, especially in border areas [22–24]. The whole system of Czechoslovak fortifications was inspired by the model of the Maginot Line, which was a system of fortifications built in France after the experience of the First World War [22,24]. As a result of the Munich Agreement (September 1938), the territory where the Czechoslovak fortification system was located, fell to the then German government [21–23,25]. The fortification system thus failed to fulfill its expected defensive purpose, as after 30 September 1938 all objects were abandoned by the Czechoslovak army and subsequently occupied by German troops [21–23,25]. At the end of WWII, some of the buildings served as strong points for the German army against the advancing Soviet army and were damaged during battle [22,24]. In the post-war period (1945–1989) the objects gradually lost their military significance [24–26]. Only a few segments of the whole Czechoslovak fortification system were renovated for the purpose of building the so-called Iron Curtain, adapted as fallout shelters, or used as storage facilities for military material [25].

Figure 1. CEFs units on the whole territory of today's Czech Republic—noticeable continuing line to Slovakia; triangle marks are CEFs units from 1936, dot marks are CEFs units from 1937–1938 [27–29].

Some of the fortification units were destroyed as a result of devastation during the liberation battles or as an obstacle to technical infrastructure (quarries, transport infrastructure, etc.) [27]. The remaining buildings were used, for example, as warehouses for various materials (e.g., fruits and vegetables, fertilizers, or sprays) or remained abandoned [30–32]. Only a few segments of the military fortification units are protected as a cultural monument and may be marked for our purposes as UA-CEFs (Used Abandoned Casemates with Enhanced Fortification) [33]. Currently, some fortification objects are offered for sale to private ownership [25,34]. We can therefore conclude that as a result of historical events, a significant part of the fortification system built in 1935–1938 became a unique type of military brownfields almost immediately after the end of WWII.

The construction of the Czechoslovakia fortifications complex in the territory of the Moravian-Silesian Region (according to the current territorial division of the Czech Republic) began in the Ostrava region. It was expected that the enemy would make the greatest offensive here. The fortification system was also intended to serve as protection for the industrial area (the so-called 'steel heart of the republic') [24,35]. The complete line of fortifications (CEFs and heavy fortifications) then continued along the border with Germany towards western and southern Bohemia [24,36]. Another line ran through southern Moravia and ended at Bratislava but was originally intended to continue further east to Košice (border with Hungary) [24,36].

The subject of this study, as an example of selected Military fortification units as specific forms of brownfields, is an A-CEF 'model 37' [27,32]. This construction is reinforced concrete with a front wall and ceiling thickness of 80 or 120 cm (normal or reinforced modification) [27,32]. From the direction of the expected enemy attack, the fortress was additionally provided with an embankment made out of boulders and covered with a layer of earth and grass, which further strengthened and camouflaged the object [22,32]. The entrance to the building consisted of a bar and one armored door (at right angles to the bar) [22,27]. Over time, five infantry types were designed in three basic levels of resistance, which could be used to protect any terrain without the need to build atypical solutions. This system simplified, accelerated, and reduced construction costs thanks to the possibility of using standardized internal equipment [21,25,27]. At the same time, the building was equipped with an entrance loophole, grenade chutes, and one or two periscopes in the ceiling of the fortress [25]. The crew was made up of 4 to 6 men, while the size of the interior space was about 8 m^2 [32].

These objects as brownfields represented by the Czechoslovak fortification units (A-CEFs) are an integral part of the Czech border post-military landscape. Therefore, we can also speak of the objects of the Czechoslovak fortifications as a phenomenon of the post-military landscape. A part of the post-military landscape with A-CEFs is shown in Figure 2.

Figure 2. Visible line of Czechoslovak fortification complex (A-CEFs) in part of Moravian-Silesian Region, on part of the section 'Milostovice'; photographed from drone.

The aim of the study is to provide a conceptual approach to A-CEFs as specific types of brownfield, while introducing and describing the roles which A-CEFs play in the landscape and how these roles relate to the definition of brownfields. For this reason, four approaches were chosen to study the function and potential of individual A-CEFs. Each of these approaches corresponds to one of the layers that can be recognized in the landscape [37]:

The first approach corresponds to the landscape layer of **genius loci** of the place—the aim of the search is stories [38]. Our aim is to determine whether or not A-CEFs fulfill this function. At first glance, it is evident that A-CEFs have lost their original function. Although they were abandoned, they can still play an important function in the hidden curriculum of the landscape. The term hidden curriculum is borrowed from the field of education. This refers to the hidden lessons of education that are taught by the school and that do not follow the official plans and intentions of the school system or teachers (as opposed to the regular "curriculum", which is the official content of education in the broader sense). The hidden curriculum of the landscape refers to activities of an unofficial character (e.g., off-trail walking, camping outside designated areas, entering bunkers), as opposed to official use (e.g., in the context of tourism, use of conventional accommodation, walking on designated hiking trails, etc.) [39]. In some cases, the hidden curriculum may also be illegal (e.g., entering abandoned mines, abandoned buildings).

The second approach corresponds to the **socio-economic sphere** of the landscape layer (human creations, their history, etc.) [37]. In our case, we select anthropogenic elements from the post-military landscape, which at first glance have lost their function and therefore meet the definition of a brownfield. On the other hand, the first approach shows that A-CEFs can still perform certain functions in the landscape (hidden curriculum). From the point of view of this approach, these objects have their historical value. Not only UA-CEFs (e.g., as museums), but also A-CEFs represent specific elements in the landscape that can perform certain functions but at the same time fall into the category of military brownfields. The problematic question of the second approach is: Are A-CEFs 'real' brownfields and is their remediation therefore necessary?

The third approach corresponds to the **functional relationship between human creations and landscape** as layers [37]. In our case, we are focusing on a landscaped enclave influenced by human activity. This enclave is directly formed by the A-CEFs or UA-CEFs themselves and their immediate surroundings. To identify the maximum possible use of anthropogenic elements in the post-military landscape is necessary to look for their function and potential. The problematic question is: What is the significance of these objects in landscape interactions?

The fourth approach corresponds to **natural conditions** (living and non-living nature, natural processes, and occurrences) as the landscape layer [40]. In our case, this involves obtaining biological data from a field survey. We look at post-military sites (A-CEFs as brownfields) not only as purely anthropogenic habitats but as semi-anthropogenic to natural habitats. Invertebrates, which are expected to be present in the interior of A-CEFs [41,42], were chosen as a model group of organisms. The problematic questions, in this case, are: Which species occur in bunkers, and what is the nature of their distribution in these objects? Also, which environmental factors may the distribution of these species depend on?

2. Materials and Methods

The Czech Republic is located in the heart of Europe, bordering Germany, Poland, Slovakia, and Austria. The Czech Republic is divided into 14 regions. The area of interest for the purposes of this paper is the border region of the Moravian-Silesian Region, where the A-CEFs or UA-CEFs under study are located. The Moravian-Silesian Region is situated in the eastern part of the territory of the Czech Republic. The northern and north-eastern part of the region borders Poland (today's border does not correspond with the pre-WWII state border), and the south-eastern part of the region borders Slovakia. The region also borders on the west and south-west with other territorial units—the Olomouc Region and part of the southern border is adjacent to the Zlín Region. The geographical situation of the Moravian-Silesian Region is shown in Figure A1.

Throughout former Czechoslovakia (1918–1938), almost 12,000 CEFs were built (or planned), as well as almost 1000 units of heavy fortifications (artillery logs and forts), which are not the subject of this study [25]. In the Moravian-Silesian region alone (in terms of the current administrative structure of the Czech Republic) 896 CEFs were planned [27]. The continuity of the whole line in the Moravian-Silesian Region is shown in Figure A1. In this figure, there is also a clear continuity with the line in the Olomouc Region. Due to a large number of these objects, only some of them were selected for the purpose of a more detailed study.

Geology and geomorphology played a significant role in the construction of CEFs [35]. It is worth noting that the Moravian-Silesian Region is covered by two geological units, namely the Bohemian Massif and the outer Western Carpathians [43,44]. The development of these two geological units is complemented by quaternary sediments whose origin is linked to continental glaciation, which left deposits of gravels and sands [43,44]. There are all types of relief from highlands and hills to lowlands in the Moravian-Silesian Region [43,44]. As the zoological survey shows, the amount of precipitation and the number of days with snow are crucial. According to Quitt [45], the lowland areas of the Moravian-Silesian Region fall into a moderately warm climatic area, while the mountain and foothill areas fall into a cold climatic area. In relation to the biogeographical classification of the Czech Republic, the territory of the Moravian-Silesian Region is part of the Central European deciduous forest province (like the vast majority territory of the Czech Republic), where parts of three subprovinces meet—the Hercynian subprovince, the Polonian subprovince and the West Carpathian subprovince (simplified in the direction from west to east) [46]. The boundary between the subprovinces is not distinct, and in a large part of the area of interest, it can be characterized as a transitional zone of mutual influence. The fauna here is relatively diverse, which is related to the geographical conditions (mixing of West Carpathian, Polonian, and Hercynian elements) [46].

2.1. Genius Loci—Layer of the Landscape: Searching for Stories of the A-CEFs

One of the functions of A-CEFs in a post-military landscape is the role they play in specific genius loci [14,47]. In order to capture this potential, a guided interview method was chosen, where respondents were asked about the stories associated with A-CEFs (including capturing the wider context associated with the place). Due to the specific topic of the research, it was necessary to approach suitable respondents who are in some way affected by the genius loci and the associated function of A-CEFs in the landscape. Priority was given to staff from organizations dedicated to leisure activities for children and young people, as well as owners of buildings and last but not least, military history clubs and historians. The initial part of the guided interview dealt with information about the respondent such as age, relationship with WWII history (work/free-time activities/education, etc.), when they first heard about this issue, or if they are visitors/owners of the bunker. As part of the guided interview, respondents were asked two sets of questions with a series of supplementary and extension questions: Q1. Do you know a story associated with A-CEFs (when it happened, where it happened, etc.)? Q2. What do you think about the story (truthfulness, authenticity)? These questions were asked in such a way as to make it clear that the focus of the research is on stories associated with A-CEFs as brownfields and not historical stories (e.g., associated with direct participants in historical events that are associated with A-CEFs serving their original purpose). Respondents were contacted either in person or via electronic communication. Guided interviews were recorded on a dictation machine in the case of in-person interviews. Subsequently, the results of the interviews were transcribed and analyzed.

2.2. Socio-Economic Sphere—Layer of the Landscape: A-CEFs as Brownfields

For the purpose of our study, the line of A-CEFs (or UA-CEFs) in the Moravian-Silesian Region which stretches from west to east and follows the current northern state border with the Republic of Poland were selected. The entire line of Czechoslovak fortifications (A-CEFs, heavy fortifications, artillery forts) in the Moravian-Silesian region is represented graphically in Figure A1. In this output, the A-CEFs units were marked according to their structural and technical condition (existing—green, destroyed—orange, initiated—purple, obliterated—blue, unbuilt—grey) and are supplemented by heavy fortifications units (red).

This region was chosen because the authors are familiar with the local terrain and because some of the fortifications were among the most completed before the start of WWII. Furthermore, at the end of WWII in 1945, they played an important role during the war between the German and Soviet armies, and after the war, they were not used (the objects on the border with Poland did not become part of the so-called Iron Curtain). They were abandoned and accessed for casual visitors in the frame of the hidden curriculum of the landscape. Due to the large number of existing A-CEFs on the territory of the Moravian-Silesian Region, the selection was made in such a way as to take into account the greatest possible heterogeneity of the selected objects. This includes altitude (highest and lowest positions of the Moravian-Silesian Region, middle positions), location within the Moravian-Silesian Region (the most eastern and the most western), and also the character of the surrounding environment (open, semi-open, and closed exterior environment).

In order to select a suitable sample of fortification units (A-CEFs) for our study, we used data from available historical military maps first depicting the line of fortifications throughout former Czechoslovakia. The original assumption was to divide the line in the territory of today's Moravian-Silesian Region into 10 groups/sections (A-J). Within each group, a selection of 5 existing A-CEFs in different environments (forest/forest edge/arc) were considered. While searching for more detailed information for the selection of the 10 groups, websites of friends of military history were found which contained, among other things, databases with more detailed information—for example, information regarding the technical status. These websites were 'The Interactive Map of Czechoslovak Fortifications 1935–1938' and 'Information on Light Fortifications 1936–1938' [27,36]. A search of these databases revealed that one of the originally intended groups A-CEFs had not been built or the fortifications had been obliterated (this group of buildings was, of course, excluded from the further investigation) Figure A2.

During the field survey, some of the buildings were inaccessible (locked or walled entrances, located on private fenced land, flooded with water, and exceptionally, could not be traced in the field). For these reasons, this methodology has been partially abandoned and the distribution of the surveyed Selected Abandoned Casemates with Enhanced Fortification (SA-CEFs) within the Moravian-Silesian Region results in a less than even distribution. It was possible to include the highest A-CEFs as well as the most eastern and the most western A-CEFs in the SA-CEFs. The selection of 39 SA-CEFs is sufficient for further data analysis and can be suitably supplemented or extended in the future. The distribution of individual SA-CEFs in the Moravian-Silesian Region is shown in Figure A3.

Subsequently, each SA-CEF was categorized according to its geomorphological location (WGS-84), climatic conditions, altitude, administrative section according to the former military administration (original military markings, military section assignment, military numbering), and current ownership. The numbering of individual units does not correspond to the order of data collection but was assigned retrospectively. Within each of the 39 SA-CEFs were in field survey semi-quantitatively detected the orientation of the entrance in relation to cardinal directions, the character of the interior environment (dry/wet/flood), the presence of organic and inorganic material in indoor spaces (none/little/lot), human use (unused/occasionally used/intensively used), accessibility of the entrance were also monitored in each unit (open/semi-open/closed) and the type of exterior environment (open/transitional/closed). The expected output is a graphical representation of the corresponding sections. These 39 SA-CEFs will be further processed not only for brownfields issues but also for the zoological survey.

2.3. Functional Relationship—Layer of the Landscape: Determination of Functional Potential of A-CEFs

The aim was to present a way of looking at these specific brownfields from the perspective of this approach of the landscape—in terms of the interactions between the objects and the surrounding environment [48]. The question is whether and how these different layers in the post-military landscape interact with each other and what significance they play in the landscape. From this perspective, in the case of the A-CEFs, it is true that they have not been studied at all yet. For this reason, the analysis was not based on empirical data, but on the methodology of the initial approach (especially field observations) to these objects.

For the purposes of our research, we have decided to consider the objects of A-CEFs in this post-military landscape in two specific ways, namely:

(a)   a semi-natural ecosystem of the external environment (consisting of the A-CEF object itself and its immediate surroundings; analogy with rock),

(b)   the semi-natural ecosystem of the A-CEF's internal environment (analogy with a cave).

This approach required a field investigation of individual A-CEFs and the definition of a model of individual objects of the post-military landscape.

We further assume that individual A-CEFs can impact us as natural elements under certain conditions. The expected output of this research is then a graphical representation with the naming of the different parts of these post-military landscape elements observed from different perspectives.

2.4. Natural Conditions—Layer of the Landscape: Zoological Survey of the Interior Environment of the SA-CEFs

The aim of the zoological survey was to carry out an inventory of invertebrate fauna in A-CEFs. As this is the first approach of this character to the SA-CEFs, the zoological survey was simplified to the extent that a) only indoor areas were studied, and b) only during the winter period (February/March 2014). This ensured that the number of potential taxa found was minimized as much as possible. Findings were expected of invertebrate species that are able to survive in similar types of environments for long periods of time (e.g., cellars, adits), but which usually live in suitable habitats outside this type of environment (so-called troglophilic or stygophilic species), as well as invertebrates that seek out similar environments for hibernation (hibernation). Probably the most representative-rich group were the expected so-called accidental guests of indoor spaces. Finds of invertebrates very closely adapted to living in underground spaces (troglobionts and stygobionts) were rather not expected.

Collecting was carried out in 39 SA-CEFs. Invertebrates of indoor spaces of the military fortification complex were studied by using conventional flashlights. Recorded species were examined on the walls and ceilings (spiders, butterflies, and moths) or on the floor under various objects like stones, remains of wood (snails, isopods). Collecting techniques as grids or kick sampling methods here were not used. In several cases, it was necessary to collect specimens for further determination by using entomological tweezers or an exhauster. Specimens were fixed in 70% alcohol or killed by vapors of ethyl acetate.

Subsequently, a partial objective of the faunistic survey was to select a suitable model group of invertebrate animals and characterize it with selected diagnostic features of zoocenoses (abundance, dominance, and frequency) [49]. Based on their abundance, we also performed inter-comparison of the SA-CEFs using multicriteria analysis. Selected independent environmental variables were included in the overall analysis. The environmental variables included altitude, humidity conditions inside the SA-CEFs, rate of human use, presence of organic material in the indoor environment, accessibility of the entrance, and the character of the SA-CEF's surrounding exterior environment. The collected data were processed using the R 4.0.5 program, calculating the similarity between the SA-CEFs using the Bray—Curtis index, the distribution of each sampling area depending on the selected environmental parameter using multivariate data analysis (MDS). SA-CEFs, where no live individuals were found, were not included in the analysis.

All data were recorded in a Microsoft Excel spreadsheet. Map outputs were processed in QGIS and zoological data and graphical outputs were processed in R 4.0.5.

3. Results

During the implementation of the research aimed at finding the functions that SA-CEFs respectively A-CEFs fulfill in the landscape, a set of results was obtained from all determined four approaches.

3.1. Genius Loci—Layer of the Landscape: Searching for Stories of the SA-CEFs

In total, 27 respondents were interviewed. Obtained data do not allow for a more general evaluation, which does not play a significant role in our case. Given that SA-CEFs represent anthropogenic elements in the post-military landscape with a specific history and therefore place with unique genius loci, our aim was to find stories that would capture this genius loci and thus essentially underline the function that A-CEFs fulfill not only in the first approach (Figure A4).

Most respondents had difficulty recognizing between stories that are historical and stories that relate to the site and its function as a brownfields site. Stories whose origins apparently date back to the post-war period were perceived by some respondents (historians, members of military history clubs) as unfounded, apparently fictional. The most frequently recurring motif in the category of unfounded stories was a military-themed plot set in the WWII period. These stories or even 'fairy tales' were looked upon with disdain by respondents familiar with the history of A-CEFs.

In general, they put themselves in the role of those who want to prevent the spread of these fallacies, and thus they were also reluctant to share them with us and thus to participate further in their spread. These were mainly history experts, but they had encountered similar 'fallacies' at a younger age before they became experts, and these 'fallacies' were often at the origin of their interest in the history of A-CEFs and similar objects.

Only when asked additional questions did these experts comment on the subject of the stories they described as fictional. These included, for example, stories concerning the existence of vast underground spaces, ammunition stores, archives, underground factories, mass graves, etc.

On the contrary, the respondents from among the leaders of clubs working with children in leisure activities had a rather positive attitude towards unfounded stories and 'fairy tales'. This is due to the fact that these stories (fabulations) present objects in a more interesting (adventurous) framework and thus fulfill different functions in troop games, troop rehearsals, or even ceremonies.

The underestimation of the dangers of amateur inspection of the interior of fortress buildings in general (ignorance of the interior construction design, and therefore basically the pitfalls in the form of shafts, wells, and various ventilation openings) was often mentioned. Furthermore, the presence of homeless people who do not hesitate to use various traps to secure their 'property', or the presence of criminal elements.

The analysis of the statements shows some connection between the respondents and the specific SA-CEFs mentioned in the stories (e.g., interactions the respondent had with the object in early childhood or especially in adolescence). Conversely, some of the statements were of a general character, i.e., a story that can be applied to any object (the recurring motif of the underground in the A-CEFs).

The above results clearly show that A-CEFs fulfill 'hidden' functions with this landscape layer which is associated with genius loci. These functions are educational, cultural, social, etc. but at the same time closely connected to the genius loci of the place and thus constitute part of the hidden curriculum of the post-military landscape. This conclusion, therefore, corresponds to our stated objectives about the role of A-CEFs as a phenomenon of the post-military landscape.

3.2. Socio-Economic Sphere—Layer of the Landscape: SA-CEFs as Brownfields

The analysis of available data shows that in the Moravian-Silesian Region a line of A-CEFs was created with a total number of 896 objects, of which 591 are still existing, 40 were destroyed (but the remains of the building are still visible), 16 were initiated but not finished, another 182 were obliterated for various reasons and 67 were planned but their construction was never started (unbuilt). The overall state of A-CEFs (including UA-CEFs) in the Moravian-Silesian Region is presented in Figure 3. This line of A-CEFs (or UA-CEFs) was also supplemented in Moravian-Silesian Region by units of heavy fortifications including artillery logs (which was not the aim of the survey). Brownfield in the built-up area of a municipality can represent an economic and social burden (Figure A4B).

The categorization into individual sections is shown in Table 1. The results obtained in the field survey (2014–2021) based on the proposed methodology (chapter) are presented in Tables A1–A5. These 39 SA-CEFs units and their selected characteristics were further studied for zoological survey purposes.

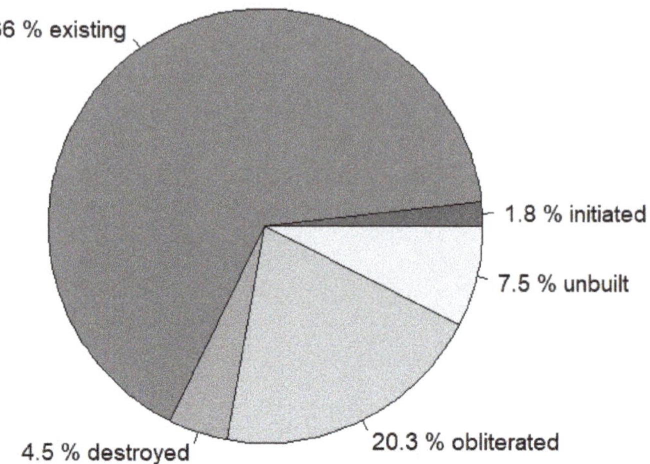

Figure 3. Percentage of the current structural and technical condition of A-CEFs (respectively UA-CEFs) in the Moravian-Silesian Region (n = 896).

Table 1. Summary table of SA-CEFs within each section for further data processing.

Section:	Total Number of SA-CEFs in Section:	Corresponding Identification of SA-CEFs in the Database
A	9	2, 3, 4, 7, 8, 9, 37, 38, 39
B	6	10, 11, 13, 14, 15, 16
C	2	5, 6
D	3	33, 34
E	2	29, 30, 31
F	5	18, 19, 20, 21, 22
G	5	17, 24, 25, 26, 27
H	1	36
I	6	1, 12, 23, 28, 32, 39

Based on our own field survey were found the following rates of human use for each SA-CEF were: 7 units were intensively used, 16 units were occasionally used and 16 units were not used—Figure 4A. The rate of human use was assessed primarily by the presence of artifacts associated with recent and repeated human presence (tables, chairs, lounge chairs, kitchen equipment, food remains, but also activities associated with efforts to restore the property to its original condition, etc.).

The rate of human use of these units is certainly related to the access restrictions. The data shows that most of these units are open and therefore accessible (26), a large proportion is accessible due to overcoming barriers (12), while only one unit of the SA-CEFs is completely closed and therefore inaccessible—Figure 4B. These were, for example, SA-CEFs in front of whose entrance was overgrown with trees, bushes, and/or partially grounded.

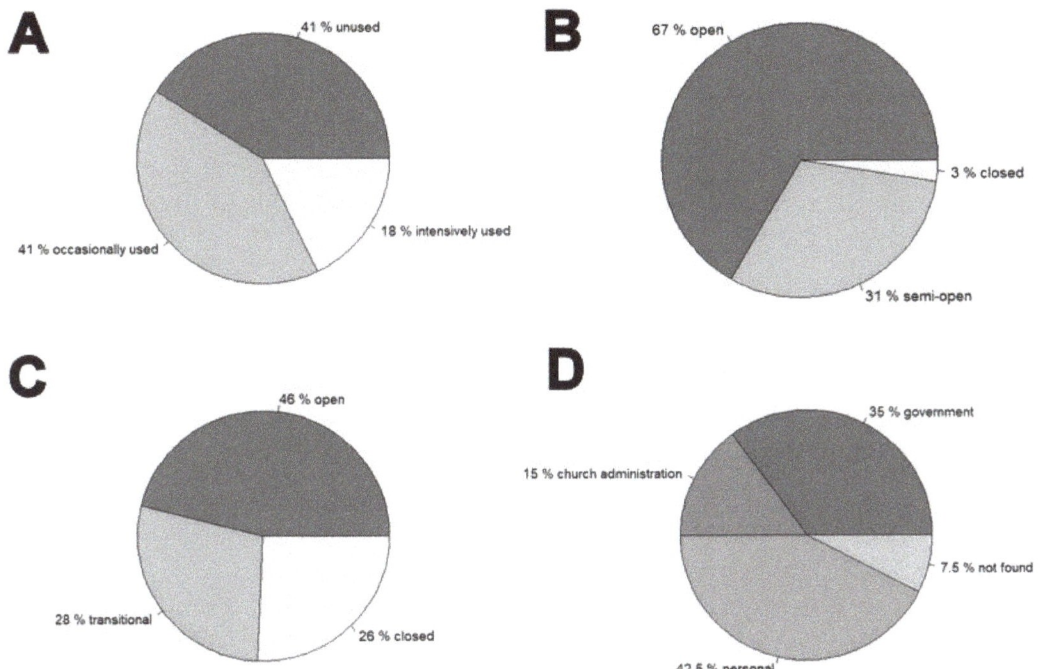

Figure 4. (**A**)—Rate of human use of indoor environment in SA-CEFs (*n* = 39); (**B**)—Entrance accessibility in SA-CEFs (*n* = 39); (**C**)—Type of SA-CEF's surrounding exterior environment (*n* = 39); (**D**)—Distribution of SA-CEF's owners (*n* = 39).

Along with the previous two factors—the rate of human use and accessibility of entrance—the location of SA-CEFs units in the landscape may also be related simultaneously. For this reason, a simple analysis of their external environment was carried out and from a total of 39 SA-CEFs, 18 were located in open landscape (meadows, pastures, fields), 10 were located in closed landscapes (forests, scrub) and the remaining 11 SA-CEFs were located in the transitional zone (forest edges, etc.)—Figure 4C.

According to the analysis of property rights, the SA-CEFs were divided into the following categories: government, church administration, personal (physical persons, joint property of married couples, associations, and cooperatives), and the last category is the 'not found' owners—where the SA-CEFs have not been entered in the land register or have an unidentified owner.

From the selected 39 SA-CEFs, only 14 units were owned by the government, and 17 were privately owned ('personal'). In addition, 6 units were owned by a church administration, and 3 units were classified as 'not found' (owner is unknown or owner was insufficiently identified)—Figure 4D. For one of the SA-CEF (number 17), it was determined that part of this SA-CEF is owned by a person and part is owned by the government—for this reason, the data set of Figure 4D is divided into 40 units.

3.3. Functional Relationship—Layer of the Landscape: Determination of Functional Potential of a-CEFs

On the basis of the analyses carried out, it can be concluded that individual SA-CEFs (respectively A-CEFs) as anthropogenic elements in the post-military landscape may represent semi-natural elements from a certain point of view. The SA-CEF (or A-CEFs) object itself may be perceived in the landscape at first sight as a 'rock', with a corresponding growth of mosses or vegetation (example shown in Figures 5 and A4B). In contrast, from the point of view of the interior environment, the A-CEFs can be seen as a 'cave', which can provide shelter for various species of organisms. This 'cave-like' environment of the A-CEFs' interior is not only characterized by a constant temperature throughout the year, but also by calcite deposits and soda straws (as they are called) on the walls and ceilings of the building as 'stalactites'. In terms of geomorphological shapes, the A-CEFs can be perceived as a concave shape that forms an unmistakable step in the terrain. From the point of view of the pedological characteristics, we can consider the object itself as an anthroposol (ceiling, embankment made out of boulders and covered with a layer of earth and grass). In Figure 5 we can observe different perspectives of view as we see the different functions of A-CEFs in the landscape.

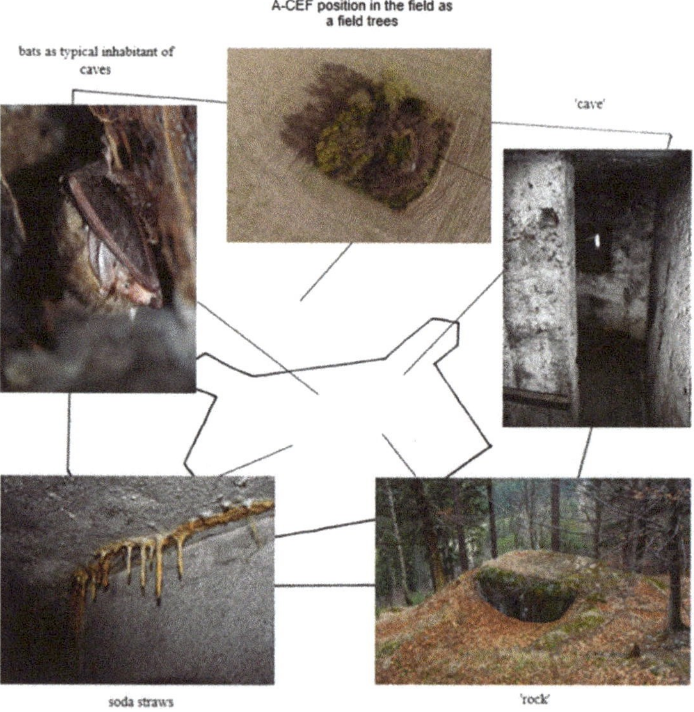

Figure 5. Different points of view of individual functions of A-CEFs.

3.4. Natural Conditions—Layer of the Landscape: Zoological Survey of the Interior Environment of the SA-CEFs

Representatives of the following taxa were found in the interiors of 39 SA-CEFs: Oligochaeta, Gastropoda (Pulmonata), Arachnida (Opiliones, Araneae, Acari, Pseudoscorpiones), Malacostraca (Isopoda), Myriapoda (Chilopoda, Diplopoda), Hexapoda (Collembola, Diplura, Orthoptera, Dermaptera, Hemiptera, Neuroptera, Coleoptera, Lepidoptera, Diptera, Hymenoptera). The presentation of these results would exceed the scope of this paper. A total of 104 species were identified. Among the taxa mentioned above, Lepidoptera (butterflies and moths), whose adults seek out similar spaces for overwintering or hibernating, and Gastropoda (land snails), which can survive in similar spaces for a long time, or are so-called accidental guests, were chosen as model groups. These two taxa have the advantage of relatively easy and unambiguous determination in the field, with the consequent possibility of determining presence/absence in the SA-CEF and estimating absolute abundance.

A total of 9 species from the model group Lepidoptera were recorded in the interior of the SA-CEFs and a total of 732 live individuals were identified (Figure A5A, Table A6).

From the model group Gastropoda, 20 species were recorded and a total of 180 live individuals were identified (Table A7). The species *Inachis io* represents the butterfly with the highest frequency (91.89%) and the highest dominance (42.76%). The land snail with the highest frequency was *Monachoides incarnatus* (63.64%) and the land snail with the highest dominance was *Helix pomatia* (24.44%) (Figure A5B).

The multicriteria analysis (MDS), which was calculated separately for butterflies and moths, and land snails, does not allow us to interpret the main directions of variability in the species data. Figure 6 shows an ordination diagram (MDS) depicting the distribution of individual SA-CEFs depending on the presence of organic material (land snails) and Figure 7 shows an ordination diagram (MDS) depicting the distribution of individual SA-CEFs depending on the nature of the surrounding environment (butterflies and moths).

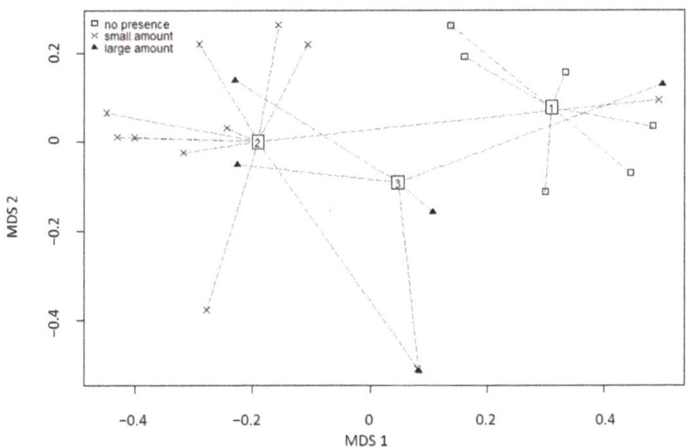

Figure 6. Ordination diagram (MDS) showing the distribution of individual SA-CEFs depending on the presence of organic material (land snails, for all SA-CEFs, without data adjustment).

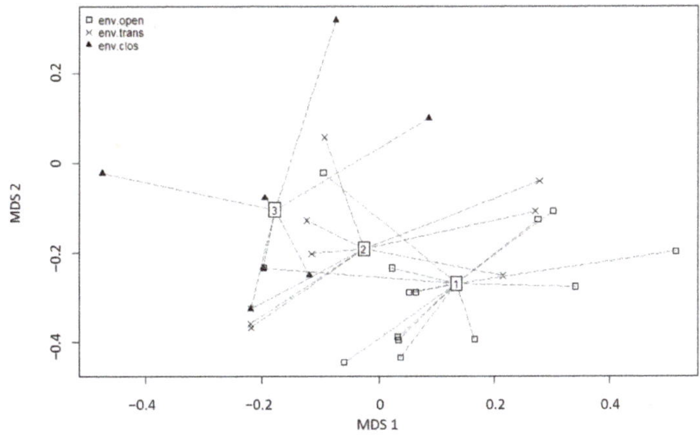

Figure 7. Ordination diagram (MDS) showing the distribution of individual SA-CEFs as a function of the nature of the surrounding environment (butterflies and moths, for all SA-CEFs, without data adjustment).

4. Discussion

From the partial results that correspond to the individual approaches (in Material and Methods), conclusions can be drawn that will have an impact on further progress in the study of A-CEFs as phenomenon elements of the Czech borderlands landscape.

4.1. Genius Loci—Layer of the Landscape: Searching for Stories of the SA-CEFs

The guided interview as a research method was not focused on the general public at this stage, but mainly at people who were expected to have some knowledge of the subject—people who deal with this in their profession (museum workers) as well as hobbyists who are involved in this field in their free time. For the guided interview, which is time-consuming, the number of respondents (27) was not very high, but for our research it is sufficient.

In terms of such a targeted guided interview focused on finding out the stories, neither the ratio of professions nor the age of the respondents is decisive in the results. It is clear from the results that the stories associated with the atmosphere of a place are absorbing regardless of age or profession (but again in a sample of people who have an established relationship with the objects). Both the verbal and written responses from respondents showed their passion for A-CEFs (respectively SA-CEFs) and their enthusiasm for our interest. This is essentially consistent with the function A-CEFs fulfill in the genius loci as a layer of landscape: A-CEFs evoke a deep emotional response in humans. In this context, it is worth quoting part of the statement of one respondent: 'The fortification has absolutely incredible genius loci if you are lucky enough to be hit by it. There are people who are not interested in fortifications, or who are uncomfortable with fortifications for some reason. Then there are the people who are interested in it, who like to read or listen to it. Many visitors leave the sites surprised at all the new things they have learned. And then there are people who, once they've been introduced to bunkers, have never been able to get turned away from them. That's something that can't be described as anything other than their diagnosis.'

These conclusions suggest the usefulness of the method (and results) of the guided interview, but at the same time open the perspective for the creation of a structured questionnaire. A structured questionnaire would be aimed at a broader public and would also take into account the attitudes of people who may not have a positive attitude towards A-CEFs, including the subsequent statistical evaluation of the data. From the results of the questionnaire survey and in the context of the above, it is necessary to appropriately define the conceptual apparatus related to the use of the term 'story', which can be perceived differently within the genius loci as the layer of the landscape (subjective experience of the visitors, etc.) and the cultural heritage (historically documented event, etc.). As already noted in the results, also fabricated stories play an important role in the search for the function of A-CEFs. Given these facts, it is necessary to recognize between different types of stories: historically based stories (historically recorded, more likely to be the accounts of direct participants), historically unfounded stories (not historical fact, but the use of this fact to locate characters, time and space), and fabricated stories (Table 2).

Table 2. The role that different story types play in genius loci and socio-economic sphere as landscape layers.

	Socio-Economic Sphere	**Genius Loci**
historically based stories	they play a key role	serve as inspiration to explore on your own; they illustrate the spirit of the place
historically unfounded stories	can play an important role	can serve as inspiration for various activities
fabricated stories	their influence is perceived as contradictory to negative	they play a key role

Although many A-CEFs appear at first sight to be abandoned (A-CEFs as brownfields), they play a role in something that is harder to grasp, and what we might call the 'hidden curriculum of the landscape'. From this perspective, they are not 'really abandoned' but only as abandoned perform social and educational functions. At the same time, fabulations (apparently fictional stories) play an important role, which is considered worthless, confusing, or even undesirable from the point of view of the cultural heritage. The fact that the casual visitors make up their own stories when interacting with A-CEFs is remarkable when looking at A-CEFs as brownfields, even though a negative phenomenon such as the Goliath effect may be associated with it (Figure A4) [50]. At this point, it is also worth highlighting that, apparently due to the occurrence of SA-CEFs in the open landscape (outside human settlements), we have not observed negative uses (e.g., squatter settlements). But this statement cannot be applied to all A-CEFs, as their very small size makes intensive use rather unlikely.

The SA-CEF units in the landscape were perceived positively by the respondents as a kind of historical milestone reminding them of an important historical stage and may also be part of 'family heritage'. Recording of negative perceptions of the objects is rather to be expected only from the results of a structured questionnaire directed to the general public. This would simultaneously capture the wider variation in the public's perception of post-military landscapes, and hence A-CEFs, for the purposes of our research, and thus also provide a valuable stimulus to a comprehensive view of the issue of A-CEFs as brownfields.

4.2. Socio-Economic Sphere—Layer of the Landscape: SA-CEFs as Brownfields

Data analysis has shown that a total of 591 units of A-CEFs and UA-CEFs are still existing in Moravian-Silesian Region (out of a total of 896 originally planned or realized). In 40 cases of A-CEFs and UA-CEFs, it is possible to trace their remains in the landscape. A total of 182 have been permanently obliterated for various reasons. It is interesting to note that a very significant number of all of them have survived and it is questionable whether this is related to difficulties in their obliteration. In certain circumstances, it may be related to their strategic importance within the Czech army for a certain period, or also to the subsequent efforts of leading figures in society (political representation) to preserve these objects as an integral part of the landscape that the Czechoslovak fortification line forms from a historical point of view. Although the study focused only on a limited set of SA-CEFs ($n = 39$), relative to the total number, it is still possible to draw some conclusions of a more general nature.

The ownership of individual A-CEFs or UA-CEFs is complicated in the former Czech Republic due to the complex evolution of property rights. In the set of SA-CEFs, we were interested in the relationship between the type of ownership and the rate of human use. Privately owned buildings were expected to have a higher rate of use, e.g., for recreational purposes or as storage facilities, or to be eventually made inaccessible to the public. The vast majority of privately owned SA-CEFs are not used—Figure 8A.

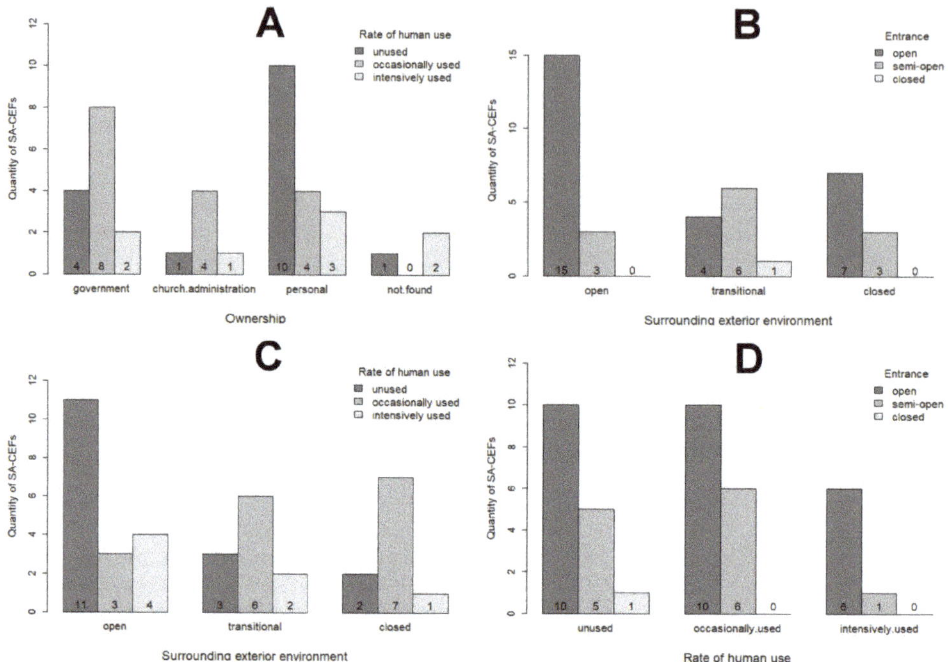

Figure 8. (**A**)—Relation between 'Ownership of SA-CEFs' and 'Rate of human use of indoor environment in SA-CEFs'. (**B**)—Relation between 'Type of surrounding exterior environment of SA-CEFs' and 'Entrance accessibility in SA-CEFs'. (**C**)—Relation between 'Type of surrounding exterior environment of SA-CEFs' and 'Rate of human use of indoor environment in SA-CEFs'. (**D**)—Relation between 'Rate of human use of indoor environment in SA-CEFs' and 'Entrance accessibility in SA-CEFs'.

When comparing the type of exterior environment and accessibility of entrance to SA-CEFs (Figure 8B), open landscape SA-CEFs are also generally more accessible. A cross-comparison of the level of human use and type of exterior environment in the SA-CEFs shows (Figure 8C) that open landscape SA-CEFs are the most visited by people. In both cases, this may be related to the possible cultivation of land in the immediate neighborhood of SA-CEFs (one reason for the earlier obliteration of SA-CEFs) versus forest management. The fact that SA-CEFs in open spaces are also more easily found by potential visitors may also be a factor to some extent. Also, when human use rates and accessibility of SA-CEFs are compared to each other (Figure 8D), it appears that logically those SA-CEFs that are also open access are used with the greatest intensity. This comparison also shows that accessibility of the entrance is not a decisive factor for the rate of human use in the sample of SA-CEFs.

4.3. Functional Relationship—Layer of the Landscape: Determination of Functional Potential of SA-CEFs

The analyses and results show that there are interactions between SA-CEFs (respectively A-CEFs) and the surrounding landscape, and SA-CEFs (respectively A-CEFs) can be viewed as semi-natural ecosystems. Viewing A-CEFs as a 'rock' or 'cave' (Figure 5 or Figure A4C) offers scope for further research to support this functional potential of SA-CEFs (respectively A-CEFs) in the landscape and further emphasize their importance. At this point, it is at least possible to consider the interior spaces of SA-CEFs (respectively A-CEFs) as overwintering or hibernating sites for some species of butterflies and moths (Results in Section 3.4, Table A6), or also as habitats for various species of invertebrates with different relations to underground spaces. The analysis carried out offers further possible directions for research (SA-CEFs, respectively A-CEFS, as anthrosols/leptosols, the embankment as anthrosols/cambisols, the walls of SA-CEFs, respectively A-CEFS, as a substrate for calciphilic species of organisms in the acidic environment of mountain forests, etc.). Finally, it is also possible to study A-CEFs from the perspective of island biogeography, where A-CEFs objects can be viewed as terrestrial islands [51]. In this case, it would be appropriate to state that A-CEFs perform ecosystem services in the landscape.

This leads us to the idea that A-CEFs as anthropogenic elements of the post-military landscape has a function (in themselves) and potential in the landscape. However, this potential may not be obvious given that these objects are viewed as classic brownfields (given the definition of brownfields). For this reason, it is appropriate to further address the question of the additional functional potential of A-CEF, which relates to its re-use, but also that it is a hidden functional potential that should be taken into account.

The historical and architectural significance of the relics of the post-military landscape (the fortress lines and their individual objects) are accepted from a global perspective. Therefore, the aim was to point out the functional significance of these objects. This functional significance of these objects can be fulfilled in the landscape itself, regardless of the hidden curriculum of the landscape (based on interaction with humans). The proposed scheme (Figure 5) shows a way of looking at these objects, or rather at their hidden potential. However, the question remains about how to incorporate this view into the definition of brownfields. It can be stated that in our case this approach does not only apply to every A-CEF (in the case of UA-CEFs, its further function is determined by its new use, re-use) but also to other brownfields that may have hidden functional potential. In the case of their re-use may be the hidden functional potential lost. The most illustrative example is the restoration of A-CEF to its wartime condition—cleaning, cleaning the surface, closing all entrances, adding paint, etc. At the same time, it is clear that given their large numbers, not every A-CEF can be restored to this dignified state (as a reminder of the willingness to defend their young homeland for some citizens). Since they are rather small in size (internal area of about 8 m^2), they can rather be used only as foundations for huts or for storing domestic crops.

4.4. Natural Conditions—Layer of the Landscape: Zoological Survey of the Interior Environment of the SA-CEFs

The largest number of hibernating butterfly species was recorded in SA-CEFs number 14 and 32, a total of 5 species out of 9 species. The largest number of hibernating butterfly individuals was also recorded in SA-CEF number 14 (62 live individuals in total). Only two SA-CEFs had no records (SA-CEFs number 13 and 22). Only 1 and 2 species were recorded in SA-CEFs number 3 and 13. SA-CEFs with such a small number of species recorded also had to be excluded from further analysis (18 SA-CEFs in total). Due to this fact and the very low total number of species found, multivariate data analysis became meaningless.

The most frequently recorded butterflies and moths (Lepidoptera) were *Inachis io* and *Scoliopteryx libatrix*. *Inachis io* from the family Nymphalidae is one of the most common butterflies of opened landscape, but forest edges too. Normally, it hibernates in heatless areas of the building (in attics, in cellars, etc.), and in some cases, it is in the hibernating places even more numerous than the other species, *Scoliopteryx libatrix* [42]. Very common is moth *Scoliopteryx libatrix* from the family Noctuidae, which prefers moist habitats (shores of streams, rivers, and lakes), it occurs also along roadsides, edges of woods, and penetrates well as into the gardens and parks. In autumn, it seeks to hibernate a variety of cavities, caves, tunnels, and cellars [52,53]. It is the most common species of all types of underground space [42].

The largest number of land snail species (Gastropoda) was recorded in SA-CEF number 37, with a total of 9 species out of a total of 20 species. The largest number of land snail individuals was also recorded in SA-CEF number 37 (42 living individuals in total). In total, no records were recorded in 17 SA-CEFs. In SA-CEFs numbers 3 and 10 were recorded only 1 and 2 species. SA-CEFs with such a small number of species recorded also had to be excluded from further analysis. Thus, for the subsequent multivariate data analysis, it was possible to use data sets relating to only 9 SA-CEFs, and thus, as in the previous case, the multivariate data analysis itself became meaningless.

The most frequently recorded species of land snail were *Monachoides incarnatus* and *Helix pomatia*. *Monachoides incarnatus* from the family Hygromiidae occurs in a variety of forest habitats. It is most commonly found in deciduous forests, in the scrub, often around streams and wet rocks. It also inhabits various secondary habitats such as quarries, gardens, or parks. *Helix pomatia* from the family Helicidae inhabits various types of habitats (especially on limestone), light deciduous forests, scrub, meadows, vegetation along streams, but also cultural areas (gardens, thickets, etc.). In both cases, these are species that can also occasionally be found in different types of underground spaces [41,54].

In addition to 9 species of butterflies and moths, and 20 species of land snails, 75 other invertebrate species from various taxonomic categories have been documented in SA-CEFs. The presence of vertebrates (e.g., bats, rodents) or their resident signs (e.g., marten droppings, bird nests) have also been recorded in SA-CEFs.

The faunistic survey carried out confirmed that the interior spaces of the A-CEFs should be used by various species of animals. Among the species found, for example, spiders (*Meta menardi* or *Metelina merianae*) are able to survive in SA-CEFs for long periods of time, as they inhabit similar habitat types [55]. Butterflies and moths use them for hibernating, and land snails represent so-called accidental guests whose occurrence in indoor spaces is related to the composition of the malacofauna in the immediate surroundings of SA-CEFs. However, some of the land snail species found may also survive in similar types of environments over the long term. These include species such as *Oxychilus cellarius* or *Limax cinereoniger* [41,56]. The occurrence of animals very closely adapted to living in underground spaces (troglobionts and stygobionts) has not been proven. The occurrence of euryvalent species of butterflies and moths (Lepidoptera) and land snails (Gastropoda) is random or in some cases quite rare in the interior of SA-CEFs.

5. Conclusions

This paper focuses on A-CEFs in the military fortification complex of the Czech borderlands landscape as specific forms of brownfields. The issue of brownfields as an environmental problem can be viewed from different angles. This corresponds to the different definitions, which are also numerous within Europe [1]. Internationally, the accepted definition is CABERNET, which is based on the CLARINET definition. According to this definition, brownfield sites are areas that have been affected by the previous use of the building, site, and surrounding land. Brownfields are abandoned and under-used, may have real or perceived contamination problems, are predominately located in built-up areas, and require intervention that would enable its continued use.

Based on this definition, the A-CEFs, as part of the military fortification complex of the Czechoslovak borderlands built before the Second World War, are clearly classified as military brownfields. However, these units are very specific compared to other military facilities (in this classification of brownfields). They were not built as an integral part of military complexes or barracks. They were built as separate defensive units, but they all form the whole of a military fortress complex copying the Czechoslovak borderlands. Most of them were built in the open landscape and nowadays form is a phenomenon of the post-military landscape. Referring to the previous definition, are A-CEFs really useless and require intervention to bring them into beneficial use?

For our research, A-CEFs may be technically brownfields, but they are also an integral part of the cultural landscape (post-military landscape) and cultural heritage. As part of cultural heritage, they can enhance social, cultural, environmental, and economic sustainability, preserving diversity and place identity. However, in processing the field survey data, it was found that A-CEFs (respectively SA-CEFs) have further, hidden, functional potential. Our research has shown that A-CEFs as brownfields interact in a specific way with humans (hidden curriculum of the landscape) and, also interact with nature, where they create suitable conditions, e.g., overwintering or hibernating invertebrates, can be considered as terrestrial islands in the landscape, 'rock', 'cave', habitat and thus can also fulfill ecosystem services. For this reason, A-CEFs cannot, therefore, be seen only as brownfields that need to be f. e. remediated or re-used. Even without such interventions, they have a function and significance in the landscape.

What is the best term to describe the hidden functional potential of A-CEFs? To find a suitable term, it is possible to start from the concept of landscape singularity. This term is used mostly in art and architecture [57,58]. Exceptionally, this term can be used to describe unique features in the landscape as landscape singularities [59]: 'Landscape singularity represents linear or point singularities in the landscape that are natural (watercourse, rock), cultural (urban line, building object) or also historical, but more often a combination of these.' Based on the above characterization of landscape singularity, it can be concluded that the term "singularity" appears to be, in part, the most appropriate name for the character of individual A-CEFs. The common and typical feature of these A-CEFs is that they are all part of the cultural heritage (although they are not listed as heritage sites, e.g., by UNESCO), but for obvious reasons, there is not enough interest in their use (a large number of them). On the other hand, in contrast to the singularity, these sites are "disturbing" in terms of the impossibility of using the site for other purposes (e.g., as arable land), they represent a specific type of brownfield. Their functional potential is hidden and can be discovered only by more detailed study. Although A-CEFs is not currently listed and may be obliterated, under certain conditions they could be left in the context of nature and landscape conservation interests. After all, the analyses and surveys conducted clearly demonstrate that SA-CEFs perform several important hidden functions in the landscape for which they cannot be seen as brownfields. The term that would best describe the hidden functional potential of A-CEFs is a hidden singularity.

The concept of the hidden singularity can be incorporated into the definition of brownfields, where we need to look at them not just as environmental problems, but as environmental opportunities (from a biocentric perspective).

From the outset, this article has challenged the definition of brownfields (in the context of A-CEFs), which is primarily based on the under-use of brownfields or their disruptive interference with the landscape. Considering the results of the research conducted on SA-CEFs, it can be concluded that the concept of hidden singularity can take into account the hidden functional potential and in this case can be included in the classification of brownfields. The hidden singularity of brownfields means that brownfields perform functions in different layers of the landscape (including the hidden curriculum), although these layers may be hidden. Nevertheless, this hidden functional potential can be identified and quantified. From an environmental point of view, the hidden singularity makes brownfields sites that do not need to be revitalized or find new uses for them. Brownfields as an opportunity for investors always assume revitalization or re-use—based on the existing definition of brownfields. In our case, we are discussing the revitalization of A-CEFs to the war form (state of 1937–1938) with a museum exhibition (we can talk about regularity). In the case of re-use, we are discussing, for example, the obliteration of the A-CEFs in order to use the land (irregularity). In the category of brownfields as an opportunity for an investor, the hidden functional potential of A-CEFs could be completely eliminated. The above classification can be applied not only to A-CEFs, but also to other types of brownfields with a similar character. We would like to address the issue of hidden singularity and the hidden curriculum of brownfields in the future.

Author Contributions: Conceptualization, J.K., A.B. and J.V.; methodology, J.K., A.B. and J.V.; software, J.K. and A.B.; formal analysis, J.K., A.B. and J.V.; investigation, J.K., A.B. and J.V.; resources, J.K., A.B. and J.V.; writing—original draft preparation, J.K., A.B. and J.V.; writing—review and editing, J.K., A.B. and J.V.; supervision, J.K.; funding acquisition, J.K. and A.B. All authors have read and agreed to the published version of the manuscript.

Funding: This research was funded by SGS grant number SP2021/113 and SP2022/122 and the APC was funded by VSB—Technical University of Ostrava—Department of Environmental Engineering,

(Faculty of Mining and Geology) and Department of Building Materials and Diagnostics of Structures (Faculty of Civil engineering).

Acknowledgments: We thank our family members for their patience and participation in the exploration of the CEFs.

Conflicts of Interest: The authors declare no conflict of interest.

Appendix A

Figure A1. Display of the line of Czechoslovak fortifications in the Moravian-Silesian Region—demonstration of graphic interpretation of the obtained data, presentation of the current structural and technical condition of the CEFs using the QGIS program.

Figure A2. Situation of A-CEFs and corresponding sections of SA-CEFs (from A to I) in the borderlands of the Moravian-Silesian Region [28,33–35].

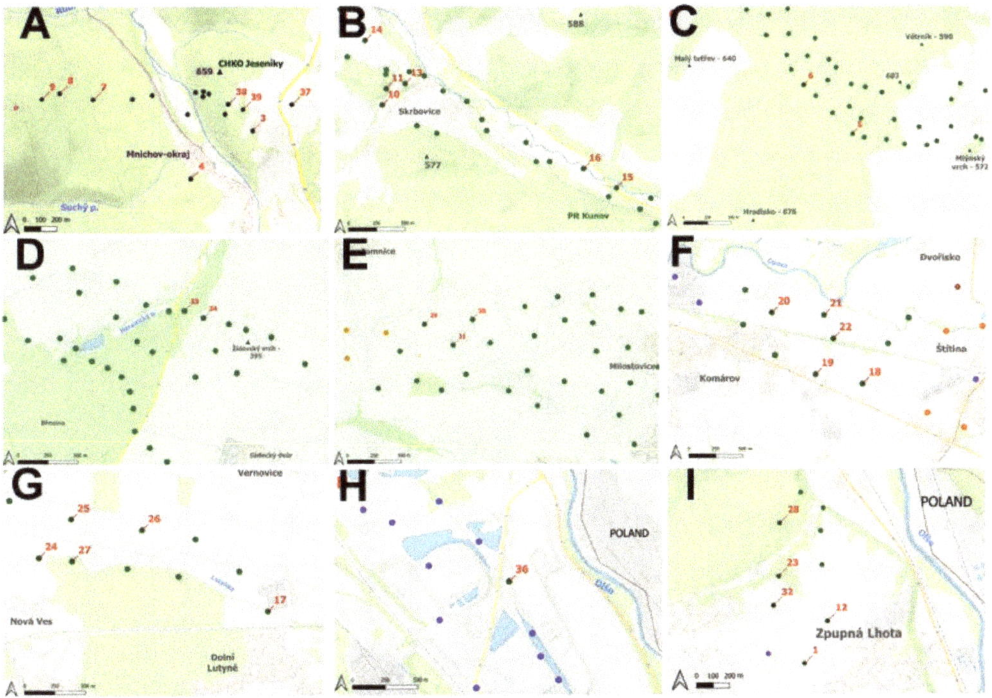

Figure A3. Selected parts of fortification line in Moravian-Silesian Region. Individual parts (**A–I**) are shown in Figure A2. Key: green—Existing A-CEF or UA-CEF; pink—Construction was initiated A-CEF; orange—destroyed A-CEF; blue—obliterated A-CEF; red number—SA-CEF.

Figure A4. (**A**): Each A-CEF object has unique genius loci (e.g., SA-CEF number 38). (**B**): A-CEFs represents a specific type of military brownfields (e.g., SA-CEF number 17). (**C**): SA-CEF number 38 is one of the existing and highest situated objects (715 m above sea level). It serves as an occasional refuge for adventurers and 'could certainly tell' many stories. (**D**): Some A-CEFs can be easily mistaken for rock (e.g., SA-CEF number 39).

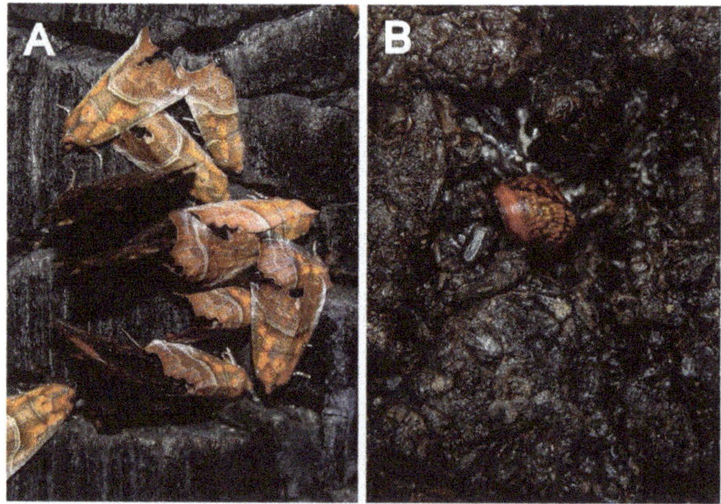

Figure A5. (**A**): A group of haralds (*Scoliopteryx libatrix*) on the walls of a bunker that was burned during World War II. We can observe orange drawings on the enclosed wings that resemble the letter 'M' (SA-CEF number 27). (**B**): Land snail *Monachoides incarnatus* hidden in a wall gap (SA-CEF number 37).

Table A1. Overview table of SA-CEFs studied objects with their description. The numbering of the individual SA-CEFs corresponds to the data in Table 1—part 1.

No.	Name:	N:	E:	Altitude:	Military Administrative Section:	Ownership:
1	XIX/432/A-140Z	49.7681898	18.5956608	284	XIX Louky	personal
2	XXI/10/E	50.1581184	17.3699108	660	XXI Mnichov	government
3	XXI/11/B1-80Z	50.1571897	17.3723533	647	XXI Mnichov	not found
4	XXI/15/E	50.1545095	17.3669212	614	XXI Mnichov	church adm.
5	VII-3/12/B1-100	49.9861001	17.5754216	607	VII-3 Hohnberg (Velký Tetřev), 3.sled	church adm.
6	VII-3/13/A-120	49.9907793	17.5681580	619	VII-3 Hohnberg (Velký Tetřev), 3.sled	church adm.
7	XXI/14/A-120	50.1589392	17.3584551	655	XXI Mnichov	church adm.
8	XXI/122/A-180	50.1593096	17.3555914	715	XXI Mnichov	church adm.
9	XXI/123/A-140	50.1589694	17.3540345	747	XXI Mnichov	church adm.
10	XI/619/E	50.0502091	17.4780221	448	XI Bretnov (Široká Niva)	government
11	XI/594/B1-80	50.0515198	17.4 785056	484	XI Bretnov (Široká Niva)	government
12	XIX/431/A-180Z	49.7704887	18.5976119	282	XIX Louky	personal
13	XI/461/B2-90	50.0519298	17.4808423	458	XI Bretnov (Široká Niva)	government
14	XI/463/A-140Z	50.0554488	17.4758975	453	XI Bretnov (Široká Niva)	government
15	X/449/A-140Z	50.0434196	17.5070920	427	X Nové Heřminovy	Government
16	X/450/A-160	50.0449989	17.5030057	416	X Nové Heřminovy	personal
17	XVI/208/A-140	49.9136298	18.4120698	201	XVI Nový Bohumín	government/ personal
18	II/16/A-120	49.9118497	17.9925046	247	II Komárov	not found
19	II/10/A-120	49.9126396	17.9864304	245	II Komárov	personal
20	II/5/A-140	49.9178202	17.9806877	237	II Komárov	personal
21	II/4/A-160	49.9176098	17.9875375	236	II Komárov	personal
22	II/7/A-120	49.9156297	17.9887584	236	II Komárov	personal

Table A2. Overview table of SA-CEFs studied objects with their description. The numbering of the individual SA-CEFs corresponds to the data in Table 1—part 2.

No.	Name:	N:	E:	Altitude:	Military Administrative Section:	Ownership:
23	XIX/600/B2-80	49.7729196	18.5934953	278	XIX Louky	personal
24	XVI/204/A-140	49.9176600	18.3854910	200	XVI Nový Bohumín	personal
25	XVI/7/A-120	49.9205797	18.3892101	194	XVI Nový Bohumín	personal
26	XVI/8/A-140Z	49.9197700	18.3973602	196	XVI Nový Bohumín	government
27	XVI/205/A-140	49.9174100	18.3893288	206	XVI Nový Bohumín	government
28	XIX/659/B1-80	49.7757991	18.5935640	285	XIX Louky	government
29	IV/911/A-160	49.9537	17.7987	331	IV Milostovice	personal
30	IV/908/A-160Z	49.9540897	17.8049218	327	IV Milostovice	personal
31	IV/912/A-120	49.95193	17.80241	330	IV Milostovice	personal
32	XIX/602/A-140	49.7713399	18.5930522	283	XIX Louky	personal
33	V/1048/A-160	49.9619599	17.7262197	368	V Sádek	government
34	V/958/A-140	49.9614186	17.7284309	373	V Sádek	personal
35	XIX/656/E	49.7799299	18.5907008	276	XIX Louky	government
36	XIX/401/A-160Z	49.8251091	18.5624953	233	XIX Louky	not found
37	XXI/6/A-180	50.15867	17.37575	646	XXI Mnichov	personal
38	XXI/2/A-140	50.1586885	17.3701837	661	XXI Mnichov	government
39	XXI/1/A-120Z	50.1583883	17.3715015	663	XXI Mnichov	government

Table A3. Table of SA-CEFs study objects with data that were obtained by field survey (2014-2021). The numbering of the individual SA-CEFs corresponds to the data in Table 1—part 1.

No.	Azimuth of Entrance:	Exterior Environment:	Interior Environment:	Organic Material in Interior:	Anorganic Material in Interior:	Rate of Human Use:	Entrance:
1	292,5/NWW	open	dry	none	none	intensively used	open
2	135/SE	closed	dry	little	none	occasionally used	open
3	157,5/SSE	transitional	dry	little	lot	intensively used	semi-open
4	180/S	closed	wet	lot	lot	intensively used	open
5	157,5/SSE	closed	flood	none	lot	occasionally used	semi-open
6	225/SW	closed	dry	lot	none	occasionally used	open
7	180/S	closed	wet	lot	lot	occasionally used	open
8	157,5/SSE	closed	dry	little	none	occasionally used	semi-open
9	157,5/SSE	transitional	dry	none	none	unused	semi-open
10	225/SW	transitional	flood	little	little	occasionally used	semi-open
11	202,5/SSW	transitional	dry	little	none	occasionally used	open
12	270/W	open	flood	none	lot	unused	open
13	270/W	transitional	dry	none	none	intensively used	open
14	225/SW	open	dry	little	none	unused	open
15	202,5/SSW	closed	wet	lot	little	unused	open
16	202,5/SSW	transitional	wet	none	little	unused	semi-open
17	202,5/SSW	open	dry	little	little	intensively used	open

Table A4. Table of SA-CEFs study objects with data that were obtained by field survey (2014-2021). The numbering of the individual SA-CEFs corresponds to the data in Table 1—part 2.

No.	Azimuth of Entrance:	Exterior Environment:	Interior Environment:	Organic Material in Interior:	Anorganic Material in Interior:	Rate of Human Use:	Entrance:
18	180/S	open	dry	none	lot	intensively used	open
19	202,5/SSW	open	wet	lot	lot	occasionally used	open
20	180/S	open	dry	little	none	unused	semi-open
21	180/S	open	dry	none	none	unused	open
22	202,5/SSW	open	wet	none	lot	unused	semi-open
23	270/W	transitional	dry	lot	little	occasionally used	open
24	225/SW	open	dry	none	lot	occasionally used	open
25	225/SW	open	flood	none	none	unused	open
26	180/S	open	dry	little	little	unused	open
27	180/S	transitional	dry	little	lot	occasionally used	open
28	247,5/SWW	closed	dry	little	lot	occasionally used	open
29	157,5/SSE	open	dry	none	none	unused	open
30	180/S	open	dry	none	none	unused	open
31	157,5/SSE	transitional	dry	none	none	unused	closed
32	270/W	open	dry	little	none	occasionally used	open
33	180/S	closed	dry	none	none	unused	open
34	202,5/SSW	open	dry	none	none	unused	open
35	247,5/SWW	closed	dry	little	little	occasionally used	semi-open
36	225/SW	open	flood	lot	little	unused	semi-open
37	202,5/SSW	open	wet	lot	little	intensively used	open

Table A5. Table of SA-CEFs study objects with data that were obtained by field survey (2014-2021). The numbering of the individual SA-CEFs corresponds to the data in Table 1—part 3.

No.	Azimuth of Entrance:	Exterior Environment:	Interior Environment:	Organic Material in Interior:	Anorganic Material in Interior:	Rate of Human Use:	Entrance:
38	225/SW	transition	dry	little	none	occasionally used	semi-open
39	180/S	transition	dry	little	none	occasionally used	semi-open

Table A6. Species of butterflies and moths (Lepidoptera) and their abundances in each SA-CEFs in 2014, F—frequency in %, D—dominance in %.

Species:		*Aglais urticae*	*Agonopterix arenella*	*Agonopterix curvipunctosa*	*Agonopterix heracliana*	*Agonopterix ocellana*	*Hypena rostralis*	*Inachis io*	*Nymphalis polychloros*	*Scoliopterix libatrix*
	1	1	~	~	~	~	2	19	~	7
	2	~	~	~	~	~	~	11	~	24
	3	~	~	~	~	~	~	24	~	12
	4	~	~	~	~	~	~	7	~	12
	5	~	~	~	1	~	~	6	~	8
	6	~	~	~	~	~	~	~	~	12
	7	~	~	~	~	~	~	2	~	11
	8	~	~	~	~	~	~	6	~	28
	9	~	~	~	1	~	1	1	~	1
	10	~	~	~	~	~	1	2	~	1
	11	~	1	~	~	~	~	8	~	7
	12	~	~	~	~	~	1	6	1	8
	14	5	~	~	3	~	3	29	~	22
	15	~	~	~	~	~	~	4	~	22
	16	~	~	~	~	~	~	7	~	8
	17	~	~	~	~	~	1	1	~	~
	18	~	~	~	~	1	7	2	~	1
	19	2	~	~	~	~	~	10	~	3
Object number:	20	1	~	~	~	~	1	1	~	1
	21	~	~	~	~	~	1	1	~	~
	23	~	1	~	~	~	~	1	~	7
	24	~	~	~	1	~	3	15	~	2
	25	~	~	~	~	~	5	8	~	7
	26	~	~	~	~	~	15	22	~	5
	27	~	~	~	~	1	3	5	~	1
	28	~	~	~	~	~	~	9	~	20
	29	~	~	~	~	~	~	6	~	~
	30	~	~	~	~	~	~	21	~	~
	31	~	~	~	1	~	27	~	~	4
	32	3	~	~	~	~	5	19	1	10
	33	~	3	1	~	~	~	~	~	2
	34	~	1	~	~	~	1	10	~	~
	35	~	~	~	~	~	~	1	~	1
	36	~	~	~	~	~	~	1	~	5
	37	~	~	~	~	~	~	7	~	18
	38	1	~	~	~	~	~	19	~	20
	39	~	~	~	2	~	~	22	~	19
Σ		13	6	1	9	2	77	313	2	309
F (%)		16.22	10.81	2.7	16.22	5.41	43.24	91.89	5.41	86.49
D (%)		1.78	0.82	0.14	1.23	0.27	10.52	42.76	0.27	42.21

Table A7. Species of land snails (Gastropoda) and their abundances in each SA-CEFs in 2014, F—frequency in %, D—dominance in %.

Species:	*Aegopinella nitens*	*Alinda biplicata*	*Arianta arbustorum*	*Arion vulgaris*	*Cepaea hortensis*	*Cochlicopa lubrica*	*Discus rotundatus*	*Discus ruderatus*	*Ena montana*	*Fruticicola fruticum*	*Helix pomatia*	*Limax cinereoniger*	*Macrogastra ventricosa*	*Monachoides incarnatus*	*Monachoides vicinus*	*Oxychilus cellarius*	*Perforatella bidentata*	*Trochulus hispidus*	*Vitrea crystallina*	*Vitrina pellucida*
2	~	~	1	~	~	~	~	~	~	~	~	~	~	1	~	~	~	~	~	~
3	~	~	1	~	~	~	~	~	~	~	~	~	~	2	~	~	~	~	~	~
7	~	~	~	~	~	~	17	1	~	~	~	~	~	~	~	~	~	~	~	~
8	~	~	~	~	~	~	~	2	~	~	~	~	~	~	~	~	~	~	1	~
10	1	~	2	~	~	~	~	~	~	~	~	1	~	3	2	~	~	~	~	~
11	~	~	3	~	~	~	~	~	~	~	~	~	~	3	~	~	~	~	~	~
14	~	~	1	~	~	~	~	~	~	~	~	~	~	~	~	~	~	~	~	~
15	~	~	2	~	1	~	~	~	3	~	~	~	~	4	~	~	1	3	~	~
17	~	~	~	~	~	~	~	~	~	1	~	~	~	1	~	~	~	~	~	~
19	~	~	~	~	1	~	~	~	~	2	3	~	~	~	~	~	~	~	~	~
20	~	~	~	1	~	~	~	~	~	3	4	1	~	~	~	~	~	~	~	~
21	~	~	~	~	~	~	~	~	~	~	7	~	~	1	~	~	~	~	~	~
22	~	~	~	~	~	~	~	~	~	3	1	~	1	1	~	1	~	1	~	~
23	~	2	~	~	~	~	~	~	~	~	~	~	~	2	~	~	~	~	~	3
25	~	~	~	~	~	3	~	~	~	~	1	~	~	~	~	~	~	~	~	~
27	~	~	~	~	~	~	~	~	~	~	1	~	~	4	~	~	~	~	~	~
29	~	~	~	~	2	~	~	~	~	~	3	~	~	~	~	~	~	~	~	~
33	~	2	~	~	~	~	~	~	~	~	3	~	~	2	~	~	~	~	~	~
34	~	~	~	~	~	~	~	~	~	~	14	~	~	~	~	~	~	~	~	~
37	1	8	11	~	~	~	2	~	~	~	7	~	~	2	~	1	~	9	1	~
38	~	3	1	~	1	~	~	~	~	~	~	~	~	1	~	~	~	~	~	~
39	~	~	~	~	~	~	~	~	~	~	~	~	~	1	~	~	~	~	~	~
Σ	2	15	22	1	5	3	19	3	3	9	44	2	1	28	2	2	1	13	2	3
F (%)	9.09	18.18	36.36	4.55	18.18	4.55	9.09	9.09	4.55	18.18	45.45	9.09	4.55	63.64	4.55	9.09	4.55	13.64	9.09	4.55
D (%)	1.11	8.33	12.22	0.56	2.78	1.67	10.56	1.67	1.67	5	24.44	1.11	0.56	15.56	1.11	1.11	0.56	7.22	1.11	1.67

The row label column is "Object number:".

References

1. Oliver, L.; Ferber, U.; Grimski, D.; Millar, K.; Nathanail, P. The scale and nature of european brownfields. In Proceedings of the CABERNET 2005—International Conference on Managing Urban Land LQM Ltd., Belfast, Northern Ireland, UK, 13–15 April 2005.
2. Matoušková, L. Military Brownfields and Their Transformations since 1989 Year. *Reg. Rozv. Mezi Teor. A Praxí* **2015**, *1*, 15–23.
3. Šilhánková, V. *Rekonverze Vojenských Brownfields*, 1st ed.; Univerzita Pardubice: Pardubice, Czech Republic, 2006.
4. Prünster, H. Report of the research project "Vallo Alpino in South Tyrol". In Proceedings of the Conferenza Vallo Alpino, Fortezza, Italy, 24 September 2021.
5. Malian, A.; Fathabad, S.A.H.; Jouzdani, M.F. Detection of Buffer changes of historical monuments and sites using remote sensing methods case study: Great wall of Gorgan. In Proceedings of the 3rd EARSeL Workshop on Remote Sensing for Archaeology 1 and Cultural Heritage Management, Ghent, Belgium, 19–22 September 2012.
6. World Heritage List. Available online: https://whc.unesco.org/en/list/ (accessed on 13 December 2021).
7. Forman, T.T.R.; Godron, M. *Krajinná Ekologie*, 1st ed.; Academia: Praha, Czech Republic, 1993.
8. Farina, A. The cultural landscape as a model for the integration of ecology and economics. *BioScience* **2000**, *50*, 313–320. [CrossRef]
9. Miko, L.; Hošek, M. *Příroda a Krajina České Republiky: Zpráva o Stavu 2009*, 1st ed.; Agentura ochrany přírody a krajiny ČR: Prague, Czech Republic, 2009; p. 2.
10. Svobodová, K. *Krajinný ráz: Krajina a Krajinný ráz ve Strategickém Plánování*, 1st ed.; Faculty of Architecture, Czech Technical University in Prague: Prague, Czech Republic, 2011; p. 22.
11. Sklenička, P. *Základy krajinného plánování*, 2nd ed.; Naděžda Skleničková: Prague, Czech Republic, 2003.
12. Moran, D.; Turner, J. The prison as a postmilitary landscape. *Soc. Cult. Geogr.* **2021**, 1–19. [CrossRef]
13. Woodwars, R. Military landscapes: Agendas and approaches for future research. *Prog. Hum. Geogr.* **2014**, *38*, 40–61. [CrossRef]
14. Hendrych, J. *Tvorba Zahrad a Krajiny*, 1st ed.; Czech Technical University in Prague: Prague, Czech Republic, 2005; p. 190.
15. Lecic, N.; Vasilevska, L. Adaptive reuse in the function of cultural heritage revitalization. In Proceedings of the Built Heritage Management and Presentation (BHM&P)—National Heritage Foundation, Nis, Serbia, 30 August–1 September 2018.
16. Demek, J. *Úvod do Krajinné Ekologie*, 1st ed.; Faculty of Science Palacký University Olomouc: Olomouc, Czech Republic, 1999.

17. Jančura, P. Aktuálne problémy starostlivosti o krajinu a investiční rozvoj. In *Aktuální Problémy Ochrany Krajinného Rázu: Sborník Příspěvků z Odborného Semináře, Prague, 12 November 2007*; Vorel, I., Kupka, J., Eds.; Centrum pro krajinu: Prague, Czech Republic, 2008.

18. Zarnetske, P.L.; Baiser, B.; Strecker, A.; Record, S.; Belmaker, J.; Tuanmu, M.-N. The interplay between landscape structure and biotic interactions. *Curr. Landsc. Ecol. Rep.* **2017**, *2*, 12–29. [CrossRef]

19. With, K.A. *Essentials of Landscape Ecology*, 1st ed.; Oxford University Press: Oxford, UK, 2019; p. 656.

20. Stepanchuk, A.; Gafurova, S.; Latypova, M. «Genius Loci» as a resource for the development of historical areas of the city. In Proceedings of the IOP Conference Series: Materials Science and Engineering, Kazan, Russia, 29 April–15 May 2020; IOP Publishing: Bristol, UK, 2020; Volume 890, pp. 1–14. [CrossRef]

21. Minimuzeum of Czechoslovak fortification 1935–1938: ČESKOSLOVENSKÉ OPEVNĚNÍ SYMBOL ODHODLÁNÍ BRÁNIT DEMOKRACII. Available online: http://www.minimuzeum.com/opevneni/ (accessed on 3 August 2021).

22. Kuchař, P. *Guide to the Hlučín-Darkovičky Czechoslovak Fortification Complex*, 1st ed.; Retis Group: Ostrava, Czech Republic, 2012; p. 27.

23. Aron, L. *Československé opevnění 1935–1938*, 1st ed.; Okresní muzeum Náchod: Náchod, Czech Republic, 1990; p. 199.

24. Natura Opava: Unplanned Use of Border Fortification Bunkers in the Hlučín and Opava Regions. Available online: http://www.natura-opava.org/opavsko/zpravy/neplanovane-vyuziti-bunkru-hranicniho-opevneni-na-hlucinsku-a-opavsku-2.html (accessed on 1 June 2021).

25. Bunkers: Fortifications of Czechoslovakia. Available online: http://www.bunkry.cz/clanek/1180 (accessed on 17 September 2021).

26. Kolář, O. *Armáda 4—Ostravská Operace 1945*, 1st ed.; Jakab: Bučovice, Czech Republic, 2019; p. 52.

27. Information about casemates with enhanced fortification. Available online: www.ropiky.net (accessed on 29 May 2021).

28. Land Registry. Available online: https://geoportal.cuzk.cz/(S(4ftdbk3h4qbe44atwzv0phie))/Default.aspx?mode=TextMeta&side=dSady_RUIAN&metadataID=CZ-CUZK-SH-V&mapid=5&head_tab=sekce-02-gp&menu=25 (accessed on 27 October 2021).

29. MAPS. Available online: www.mapy.cz (accessed on 13 September 2021).

30. Břeclavský Deník: Klub Vojenské Historie Sídlí na Pohansku. Available online: https://breclavsky.denik.cz/zpravy_region/pohansko_bunkry20070525.html (accessed on 11 October 2021).

31. Česká Televize: Během několika let Vystavělo Československo Systém Pevností. Měly Zbrzdit Němce, než Přijdou Na Pomoc Spojenci. Available online: https://ct24.ceskatelevize.cz/domaci/2588257-bunkry-na-hranicich-pred-80-lety-marne-cekaly-na-vyzyvatele-dnes-lakaji-turisty (accessed on 9 October 2021).

32. Oživlý Svět Technických Památek: Betonové Pevnosti. Available online: http://www.technicke-pamatky.cz/sekce/38/betonov-pevnosti/ (accessed on 3 September 2021).

33. National Monument Institute: Monument Catalogue. Available online: https://pamatkovykatalog.cz/ (accessed on 11 September 2021).

34. Ministry of Defence & Armed Forces of the Czech Republic: Offer of Immovable Assets. Available online: http://www.onnm.army.cz/index.php?id=21&zobr=nab&up=&typ=bs (accessed on 16 October 2021).

35. Durčák, J. *Opevňování Ostravska v letech 1935 až 1938*; AVE—Informační Centrum Opavska: Opava, Czech Republic, 1995; p. 88.

36. Interactive Map of CZ Fortifications. Available online: http://mapa.opevneni.cz/ (accessed on 15 June 2021).

37. László, M.; Kocicka, E.; Kočický, D.; Diviaková, A.; Izakovièová, Z.; Špinerová, A.; Miklosova, V. Landscape as a Geosystem. In *Landscape as a Geosystem*; Springer: Cham, Switzerland, 2018; pp. 11–42. [CrossRef]

38. Kolejka, J. Postindustrial landscape of the Czech Republic—available facts, documentation and possibility of typization. In Proceedings of the Geografie Pro Život ve 21 Století: XXII. Sjezd České Geografické Společnosti, Ostrava, Czech Republic, 31 August–3 September 2010; University of Ostrava: Ostrava, Czech Republic, 2010; pp. 125–130.

39. Kupka, J.; Pohunek, J. Krajina po těžbě břidlice očima trampů a táborníků aneb o brownfieldech jinak (Nízký Jeseník). *Studia Ethnol. Pragensia* **2017**, *2*, 134–145.

40. Lipský, Z.; Demková, K.; Skaloš, J.; Kukla, P. The influence of natural conditions on changes in landscape use: A case study of the lower podoubraví region (Czech Republic). *Ekológia* **2011**, *30*, 133–140. [CrossRef]

41. Dvořák, L. Gastropods in subterranean shelters of the Czech Republic. *Malacol. Bohemoslov.* **2005**, *4*, 10–16. [CrossRef]

42. Dvořák, L. Notes on hibernation of Lepidoptera species in underground shelters of the Bohemian Forest and of West Bohemia. *Silva Gabreta* **2000**, *5*, 167–178.

43. Demek, J.; Culek, M.; Mackovčin, P.; Balatka, B.; Buček, A.; Cibulková, P.; Čermák, P.; Dobiáš, D.; Havlíček, M.; Hrádek, M.; et al. *Hory a nížiny. Zeměpisný lexikon ČR*, 1st ed.; AOPK ČR: Brno, Czech Republic, 2006; p. 580.

44. Bína, J.; Demek, J. *Z nížin do hor: Geomorfologické Jednotky České Republiky*; Academia: Prague, Czech Republic, 2012; p. 343.

45. Quitt, E. *Climatic Regions of Czechoslovakia*, 1st ed.; Geografický Ústav ČSAV: Brno, Czech Republic, 1971; p. 73.

46. Culek, M. *Biogeografické Členění České Republiky*, 1st ed.; Enigma: Prague, Czech Republic, 1996; p. 347.

47. Cílek, V.; Ložek, V.; Mudra, P.; Šprynář, P. *Vstoupit do Krajiny: O Přírodě a Paměti Středních Čech*, 1st ed.; Dokořán: Prague, Czech Republic, 2004; p. 110.

48. Bastian, O. Landscape Ecology–towards a unified discipline? *Landsc. Ecol.* **2001**, *16*, 757–766. [CrossRef]

49. Losos, B. *Ekologie živočichů*, 1st ed.; SPN: Prague, Czech Republic, 1980; p. 307.

50. Fine, G.A. The Goliath Effect: Corporate dominance and mercantile legends. *J. Am. Folk.* **1985**, *98*, 63–84. [CrossRef]

51. Kovář, P. *Geobotanika: Úvod do Ekologické Botaniky*, 1st ed.; Karolinum: Prague, Czech Republic, 2001; p. 104.

52. Macek, J.; Laštůvka, Z.; Beneš, J.; Traxler, L. *Motýli a Housenky Střední Evropy IV. Denní motýli*, 1st ed.; Academia: Prague, Czech Republic, 2015; p. 539.

53. Macek, J.; Procházka, J.; Traxler, L. *Motýli a Housenky Střední Evropy. Noční Motýli II.: Můrovití*, 1st ed.; Academia: Praha, Czech Republic, 2008; p. 492.
54. Horsák, M.; Juřičková, L.; Picka, J. *Měkkýši České a Slovenské Republiky*, 1st ed.; Kabourek: Zlín, Czech Republic, 2013.
55. Růžička, V. Pavouci v jeskyních České republiky. *Speleo* **2007**, *49*, 14–19.
56. Riedel, A. Die in West-Paläarktis unterirdisch lebenden Zonitidae sensu lato (Gastropoda, Stylommatophora). *Fragm. Faun.* **1996**, *39*, 363–390. [CrossRef]
57. Laboy, M.M. A Conceptual Space for Architecture: Singularity Through Ecological Contingency. In Proceedings of the Fall Conference: Between the Autonomous & Contingent Object, Syracuse, NY, USA, 8–10 October 2015; pp. 231–238.
58. Andrew Sargeant: Landscape Singularity: Innovative Tools for Design and Engagement. Available online: https://www.cudc.kent.edu/andrew-sargeant-032021 (accessed on 13 December 2021).
59. Agentura Ochrany Přírody a Krajiny České Republiky—Český Ráj. Available online: http://ceskyraj.ochranaprirody.cz/res/archive/324/039802.pdf?seek=1476181256 (accessed on 13 December 2021).

 *land*

Article

Land Use Optimization and Simulation of Low-Carbon-Oriented—A Case Study of Jinhua, China

Shiqi Huang [1], Furui Xi [2,*], Yiming Chen [1], Ming Gao [1], Xu Pan [1] and Ci Ren [1]

[1] College of Geoscience and Surveying Engineering, China University of Mining & Technology (Beijing), Beijing 100083, China; hsq@student.cumtb.edu.cn (S.H.); cym@student.cumtb.edu.cn (Y.C.); gm@student.cumtb.edu.cn (M.G.); px@student.cumtb.edu.cn (X.P.); rc@student.cumtb.edu.cn (C.R.)

[2] China Institute of Geo-Environment Monitoring, Beijing 100081, China

* Correspondence: xifurui@mail.cgs.gov.cn; Tel.: +86-10-83473353

Abstract: Land-use change is an important contributor to atmospheric carbon emissions. Taking Jinhua city in eastern China as an example, this study analyzed the effects on carbon emissions by land-use changes from 2005 to 2018. Then, carbon emissions that will be produced in Jinhua in 2030 were predicted based on the land-use pattern predicted by the CA-Markov model. Finally, a low-carbon optimized land-use pattern more consistent with the law of urban development was proposed based on the prediction and planning model used in this study. The results show that (1) from 2005 to 2018, the area of land used for construction in Jinhua continued to increase, while woodland and cultivated land areas decreased. Carbon emissions from land use rose at a high rate. By 2018, carbon emissions had increased by 1.9 times compared to 2015. (2) During the 2010–2015 period, the total concentration of carbon emissions decreased due to decreases in both the rate of growth in construction land and the rate of decline in a woodland area, as well as an adjustment of the energy structure and the use of polluting fertilizer and pesticide treatments. (3) The carbon emissions produced with an optimal land-use pattern in 2030 are predicted to reduce by 19%. The acreage of woodland in Jinhua's middle basin occupied by construction land and cultivated land is predicted to reduce. The additional construction land will be concentrated around the main axis of the Jinhua-Yiwu metropolitan area and will exhibit a characteristic ribbon-form with more distinct clusters. The optimized land-use pattern is more conducive to carbon reduction and more in line with the strategy of regional development in the study area. The results of this study can be used as technical support to optimize the land-use spatial pattern and reduce urban land's contribution to carbon emissions.

Keywords: land-use change; carbon flow; CA–Markov; low-carbon optimization

Citation: Huang, S.; Xi, F.; Chen, Y.; Gao, M.; Pan, X.; Ren, C. Land Use Optimization and Simulation of Low-Carbon-Oriented—A Case Study of Jinhua, China. *Land* **2021**, *10*, 1020. https://doi.org/10.3390/land10101020

Academic Editors: Dong Jiang, Jinwei Dong and Gang Lin

Received: 26 August 2021
Accepted: 24 September 2021
Published: 28 September 2021

1. Introduction

As human social and economic activities consume large amounts of resources, the concentration of greenhouse gas emissions produced is growing rapidly. The resulting climate warming problem has become a global issue that is critically related to the sustainable development of human society. The International Energy Agency (IEA) [1] showed that, in the past 30 years, the total carbon emissions produced in China have continued to rise. By 2018, the total carbon emissions had reached 9.5 billion tons. This concentration is still rising, albeit at a low speed, so the peak value of carbon emissions has not yet been reached. Therefore, in September 2020, at the United Nations General Assembly, China renewed its commitment to increase its intended nationally determined contributions and strive to achieve peak carbon dioxide emissions by 2030. "To widely promote green ways of working and living and to reduce carbon emissions placidly after peaking" [2] was listed as one of the important Chinese economic and social development goals for the 14th Five-Year period.

As the natural carrier of carbon emissions from terrestrial ecosystems and the spatial carrier of emissions from human economic and social carbon activities [3], land use is closely related to atmospheric carbon emissions. The fifth assessment report of the Intergovernmental Panel on Climate Change (IPCC) shows that land-use change accounts for about one-third of manmade carbon emissions, and land-use changes play an important role in the increase in the global atmospheric carbon dioxide concentration [4]. Choices regarding land-use structure by humans significantly impact atmospheric greenhouse gas emissions [5]. Determining how to adjust spatial patterns of land use to achieve carbon emission reduction targets is currently a research hotspot. Houghton et al. [6] summarized and analyzed 13 methods for calculating net carbon emissions caused by land-use changes. Lai [3] researched the mechanisms, characteristics, and factors contributing to the production of carbon emissions due to land-use changes in China over the past 20 years. Meanwhile, a comprehensive inventory of carbon emissions produced through land use based on land use type was developed, and a linear programming model was used to optimize land use planning to reduce carbon emissions. Liu et al. [7] constructed the LUCC (Land Use and Land Cover Change)-climate-ecosystem research system to study the effects of land-use changes on the global climate, and their results can help to determine the impacts of large-scale LUCC factors on climate change through the initiation of a budget change regarding greenhouse gases and through changes in land surface and atmosphere processes. Zhu et al. [8] developed a land-use and land-cover transfer matrix based on digital land use maps and then assigned a carbon density score to each land-use type. Finally, they quantified the spatio-temporal patterns of LUCC carbon emissions produced from 1970 to 1990 and from 1990 to 2010 in Zhejiang Province. Yang et al. [9] researched the characteristics of spatio-temporal variation in carbon storage for different land-use spatial patterns using two different scenarios in the Yellow River Basin based on the CA–Markov model and the carbon storage module of the InVEST model.

Most studies in this field have focused on analyzing and evaluating the effects of carbon emissions produced with various land-use patterns. Li et al. [10] analyzed the effects of land use patterns on carbon emissions in Jiangsu province. Zhao et al. [11] measured and analyzed the effects of land uses on carbon emissions and their spatial patterns in the Hunan province. Sun et al. [12] estimated the carbon emission intensity on land-use patterns of 31 provinces in China, revealing its rules of spatial-temporal evolution and discussing its spatial association at provincial scale with the methods of spatial autocorrelation. The law of carbon emissions produced with land-use changes and existing problems in different areas have been revealed clearly. However, a few studies have forecasted the variation trend in future carbon emissions from the perspective of the spatiotemporal pattern of land use and predicted the optimized land-use pattern with the consideration of planning objectives.

This study used Jinhua City as an example. The aim here was to provide a theoretic basis for optimization of the territorial spatial pattern and to help carbon dioxide emissions reach their peak:

(1) the carbon emissions produced from land use and the transfer matrix of land use were calculated. Based on this, an explicit carbon flow model was used to calculate carbon transfer between different land-use types. It contributed to revealing the spatio-temporal variation rules of land-use carbon emissions.

(2) we simulated the land-use spatial pattern that will be present in Jinhua under a natural change scenario in 2030 using the CA–Markov model and predicted the carbon emission change characteristics. Based on this, suggestions for carbon emission reduction from the perspective of land-use change can be proposed.

(3) the low-carbon quantitative structure obtained from multiple linear programming models and optimization-oriented driving factors were added to the FLUS model to optimize the land-use spatial pattern in Jinhua in 2030. The simulation results meet both the requirements of urban development and carbon emission reduction, providing a reference for the future green development of land use in Jinhua city.

2. Study Area

This study used Jinhua City, which is located at the intersection of Dongyang River, Wuyi River, and Jinhua River (Figure 1), as an example. Jinhua is adjacent to Taizhou in the east, Lishui in the south, Hengzhou in the west, and Shaoxing and Hangzhou in the north. Jinhua is surrounded by three mountains and contains three rivers. Most of the mountains in Jinhua are 500–1000 m high, distributed on the north and south sides. The terrain is high in the north and south while low in the middle. The total land area of Jinhua was 10,942 km^2. The population was approximately 0.71 million [13], and its GDP was CNY 4703.95 billion in 2020 [14]. As one of the important bases for advanced manufacturing industries south of the Yangtze River Delta and the central city of the central Zhejiang urban agglomeration, Jinhua's economy has been developing rapidly and relies on large-scale professional trading markets such as Yiwu International Trade City and Yongkang Hardware Technology City. National strategies, such as the integrated development of the Yangtze River Delta, have brought unprecedented opportunities for development here. However, in recent years, problems such as the extensive utilization of land resources, unscientific spatial layout, and low-performance level have emerged, leading to a massive expansion of construction land to provide land guarantees for economic construction. As of 2020, the total urban and rural land area exceeded that outlined in the planning objectives, and the per capita land acreage far exceeded the upper limit set by the national standard. A series of ecological and environmental problems have also emerged, including air pollution and intensification of the urban greenhouse effect. Therefore, the aim of this study was to explore the law of land-use carbon emissions in Jinhua City and to predict and optimize the future spatial land-use pattern. The research results could have significant benefits in terms of providing suggestions for reaching the carbon peak, which is vital for promoting progress in ecological civilization in Jinhua City.

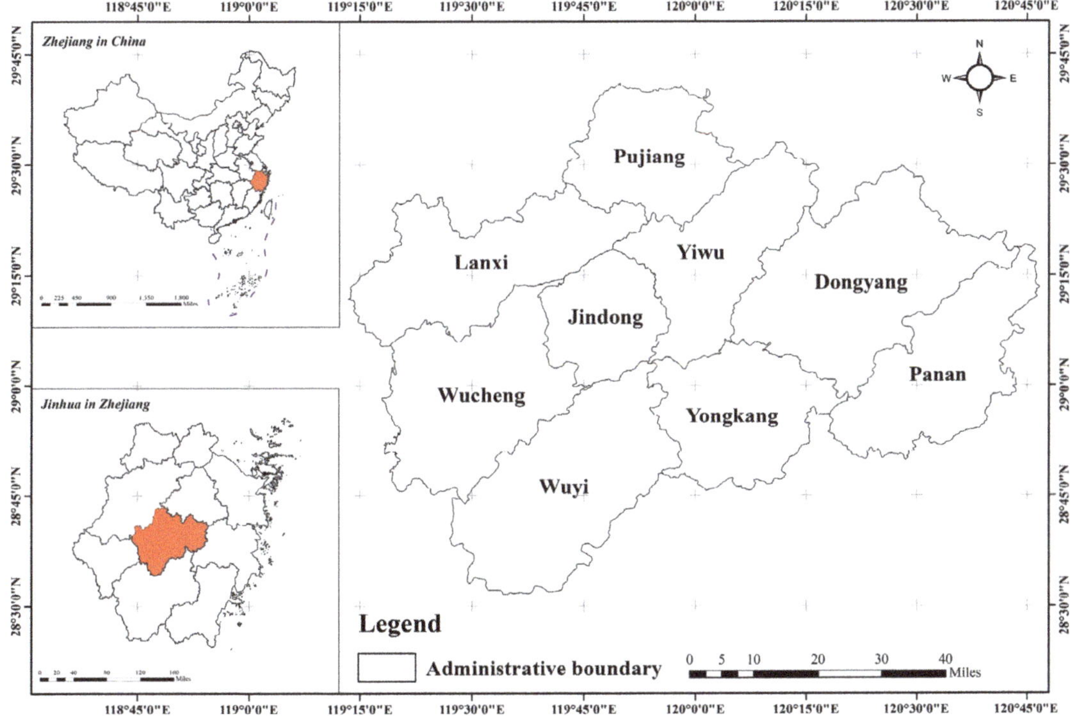

Figure 1. Location and topography of the study area.

3. Materials and Method

3.1. Data Acquisition

The data sets used in this study were as follows: (1) land use and cover raster images (30 m) in 2005, 2010, 2015, and 2018 from Data Center for Resources and Environmental Sciences, Chinese Academy of Sciences (RESDC) [15], which were reclassified into six categories of cultivated land, woodland, grassland, water area, construction land, and unused land based on the Land-Use and Land-Cover Change (LUCC) classification system; (2) population density data in 2018 from Worldpop project data sets and Gross Domestic Product (GDP) distribution data in 2015 from the Resource and Environment Data Center of the Chinese Academy of Sciences; and (3) data on energy consumption and agricultural material use including pesticide, chemical fertilizer, and agricultural film from 2005–2018 from Jinhua Statistical Yearbooks of corresponding years. The calculation parameters related to energy consumption were taken from the 2006 IPCC National Greenhouse Gas Inventory Guidelines and the General Principles for the Calculation of total Production Consumption.

3.2. Methodology

The aim of this study was to contribute to optimizing the spatial pattern of land use under the guidance of low carbon. We used the carbon emissions estimation model and land-use change forecast model. With the consideration of low-carbon objectives and strategies for regional development, the optimized land-use spatial pattern in 2030 in Jinhua was obtained (Figure 2).

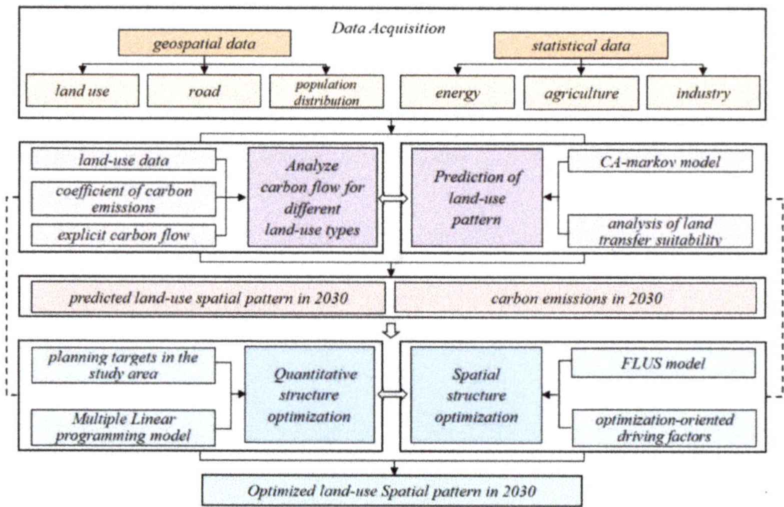

Figure 2. The technology roadmap of this study.

3.3. Accounting Model Used to Assess "Carbon Flow"

3.3.1. Calculation of Carbon Emissions

(1) To determine the carbon emissions produced from construction land, this study mainly considered the carbon emissions produced through energy consumption by the industrial and mining enterprises and the transportation industry. In accordance with the reference method proposed in the 2006 IPCC Guidelines for National Greenhouse Gas Inventories, the carbon emissions from construction land were calculated as follows:

$$Ch = \sum_i Qh_i \cdot NCV_i \cdot CC_i \cdot COF_i \cdot \frac{44}{12} \cdot \frac{1}{1000} = \sum_i SC_i \cdot F_i \tag{1}$$

$$SC_i = Qh_i \cdot CS_i \tag{2}$$

where SC_i represents the quantity of standard coal converted from every kind of energy source ($kgce$); Qh_i represents the consumption of every kind of energy (kg); and CS_i represents the standard coal conversion coefficient from each energy source ($kgce/kg$),

$$F_i = CC_i \cdot COF_i \cdot \frac{44}{12} \cdot \frac{1}{1000} \cdot 29.3076 \tag{3}$$

where F_i represents the standard coal emission factor ($kgCO_2/kgce$); CC_i represents the carbon content of fossil energy (kC/GJ); COF_i represents the carbon oxidation factor; $\frac{44}{12}$ is the carbon content per molecule CO_2($kgCO_2/kg$); $\frac{1}{1000}$ is the unit conversion factor; and 29.3076 ($MJ/kgce$) is the low calorific value of standard coal per kg according to the general rules of the GB/T2589-1990 comprehensive energy consumption calculation.

For cultivated land, because carbon emissions are produced over a long time scale, changes in crop yield have little impact on carbon absorption and emissions by the ecosystem [16]. Thus, only human-induced carbon emissions from agricultural activities were considered. This included carbon emissions caused by the powering of agricultural machinery, through the use of fertilizers and pesticides, and through agricultural irrigation. In accordance with Tristram O West [17] and the Institute of Agricultural Resources and Eco-Environment Performance of Nanjing University, the carbon emissions produced from cultivated land were calculated as follows:

$$E_b = U_{fN} \cdot A_N + U_{fP} \cdot A_P + U_{fK} \cdot A_K + U_m \cdot B + U_p \cdot C + U_a \cdot D + S_i \cdot E + S \cdot F \tag{4}$$

where U_{fN} represents the consumption of nitrogen fertilizer; U_{fP} represents the consumption of phosphate fertilizer; U_{fK} represents the consumption of potassium fertilizer; $A_N = \frac{857.54 \ kgC}{Mg}$, $A_P = \frac{165.09 \ kgC}{Mg}$ and $A_K = 120.28 \ kgC/Mg$ are conversion coefficients of types of fertilizer; U_m represents the total power needed for agricultural machinery; U_p represents the consumption of pesticides; U_a represents the consumption of agricultural film; U_a represents the area of agricultural irrigation; S represents the area of crop planting; and $B = 0.18 \ kgC/kwh$, $C = 4907.25 \ kg/Mg$, $D = 5.18 \ kg/kg$, $E = 266.48 \ kgC/ha$, $F = 16.47 \ kgC/ha$ are conversion coefficients.

(2) Carbon emissions from other land-use types were calculated using the following direct carbon emission coefficient method:

$$C_i = S_i \times k_i \tag{5}$$

where C_i represents the carbon emissions produced from different land-use types; S_i represents the area taken up by each land-use type; and k_i represents the carbon emission coefficient of each land-use type.

In terms of the carbon emission coefficients of woodland and grassland, based on the fact that the main vegetation type present in Jinhua is evergreen coniferous forest mixed with a small amount of evergreen broad-leaved forest [18], we referred to the relational expression $CSE = carbon \ sink/NPP$ from the research results of Fang et al. [19] and Lai [3] and selected −0.374 tC/ha·a as the carbon emission coefficient for woodland in Jinhua and −0.021 t/ha·a as the carbon emission coefficient of grassland.

Regarding the water carbon emission coefficient, according to the research results of Duan et al. [20], the carbon sink capacity of the lake wetland in the Eastern Lake Area where Jinhua is located is −0.567 tC/ha·a. According to the research results of Lai Li, the carbon sink capacity of the tidal flat is −0.236 tC/ha·a. The secondary classification of waters in the LUCC classification system includes lakes, rivers, canals, reservoirs, ponds, tidal flats, and beaches. For comprehensive consideration of the carbon sink capacity of different types of water and the average carbon sink capacity of waters across the country, the carbon sink coefficient of the study was taken as −0.28 tC/ha·a [21].

Regarding the carbon emission coefficient of unused land, unused land includes sandy land, Gobi, saline-alkali land, marshland, bare land, and bare rock land. Among these, the carbon emissions are low except for swamps, which have a strong carbon sink capacity. In accordance with the research results of Lai [3], the carbon sink coefficient was taken as -0.005 tC/ha·a.

3.3.2. Analysis of Carbon Flow for Different Land-Use Types

In order to better explore the law of carbon transfer between different land-use types and evaluate the impacts of carbon emissions from land-use changes more accurately, an explicit carbon flow model was introduced [22]:

$$f_{ij} = \Delta W \cdot \Delta S \tag{6}$$

$$\Delta W = W_j - W_i = \frac{V_j}{S_j} - \frac{V_i}{S_i} \tag{7}$$

where f_{ij} represents the transfer of carbon from land type j to land type i; ΔW is defined as the carbon metabolism density during the process of conversion from land type j to land type i; ΔS represents the acreage over which the conversion occurs ; W_j and W_i represent the carbon flow density of land type j and the land type i, respectively, which is equivalent to the land carbon emission coefficient. V_j and V_i represent the carbon flow of land type j and the land type i, respectively, which is equivalent to the land carbon emissions. If $\Delta W > 0$, the land type conversion is a positive process with an increase in the carbon sink. If $\Delta W < 0$, the land type transfer is a negative process with a reduction in the carbon sink or an increase in carbon emissions.

3.4. CA–Markov Model

3.4.1. Principles of the CA–Markov Model

The Markov chain model is a Markov model with a discrete time and state. It is often used to analyze the law of land-use changes. Cellular Automaton (CA) can be used to simulate the process of natural changes with the influence of the surrounding space. In this study, CA was utilized to simulate changes in any pixels in the land-use image affected by the state of themselves and their neighboring pixels.

The traditional Markov model has difficulty predicting spatial changes in land-use patterns. The CA model focuses on partial interactions between cells, which has obvious limitations [23]. The CA–Markov model has the advantages of both the Markov model and the Cellular Automata, as spatial neighboring elements and rules of spatial distribution conversion are added to the Markov chain model. It can be used to predict spatial changes of the land-use pattern with full consideration of spatial parameters.

3.4.2. Prediction Process Used by the CA–Markov Model

The CA–Markov module in IDRISI systematically integrates the Markov chain model with the CA model. The Multi-Criteria Evaluation module can be used for the prediction of land-use spatial patterns and the analysis of land transfer suitability. We simulated the land-use pattern with the following steps:

(1) First, earlier and later land-use images were input into the Markov module to construct a probability matrix of conversion between different land-use types.
(2) Then, we analyzed the land transfer suitability. This study mainly analyzed the suitability of the spatial distribution of construction land, because the expansion of construction land is fast and is greatly affected by human factors. First, we selected the driving factors. The driving factors include constraints and influencing factors. Constraints are expressed in the form of Boolean maps in which areas to be excluded from consideration have the value 0 and those to be considered have the value 1. Here we set existing construction land, water, and nature reserve as constraints. Water includes lake, river, beach, reservoir, and pit. Nature reserve includes ecosystem

type reserves, biological species reserves, and natural heritage reserves. Influencing factors enhance or detract from the suitability of a location for the objective. All influencing factor images must be measured on the same scale. In this study, the scale 0–255 is used where 0 is not at all suitable and 255 is perfectly suitable. Then the weights of the influencing factors are produced follows the logic under the Analytical Hierarchy Process (AHP). That is, we should provide a series of pairwise comparisons of the relative importance of factors to the suitability of pixels for the activity being evaluated. (Table 1). Second, the constraints were transformed into a binary image. That is, areas with a construction land spatial distribution probability of zero were assigned a value of 0. The factors were transformed into standard data in the range of 0–255 using the Fuzzy module, and the fuzzy set membership function was applied to evaluate the effects of factors on suitability. Finally, the MCE module was used to generate a construction land transition suitability image.

(3)   Finally, the land-use spatial pattern in 2030 was predicted. We selected default 5 × 5 filters as the neighborhood scale, which is suitable for most simulation processes to ensure accuracy and also avoid running the simulation too far into the future. Then the number of iterations was set to 8. Land-use images from 2000 and 2010 were regarded as earlier and later images, respectively. We input the land-use transfer matrix and Transition suitability image collection so that we obtained a predicted land-use image for 2018. Then, the Crosstab module was applied to compare this with the actual image of land use in 2018 to get a kappa coefficient of 0.9243, indicating that the simulation effect was good. Therefore, the actual land-use image in 2018 could be served as the early image to simulate the land-use image in 2030 using the method above.

Table 1. Driving factors identified in the suitability analysis of construction land.

Constraints	Influencing Factors	Weights
Existing construction land	Population	0.1672
Water	GDP	0.1839
Nature reserve	DEM	0.0712
	slope	0.0661
	Distance to construction land	0.0831
	Distance to roadway	0.1387
	Distance to the motorway	0.1500
	Distance to railway	0.1397

3.5. Low-Carbon Optimization Model of Land Use

3.5.1. Multiple Linear Programming Model

Linear programming is a mathematical method that is used to solve the extremum of a linear objective function under constrained conditions. It provides a scientific basis for the rational utilization of limited resources. In this study, the carbon emission calculation model was added to the multivariate linear programming model to achieve the goal of minimizing carbon emissions. The constraint conditions were set according to the actual situation in the study area. The general linear programming model used was as follows:

$$F(x) = \sum_{i=1}^{6} k_i X_i \tag{8}$$

where $F(x)$ represents the total carbon emissions; k_i represents the carbon emission intensity of different land types; and X_i represents the decision variable, that is, the acreage of different land types.

With the goal of minimizing carbon emissions from land use in Jinhua by 2030, a multiple linear programming model was built. Cultivated land, woodland, grassland, water area, construction land, and unused land were selected as decision variables X_i, and the carbon emission coefficient from land use in 2018 was selected as k_i. The planning model developed was as follows:

$$min = 365.35 \cdot X_1 - 374 \cdot X_2 - 21 \cdot X_3 - 280 \cdot X_4 + 261853.76 \cdot X_5 - 5 \cdot X_6 \tag{9}$$

For the selection of decision variable constraints, we mainly referred to the Fourteenth Five-Year Plan and Long-Range Objectives Through the Year 2035 (hereinafter referred to as the Outline), National Land Planning Outline 2016–2030 (hereinafter referred to as the Plan) with consideration of Jinhua's socio-economic development and status of land-use. The specific constraint conditions are as follows:

(1) With the rapid development of the economy and the acceleration of urbanization in Jinhua, it is inevitable that the areas of cultivated land and woodland will reduce due to the expansion of construction land. So, it is necessary to strengthen the protection of cultivated land resources, strictly maintain the red line of cultivated land, and strictly control the use of cultivated land for non-agricultural construction. The national cultivated land reserves in 2020 and 2030 should not be less than 1.865 billion mu and 1.825 billion mu, respectively. Therefore, the annual rate of decrement in cultivated land should not exceed 0.2%.

(2) Since 2005, the area of woodland in Jinhua was in a state of negative growth. In order to prevent woodland from becoming occupied without limit, the annual rate of decrement in woodland should not be greater than the rate of decrement over the last decade (0.31%), which was set as the lower limit. In addition, the Plan points out that the national forest coverage rate is predicted to increase from 21.66% in 2015 to more than 24% by 2030. To achieve this goal, the annual rate of growth in woodland areas needs to reach 1%. This represents the ideal state of land conversion in Jinhua, so it was set as the upper limit.

(3) The acreage of water, grassland, and unused land have been decreasing to varying degrees over the last 10 years. The Plan points out that it is necessary to strengthen the protection and restoration of water source conservation areas to consolidate the achievements related to the grain for green and the return of farmland to lake strategies. Therefore, the annual rate of deceleration in grassland, water, and unused land areas should not exceed the deceleration rate from the past decade.

(4) In the past decade, the annual GDP growth rate in Jinhua was close to 10%, and the annual population growth rate over the past ten years was 2.78% according to the seventh census. This led to a rapid expansion of construction land. The Outline points out that we will continue to promote economic construction. Thus, the area of construction land will inevitably continue to increase in the future. At the same time, intensive and economic use of land should be strengthened. The Plan points out that the land-use intensity in 2030 should be restricted to 4.62% based on a level of 4.02% in 2015. Thus, the growth rate of the construction land area should be capped at 0.93% with consideration of the socio-economic situation present in Jinhua.

According to the analysis above, we set the following constraint conditions for the decision variables set:

$$\sum_{i=1}^{6} X_i = 1092199.05, \; X_i \geq 0 \tag{10}$$

$$(1 - 0.2\%)^{12} \cdot 292914.63 < X_1 \tag{11}$$

$$(1 - 0.31\%)^{12} \cdot 655048.71 < X_2 \leq (1 + 1.0\%)^{12} \cdot 655048.71 \tag{12}$$

$$(1 - 0.45\%)^{12} \cdot 21957.48 \leq X_3 \tag{13}$$

$$(1 - 0.02\%)^{12} \cdot 15286.59 \leq X_4 \tag{14}$$

$$(1 + 0.5\%)^{10} \cdot 92522.52 < X_5 \leq (1 + 0.93\%)^{12} \cdot 92522.52 \tag{15}$$

$$(1 - 0.61\%)^{12} \cdot 363.42 \leq X_6 \tag{16}$$

3.5.2. Low-Carbon Space Optimization

The FLUS model is used to simulate changes and future patterns of land use, which can aid in spatial optimization and assist decision-making [24]. The model uses the artificial neural network (ANN) to calculate the transition probability of each land type and then integrates the system dynamics model (SD) and the Cellular Automaton model (CA) for further analysis. The self-adaptive inertia and competition mechanism are used in the CA model to deal with complex competition and interactions between different land-use types [25]. In this study, the low-carbon land-use allocation plan was used as the quantitative target for land-use type change in the FLUS model. After that, with the consideration of planning targets in the study area's territorial space, we added the following optimization-oriented driving factors (Table 2): First, the probability of land around the main axis of the Jinhua-Yiwu metropolitan area converting to construction land was improved by taking the Jinhua-Yiwu Expressway as a driving factor in order to promote the intensive use of land resources. Second, the hilly area in the central basin was taken as a constraint in order to create a green ecological zone in the central basin area. Through the above method, the spatial pattern of land use after low-carbon optimization was simulated.

Table 2. Optimization-oriented driving factors.

Optimization-Oriented Driving Factors	Description
Jinhua-Yiwu Expressway	To improve the probability of land around the main axis of the Jinhua-Yiwu metropolitan area converting to construction land
Hilly area in the central basin	To reduce the probability of the green ecological zone in the central basin area

4. Results

4.1. Spatio-Temporal Patterns of Land Use in Jinhua

According to the land-use images from 2005 to 2018 (Figure 3), transfer matrix and area changes of different land-use types in Jinhua City were obtained (Tables 3–6). During the 2005–2018 period, the main land-use type present in Jinhua City was woodland, followed by cultivated land. During three periods, 2005–2010, 2010–2015, and 2015–2018, the total land transfer area gradually increased, reaching a maximum of 58,554 ha in the third period. This area was much larger than that present in the first two periods, indicating that the degree of land use evolution in the study area has intensified in recent years. A total of 34.6% of the converted land area represented the transfer-out of construction land, and 31.9% represented the transfer-in of cultivated land. Other land types also changed to different degrees. The acreage of cultivated land continued to decrease at a relatively stable rate, and the proportion converted to woodland increased, while the proportion converted to construction land decreased, thanks to the Grain for Green strategy. The acreage of woodland continued to decrease, with the decrement rate being the smallest in 2010–2015 and the largest in 2015–2018: 0.03% and 0.61% respectively. The area converted from woodland to cultivated land gradually increased. The acreage of construction land continued to grow rapidly. More than 80% of new construction land came from the transfer of cultivated land. The growth rate of construction land declined slightly from 2010 to 2015, but after 2015, it entered a period of rapid expansion and the proportion of woodland occupation increased correspondingly. The acreage of water and grassland changed from positive growth to negative growth after 2010 with increases in the decrement rates. As a whole, cultivated land and woodland gradually reached a balance between occupation and compensation from 2005 to 2018. However, the expansion rate of construction was high, which seriously reduced the space of the areas absorbing carbon such as woodland, grassland, and water.

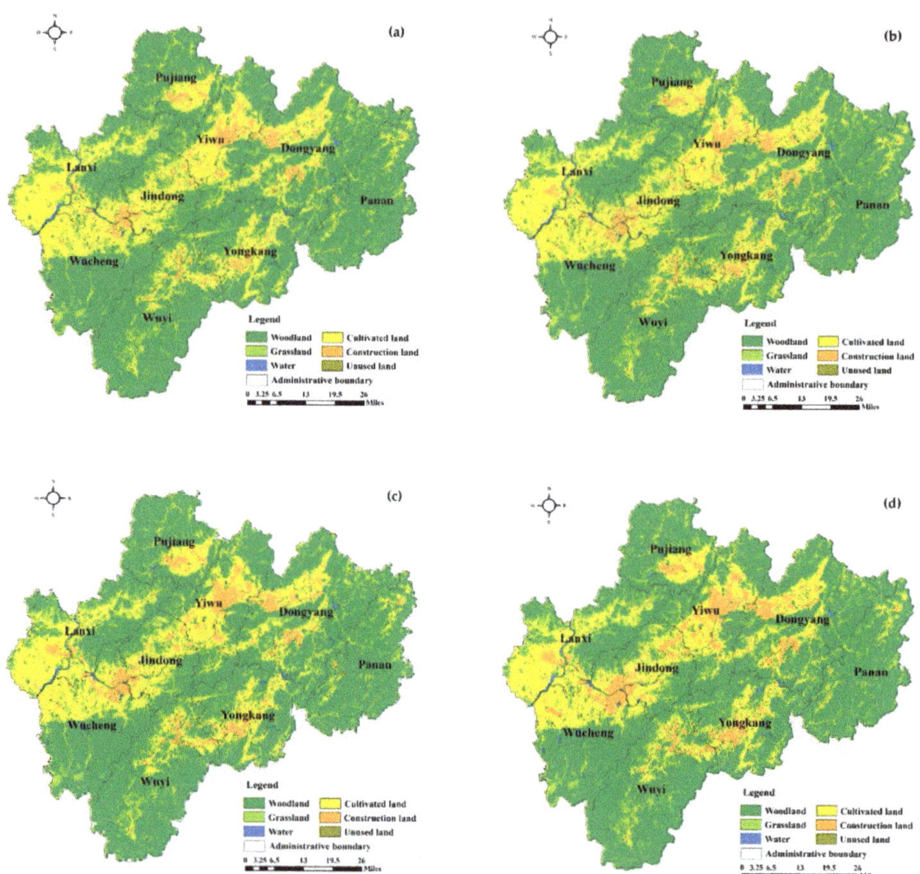

Figure 3. Spatial distribution of Jinhua land-use (**a**) 2005; (**b**) 2010; (**c**) 2015; (**d**) 2018.

Table 3. Land-use transfer matrix for 2005–2010 (km^2).

Land-Use Types	CuL	WL	GL	W	CoL	UL
CuL	3096.3	14.9	1.0	7.0	92.8	0.0
WL	20.2	6575.5	4.8	0.6	14.3	0.1
GL	0.3	1.0	212.8	0.2	0.8	0.0
W	1.7	0.3	0.0	144.1	0.6	0.0
CoL	3.2	0.8	0.0	0.2	583.3	0.0
UL	0.0	0.0	0.3	0.0	0.1	3.7

Note CuL represents Cultivated land; WL represents Woodland; GL represents Grassland; W represents Water; CoL represents Construction land; UL represents Unused land.

Table 4. Land-use transfer matrix for 2010 to 2015 (km^2).

Land-Use Types	CuL	WL	GL	W	CoL	UL
CuL	3003.6	26.5	0.9	2.2	88.3	0.0
WL	23.3	6559.3	1.7	0.7	7.5	0.1
GL	0.9	2.2	215.0	0.0	0.8	0.0
W	2.4	0.9	0.0	147.5	1.2	0.0
CoL	8.9	1.5	0.1	0.3	681.2	0.0
UL	0.0	0.1	0.0	0.0	0.0	3.7

Table 5. Land-use transfer matrix for 2015 to 2018 (km^2).

Land-Use Types	CuL	WL	GL	W	CoL	UL
CuL	2742.5	128.4	4.4	9.2	44.6	0.0
WL	124.3	6398.1	13.7	5.1	8.8	0.4
GL	4.0	16.7	198.1	0.2	0.6	0.0
W	14.0	5.0	0.2	131.1	2.5	0.0
CoL	154.3	41.9	1.2	5.1	722.5	0.2
UL	0.0	0.5	0.0	0.0	0.0	3.1

Table 6. Area changes for different land-use types from 2005 to 2018.

Land-Use Types	2005–2010		2010–2015		2015–2018	
	Variable-Area (km^2)	Rate (%)	Variable-Area (km^2)	Rate (%)	Variable-Area (km^2)	Rate (%)
CuL	−90.4	−2.8	−82.8	−2.6	−117.2	−3.8
WL	−23.7	−0.4	−1.9	0.1	−163.1	−2.4
GL	4.3	1.9	−1.3	−0.6	−6.8	−3.0
W	5.4	3.6	−1.4	−0.9	1.1	0.7
CoL	104.8	17.8	87.4	12.6	145.1	18.6
UL	−0.4	−8.6	0.0	−1.3	−0.1	−3.5

4.2. "Carbon Flow" of Land-Use in Jinhua

Through calculations, the land-use carbon sinks, carbon sources, and total carbon emissions from Jinhua City from 2005 to 2018 were determined (Table 7). From 2005 to 2018, the total carbon emissions first decreased and then increased. The growth rate reached up to 121.3% from 2005 to 2010, and there was slight negative growth from 2010 to 2015 when carbon emissions reduced by 4.3%. The growth rate of carbon emissions from 2015 to 2018 was 35.7%, a significant slow down when compared with the first stage. As for the main carbon sources, the average carbon emission intensities of cultivated land and construction land were 6.1% and 13.9% lower, respectively, in 2018 compared to 2005.

Table 7. Carbon emissions from different land-use types from 2005 to 2018 (MgC).

Years	CuL	WL	GL	W	CoL	UL	Carbon Source	Carbon Sink	Total
2005	135.0	−252.0	−0.5	−4.1	8507.4	0.0	8642.5	−256.7	8385.8
2010	142.5	−251.2	−0.5	−4.3	18,671.0	0.0	18,813.5	−255.9	18,557.6
2015	125.2	−251.1	−0.5	−4.2	17,885.3	0.0	18,010.5	−255.8	17,754.7
2018	107.0	−245.0	−0.5	−4.3	24,227.3	0.0	24,334.4	−249.7	24,084.6

Through the explicit carbon flow model, the law of carbon transfer between different land-use types in Jinhua from 2005 to 2018 was determined (Table 8, Figure 4). The results show that the carbon flow was negative in all three periods, which had an adverse effect on the reduction of regional carbon emissions. The positive carbon flow achieved in the process of land transfer has mainly come from the carbon sink formed by the conversion of construction land to other land types. This accounts for more than 98% of the positive carbon flow. Secondly, the positive carbon flow caused by the transfer of construction land continued to grow at a high rate. This indicates that the restoration of brownfield land and the construction of green mines in Jinhua have made a difference to ecological civilization construction. Negative carbon flow has mainly come from the carbon emissions generated by the conversion of other land-use types to construction land. During the period from 2010 to 2015, due to using clean energy sources instead of high-pollution and high-carbon emission energy sources and the treatment of pesticide and fertilizer pollutants, the carbon emission intensity of construction land and cultivated land was under control. Thus, the level of negative carbon flow was low and reached a minimum value. After that, the negative carbon flow started to increase at a high speed. Overall, during the 2005–2018 period, although the positive carbon flow gradually increased, the amount of negative carbon flow was always much greater than the amount of positive flow, indicating that the problem of extensive construction land expansion and the corresponding increase in negative carbon flow still existed. Thus, the spatial pattern of land use was not yet optimal. This shows that if the land-use spatial pattern continues to develop according to the current evolution law, the imbalance in regional carbon metabolism will be further aggravated.

Table 8. Carbon flow from 2005 to 2018.

The Direction of Carbon Flow	2005–2010 (tC·a^{-1})	2010–2015 (tC·a^{-1})	2015–2018 (tC·a^{-1})
Cultivated land to Woodland	1135.9	2204.2	10,161.3
Cultivated land to Grassland	42.2	41.8	185.5
Cultivated land to Water	470.6	160.8	1012.3
Cultivated land to Construction land	−2,497,282.7	−958,883.8	−4,034,420.0
Cultivated land to Unused land	0.2	0.5	1.2
Woodland to Cultivated land	−1669.8	−1902.0	−9512.0
Woodland to Grassland	−168.1	−60.4	−591.1
Woodland to Water	−5.6	−7.2	−48.2
Woodland to Construction land	−387,023.8	−82,125.8	−1,098,230.4
Woodland to Unused land	−4.0	−2.1	−18.4
Grassland to Cultivated land	−14.8	−41.4	−171.1
Grassland to Woodland	37.0	78.5	485.4
Grassland to Water	4.7	0.5	4.4
Grassland to Construction land	−20,450.4	−8776.1	−31,322.8
Grassland to Unused land	0.0	0.0	0.0
Water to Cultivated land	−123.1	−176.9	−589.1
Water to Woodland	3.0	8.4	51.9
Water to Grassland	−0.1	−1.1	−3.9
Water to Construction land	−17,241.0	−13,067.4	−133,833.7
Water to Unused land	0.0	0.0	0.0
Construction land to Cultivated land	96,500.0	239,242.5	484,289.8
Construction land to Woodland	22,998.5	40,784.5	96,445.3
Construction land to Grassland	766.6	2183.3	6685.1
Construction land to Water	6330.2	7260.2	27,728.9
Construction land to Unused land	0.0	48.5	117.8
Unused land to Cultivated land	−0.2	−0.7	−1.5
Unused land to Woodland	1.4	3.6	16.0
Unused land to Grassland	0.5	0.0	0.0
Unused land to Water	0.0	0.0	0.0
Unused land to Construction land	−3857.0	−58.9	−5043.4
Positive carbon flow	128,290.5	292,017.3	627,184.8
Negative carbon flow	−2,927,840.4	−1,065,103.6	−5,313,785.5

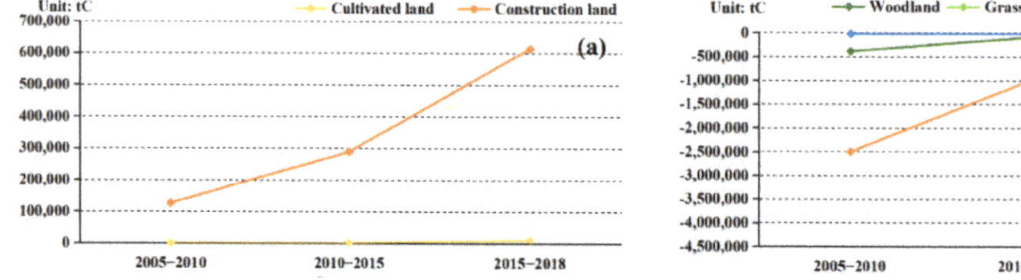

Figure 4. Trends of (**a**) positive and (**b**) negative carbon flow from 2005 to 2018.

4.3. Prediction of the Land-Use Spatial Pattern

The predicted pattern of land use for 2030 shows that if the original land-use evolution law remains unchanged, the acreage of forest land will decrease by 53,586 ha and the acreage of construction land will increase by 25,719 ha compared to 2018. The acreage of cultivated land, grassland, and water areas is also predicted to increase by 20,796, 3656, and 1637 ha, respectively. The construction land area changed the most, reaching a growth rate of 29%, while the rate of decrement in the woodland area reached 8.2%. According to the land use transfer matrix, 28% of the reduced woodland area was converted into construction land and 61% was converted into cultivated land. A total of 47% of the increase in construction land came from cultivated land and 52% came from the conversion of woodland. As for the spatial pattern of land use, the predicted results for 2030 (Figure 5) show that the majority of new construction land is likely to be concentrated in the flat central corridor, mainly due to a radiant expansion of the existing construction land to the cultivated land around it. In the east and south, Yongkang and Dongyang have relatively larger areas of new construction land. The intensity of land development on the secondary development axis of the central Zhejiang urban agglomeration is greater.

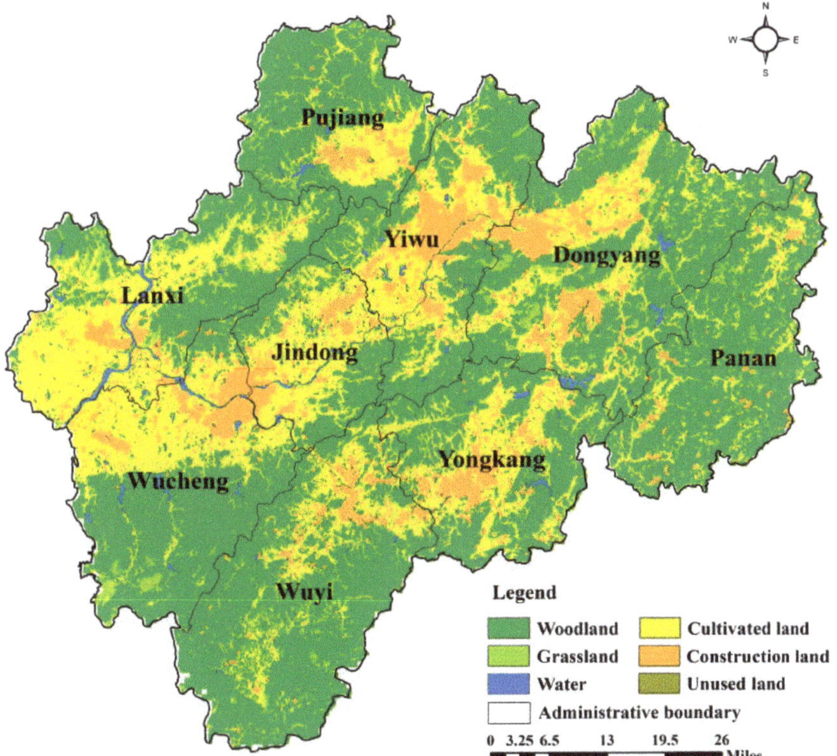

Figure 5. Image of predicted land-use in 2030.

Based on the forecasted results for the land-use pattern in 2030, carbon emissions were predicted. The carbon sources are expected to increase by 29.6% by 2030. Carbon sinks are predicted to reduce by 7.8%, and total carbon emissions are predicted to increase by 30% compared to 2018. Compared with the 2005-2018 period, the growth rate of carbon emissions is predicted to drop significantly but still remain at a relatively high level. With the gradual decrease in the average carbon emissions intensity of construction land after 2015, the continued rapid growth of carbon emissions will be largely due to the unreasonable spatial layout of land use, which needs to be adjusted and optimized.

4.4. Optimization of the Land-Use Spatial Pattern

Through the linear programming model, we determined the land-use structure plan after low-carbon optimization (Table 9). In terms of quantity, in the optimized scheme, the construction land area is reduced by 18% compared with the predicted scheme, and the grassland, water area, and unused land areas reduce in accordance with the law of land use evolution in Jinhua. The woodland area increases by 11%, indicating that improving the forest coverage is an important way to increase carbon sinks to achieve carbon reduction and promote the green development mechanism. As for the direction of transfer, 59% of the additional construction land from 2018 to 2030 comes from cultivated land and 38% comes from forestland. In the optimized land-use pattern, the proportion of construction land converted from woodland is smaller than in the predicted result.

Table 9. Carbon emissions in 2030 for the predicted and optimized land-use patterns.

Land-Use Types	Prediction for 2030		Optimization in 2030	
	Area (km^2)	Carbon Emission (MgC)	Area (km^2)	Carbon Emission (MgC)
Cultivated land	3137.1	114.6	2859.6	104.5
Woodland	6014.6	−224.9	6716.2	−251.2
Grassland	256.1	−0.5	208.0	−0.4
Water	169.2	−4.7	152.5	−4.3
Construction land	1200.4	31,433.4	982.3	25,721.6
Unused land	3.6	0.0	3.4	0.0

The quantitative targets and driving factors for low-carbon optimization were added to the FLUS model to determine the optimized land-use spatial pattern for Jinhua in 2030 (Figure 6). The additional construction land area was shown to mainly come from the radial expansion of existing construction land in the urban areas of Jinhua and Yiwu, but it was more directional compared with the forecasted expansion in all directions. The proportion of newly added construction land in the northeast of Jinhua city and the southwest of Yiwu city of the central corridor area was found to increase, which is consistent with the east–west expansion strategy of Jinhua and presents more obvious ribbon-form and cluster characteristics. Compared with the forecasted results, the expansion of the construction land area around Yongkang and Dongyang was predicted to decrease. The construction land areas in the southwest of Jinhua and the northeast of Yiwu were shown to decrease and concentrate more around the main axis of the Jinhua-Yiwu metropolitan area. They will be distributed around the Jinhua-Yiwu Expressway, which will contribute to the addition of production factors to the traffic arterial line. The green ecological zone in the central basin will be less affected by the expansion of construction land, and the acreage of cultivated land will decrease, and the acreage of woodland will increase. Total carbon emissions from the optimized land-use pattern will reduce by 19% compared with the value shown by the predicted results. Carbon sinks will increase by 11%. Carbon sources will reduce by 18%. In the scenario of low-carbon optimization, the growth rate of land-use carbon emissions is predicted to be only 6.2% from 2018–2030, lower than that of each period in the past.

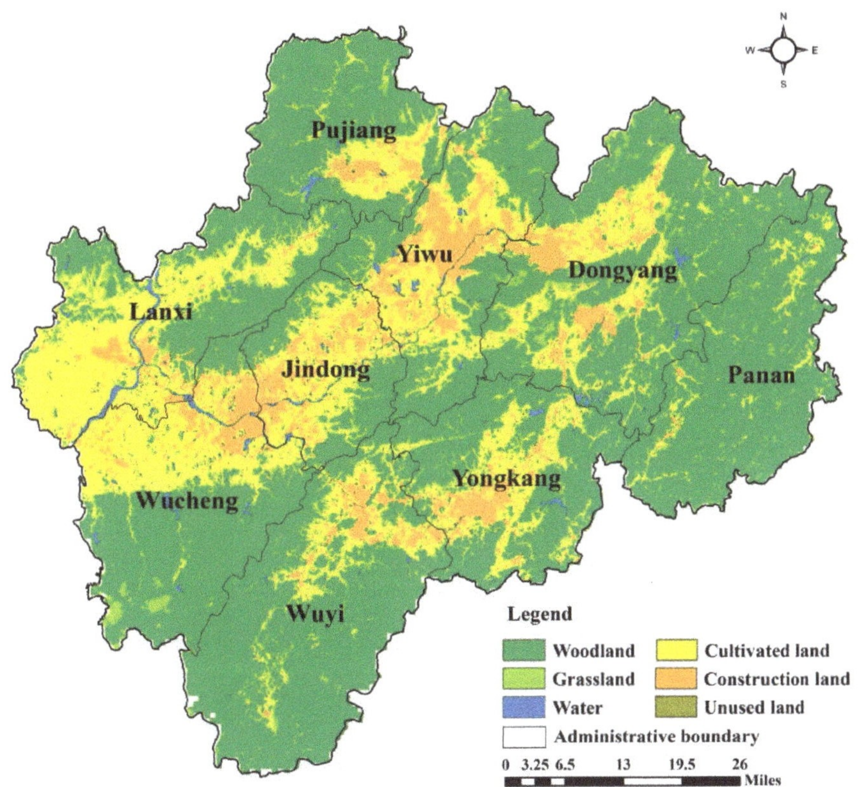

Figure 6. Optimized land-use spatial pattern in 2030.

5. Discussion

As the problems of the greenhouse effect and global warming are becoming increasingly serious, reducing carbon emissions has become an international consensus. Thus, in this paper, we proposed suggestions for low-carbon optimization of the land-use spatial distribution pattern based on an analysis of the effect of carbon emissions from land-use changes and the prediction of future tendencies in Jinhua. Many studies have explored the characteristics of land-use carbon emissions through coefficients of carbon emissions of land use in different regions and analyzed the existing problems in the land-use pattern [26–29]. Compared with previous studies, this study gives more practical advice on carbon emission reductions and proposed an optimized land-use spatial pattern using a multivariate linear programming model and a FLUS model with local driving factors.

In this study, we selected carbon emission coefficients in accordance with the actual situation in the study area. For instance, the carbon emission coefficient of woodland was selected with consideration of the carbon sink of the main forest vegetation types and the carbon emissions produced through firewood collection, forest fires, and HWP emissions of forest products in Zhejiang Province. This is more accurate than using the average coefficient at the national scale. However, the temporal effects of carbon emissions from changes in the age of stands, the site index, and the stand structure were still neglected. In terms of calculation of the carbon flow, we just considered natural and manmade carbon emissions but did not consider changes in carbon storage in land ecosystems. This may have caused the results to deviate from the actual situation. In addition, land use was divided into six types based on the first LUCC classification system, but the differences in carbon emissions intensity among different land-use types under the secondary classification were not considered. Thus, the spatial distribution of carbon emissions was not accurately described. Additionally, without driving factors, we were only able to perform pure mathematical simulations through the CA–Markov and FLUS models. The results do not reflect the influences of social and economic factors on the evolution of the land-use pattern well. Thus, in this study, driving factors, such as population, economy, transportation, and terrain were added to the models, and the impact of planning policies in the study area was also considered. The optimized simulation result in 2030 in Jinhua shows that in order to maintain stable economic performance under the premise of strictly controlling the intensity of land development, we should not only aim to limit the rate of construction land growth in terms of quantity but also to promote the construction of the Jinhua-Yiwu New Area and Jinhua-Lanxi Innovation City guided by the development of axial belts and a group layout. This will give full play to the connectivity of land development axes and promote the compound use of land and the intensive use of resources from the perspective of the spatial layout.

In future research, the carbon emission coefficient should be modified to improve the accuracy of the accounting results. Additionally, the impacts of land-use change on both carbon emissions and carbon storage should be considered. Our prediction models could be improved by refining the land-use classification so that the method can be applied to small-scale research more accurately in the future.

6. Conclusions

In this study, we analyzed the law of carbon emissions from land-use change in Jinhua City and explored a land-use optimization model under the guidance of achieving low carbon emissions. The specific conclusions are as follows:

(1) From 2005 to 2018, the land-use structure of Jinhua changed significantly. On the whole, the acreage of construction land continued to grow. The acreage of woodland and cultivated land areas continued to decrease. The acreage of grassland and water areas first increased and then decreased. Most of the woodland was converted to cultivated land, and most of the cultivated land was converted to construction land. More than 80% of the construction land was formed through transfer from cultivated land. In the process of land-use change, in 2018, the total carbon emissions were 1.9 times greater than in 2005. There was a trend for positive carbon flow to increase, but negative carbon flow was always far greater than positive carbon flow, that is, the carbon metabolism of land use was far from balanced.

(2) During the period from 2010 to 2015, carbon emissions in Jinhua showed negative growth. At this stage, the rate of decrement in woodland and the rate of growth in construction land were the smallest among the three periods. Accordingly, positive carbon flow reached its highest value and negative carbon flow reached its lowest value, indicating that changes in carbon flow due to land-use changes are closely related to the intensity of the evolution of the land-use pattern. In addition, thanks to the adjustment of the energy consumption structure and the treatment of fertilizer and pesticide pollutants in Jinhua after 2010, the average carbon emission intensity of construction land and cultivated land reduced, making the total carbon emissions of this stage somewhat lower than in the previous period.

(3) The land-use prediction results show that on the basis of the law of original land-use evolution in Jinhua, by 2030, the acreage of construction land will have increased by 25,719 ha and the woodland area will have reduced by 53,586 ha. Carbon emissions are expected to increase by 30% compared to 2018. In the low-carbon optimization scenario of land use, the expansion of construction land in 2030 will be restricted, while the area of woodland will increase. Total carbon emissions will be reduced by 19% compared with the predicted results. In terms of the spatial pattern, the additional construction land present in the optimized land-use pattern will be concentrated around the main axis of the Jinhua-Yiwu metropolitan area, consistent with the strategy of focusing on developing the Jinhua-Yiwu metropolitan area. It is helpful to

promote land agglomeration development and the efficient allocation of resources. Therefore, this study proposed a macroscopic spatial land-use pattern that contributes to carbon emission reduction with the consideration of regional planning objectives and economic development.

Author Contributions: S.H. and F.X. contributed to all aspects of this work. F.X. and Y.C. conducted data analysis. S.H. and M.G. wrote the main manuscript text. X.P. and C.R. gave some useful comments and suggestions to this work. All authors have read and agreed to the published version of the manuscript.

Funding: This work was supported by the National Natural Science Foundation of China (Grant No. 41907402) and Beijing Training Program of Innovation and Entrepreneurship for Undergraduates (Grant No. 202011413074).

Institutional Review Board Statement: Not applicable.

Informed Consent Statement: Not applicable.

Data Availability Statement: The data presented in this study are available on request from the author.

Conflicts of Interest: The authors declare no conflict of interest.

References

1. International Energy Agency. *Global Energy & CO$_2$ Status Report: The Latest Trends in Energy and Emissions in 2018*; International Energy Agency: Paris, France, 2019.
2. Political Bureau of the Central Committee of the CPC. The 14th Five-Year Planfor National Economic and Social Development and the Long-Range Objectives through the Year 2035. In Proceedings of the 19th CPC Central Committee, Beijing, China, 26 October 2020.
3. Lai, L.; Huang, X.J. *Carbon Emission Effect of Land Use in China*, 1st ed.; Nanjing University Press: Nanjing, China, 2010; pp. 3–65.
4. Stuiver, M. Atmospheric carbon dioxide and carbon reservoir change. *Science* **1978**, *199*, 253–258. [CrossRef]
5. Feddema, J.J.; Oleson, K.; Bonan, G. How important is land cover change for simulating future climates. *Science* **2005**, *310*, 1674–1678. [CrossRef] [PubMed]
6. Houghton, A.; House, J.I.; Pongratz, J. Carbon emissions from land use and land-cover change. *Biogeosciences* **2012**, *9*, 5125–5142. [CrossRef]
7. Liu, J.Y.; Shao, Q.Q.; Yan, X.D. An overview of the progress and research framework on the effects of land use change upon global climate. *Adv. Earth Sci.* **2011**, *26*, 1015–1022.
8. Zhu, E.Y.; Deng, J.S.; Zhou, M.M. Carbon emissions induced by land-use and land-cover change from 1970 to 2010 in Zhejiang, China. *Sci. Total Environ.* **2018**, *646*, 930–939. [CrossRef]
9. Yang, J.; Xie, B.P.; Zhang, D.G. Spatio-temporal evolution of carbon stocks in the Yellow River Basin based on InVEST and CA-Markov models. *Chin. J. Eco-Agric* **2011**, *29*, 1018–1029.
10. Li, Y.; Huang, X.J.; Zhen, F. Effects of land use patterns on carbon emission in Jiangsu Province. *Trans. CSAE* **2008**, *24* (Suppl. 2), 102–107.
11. Zhao, X.C.; Zhu, X.; Zhou, Y.Y. Effects of land uses on carbon emissions and their spatial-temporal patterns in Hunan Province. *Acta Sci. Circumstantiae* **2013**, *333*, 941–949.
12. Sun, H.; Liang, H.M.; Chang, X.L. Land Use Patterns on Carbon Emission and Spatial Association in China. *Econ. Geogr.* **2015**, *35*, 155–162.
13. Zhejiang Provincial Statistics Bureau. *Statistical Bulletin of the 7th Population Census of Zhejiang Province*; Zhejiang Provincial Statistics Bureau: Zhejiang, China, 2021.
14. Jinhua Municipal Statistics Bureau. *Statistical Yearbook of Jinhua in 2020*; Jinhua Municipal Statistics Bureau: Jinhua, China, 2021.
15. Liu, J.Y.; Liu, M.L.; Deng, X.Z. The land use and land cover change database and its relative studies in China. *J. Geogr. Sci.* **2002**, *12*, 275–282.
16. Li, K.R. *Land Use Change, Net Greenhouse Gas Emissions and Terrestrial Ecosystem Carbon Cycle*, 1st ed.; Meteorological Publishing House: Beijing, China, 2000; p. 260.
17. West, T.O.; Marland, G.A. Synthesis of Carbon Sequestration, Carbon Emissions, and NetCarbon Flux in Agriculture: Comparing Tillage Practices in the United States. *Agr. Ecosyst. Environ.* **2002**, *91*, 217–232. [CrossRef]
18. Piao, S.L.; Fang, J.Y.; Zhu, B. Forest biomass carbon stocks in China over the past 2 decades: Estimation based on integrated inventory and satellite data. *J. Geophys. Res.* **1984**, *110*. [CrossRef]
19. Fang, J.Y.; Guo, Z.D.; Piao, S.L. Estimation of terrestrial vegetation carbon sequestration in China from 1981 to 2000. *Sci. China Earth Sci.* **2007**, *6*, 804–812.
20. Duan, X.N.; Wang, X.K.; Lu, F. State Lab of Urban and Regional Ecology, Research Center for Eco-Environmental Sciences. *Acta Ecol. Sin.* **2008**, *2*, 463–469.

21. Xu, G.Q.; Liu, Z.Y.; Jiang, Z.H. Decomposition Model and Empirical Study of Carbon Emissions for China, 1995–2004. *Chin. J. Popul. Resour. Environ.* **2006**, *16*, 158–161.
22. Xia, C.Y.; Li, Y.; Ye, Y.M. Analyzing urban carbon metabolism based on ecological network utility: A case study of Hangzhou City. *Acta Ecol. Sin.* **2018**, *38*, 73–85.
23. Yang, G.Q.; Liu, Y.L.; Wu, Z.F. Analysis and Simulation of Land-Use Temporal and Spatial Pattern Based on CA-Markov Model. *Geomat. Inf. Sci. Wuhan Univ.* **2007**, *5*, 414–418.
24. Cao, S.; Jin, X.B.; Yang, X.H. Coupled MOP and GeoSOS-FLUS models research on optimization of land use structure and layout in Jintan district. *J. Nat. Resour.* **2019**, *34*, 1171–1185.
25. Liu, X.P.; Xun, L.; Xia, L.A. Future land use simulation model (FLUS) for simulating multiple land use scenarios by coupling human and natural effects. *Landsc. Urban. Plan.* **2017**, *168*, 94–116. [CrossRef]
26. Shi, H.X.; Mu, X.M.; Zhang, Y.L. Effects of Different Land Use Patterns on Carbon Emission in Guangyuan City of Sichuan Province. *Bull. Soil Water Conserv.* **2012**, *32*, 102–106.
27. Su, Y.L.; Zhang, Y.F. Study on Effects of Carbon Emission by Land Use Patterns of Shaanxi Province. *J. Soil Water Conserv.* **2011**, *25*, 152–156.
28. Li, P.; Li, J.T.; Liang, Y.H. Analysis of Land Use/Cover Change and Carbon Effect in Hebi in Recent 30 Years. *Ecol. Econ.* **2019**, *35*, 121–128.
29. Jing, Y.; Zuo, L.L.; Peng, W.F. Spatial-temporal Change Analysis of Carbon Emission from Land Use in Mianyang the Case Study of Mianyang. *Abbreviated J. Name* **2021**, *44*, 172–185.

 *land*

Article

Identification of Potential Land-Use Conflicts between Agricultural and Ecological Space in an Ecologically Fragile Area of Southeastern China

Jing Zhang [1,2,3], Yan Chen [1], Congmou Zhu [1], Bingbing Huang [1] and Muye Gan [1,*]

[1] Institute of Applied Remote Sensing and Information Technology, College of Environmental and Resource Sciences, Zhejiang University, Hangzhou 310058, China; zj1016@zju.edu.cn (J.Z.); _chenyan@zju.edu.cn (Y.C.); congmouzhu1993@zju.edu.cn (C.Z.); 22014172@zju.edu.cn (B.H.)

[2] Key Laboratory of Urban Land Resources Monitoring and Simulation, Ministry of Natural Resources, Shenzhen 518000, China

[3] Technology Innovation Center for Land Spatial Eco-restoration in Metropolitan Area, Ministry of Natural Resources, Shanghai 200003, China

* Correspondence: ganmuye@zju.edu.cn

Abstract: In the context of ensuring national food security, high-intensity agricultural production and construction activities have aggravated the conflicts between agricultural and ecological spaces in ecologically fragile areas, which have become one of the most important factors hindering regional sustainable development. This study took Lin'an District, a typical hilly region of southeastern China, as an example. By constructing a landscape ecological risk evaluation model, land-use conflicts between agricultural and ecological spaces were identified, spatial autocorrelation and topographic gradient characteristics were analyzed, and land-use conflict trade-off mechanisms were proposed. During 2008 and 2018, the degree of land-use conflict in Lin'an District displayed an increasing trend, and the proportion of severe conflicts increased obviously. Slope is the main factor affecting land-use conflicts in a hilly region and shows a negative correlation, mainly because areas with flat terrain are more conducive to human activities. Based on the characteristics of land-use conflicts in Lin'an District, conflict trade-off mechanisms were proposed to provide a theoretical basis and practical support for land-use conflict management. Our study provides scientific evidence for sustainable land-use planning and ecological management in ecologically fragile areas.

Keywords: land use conflict; agricultural space; ecological space; ecological fragile area

Citation: Zhang, J.; Chen, Y.; Zhu, C.; Huang, B.; Gan, M. Identification of Potential Land-Use Conflicts between Agricultural and Ecological Space in an Ecologically Fragile Area of Southeastern China. *Land* **2021**, *10*, 1011. https://doi.org/10.3390/land10101011

Academic Editors: Dong Jiang, Jinwei Dong and Gang Lin

Received: 9 August 2021
Accepted: 24 September 2021
Published: 26 September 2021

Publisher's Note: MDPI stays neutral with regard to jurisdictional claims in published maps and institutional affiliations.

1. Introduction

Mountains, water, forests, fields, lakes and grass are communities shared by all life on Earth [1]. This fact emphasizes the inseparable interactions between the agricultural and ecological elements of landscapes [2]. However, In order to guarantee food security, the Chinese government requires that a certain amount of arable land must be maintained [3]. At the same time, in order to guarantee the ecological security of the land, it also requires that areas with important ecological functions be included in the protection [4]. These tensions, if not managed properly, can accentuate land-use conflicts between agricultural space and ecological space, and the conflicts caused by land resource shortages and single-use lands are increasing, especially in ecologically fragile areas. When land-use management in a certain region fails to integrate economic development, food security and ecological protection, imbalances in land use structure and regional landscape patterns emerge. These imbalances manifest as spatial conflicts caused by land users competing for land resources out of different interests [5]. Currently, more than 50 countries have pledged to protect 30% of the planet's land and sea area by 2030. Therefore, it is significant to study and identify land-use conflicts in agricultural and ecological spaces to coordinate human–land balanced relationships for green, coordinated and sustainable regional development.

Land-use conflict mainly refers to the contradiction between interest groups in the process of land use due to the multiple functional demands of social development or the industry objectives and the resulting functional utilization conflicts [6]. Early studies on land-use conflicts mainly focused on qualitative analysis. The commonly used qualitative analysis methods include the participatory survey method [7–9], logical framework method [10,11], game theory analysis method [12–14] and other methods. The application of the participatory survey method in domestic outdoor surveys began in the 1990s, which is characterized by a high degree of participation and acceptance by the population, and the use of the participatory survey method can analyze land-use conflicts. The participatory survey method can analyze the emergence and resolution of land use conflicts from internal mechanisms [15]. The logical framework approach is an evaluation method proposed by the United States in the 1970s, which in essence shows the causal links between things, with particular emphasis on the role of indicators of project objectives and the initial participation of stakeholders. The logical framework approach enables the initial measurement and hypothesis testing of land-use conflicts [16]. Game theory focuses on the decision-making behavior of decision makers as "rational people" to maximize their own benefits in the process of land use and thus explains the underlying mechanism of land-use conflicts [17]. Although qualitative analysis methods can help us understand land-use conflict mechanisms and resolution strategies, it is difficult to measure land-use conflicts quantitatively [18].

In recent years, many scholars with different professional backgrounds and research perspectives have conducted extensive research on the identification of land use conflicts, and some quantitative analysis methods have been proposed. Commonly used quantitative analyses include pressure-state-response (PSR) [19,20], statistical regression analysis [21,22], multicriteria evaluation [23,24], and spatial analysis [7,25]. The PSR approach is essentially used to show the linkage between humans and nature, and it is used in the assessment of ecological safety and sustainable use of land to clearly understand the occurrence, development and transformation of things in a complex and changing environment and has basic systemic and integrity characteristics [26]. Statistical regression analysis is used to interpret the structural and quantitative evolution of land use and to make reasonable assumptions about the drivers that may affect changes in the future [27,28]. The multicriteria evaluation method is more flexible than the above methods and mainly focuses on multicriteria decision making. The resolution of land-use conflicts is a long-term and comprehensive process, and the use of this method can play a better role in mitigating land use conflicts [29,30]. The spatial analysis method diagnoses the spatial distribution pattern of land-use conflicts in the region by establishing land-use conflict data, identifying land use conflicts, and displaying land use conflict results on land use-related data through the GIS (geographic information system) function [31,32]. This method can quickly achieve the quantification and precision of land-use conflicts and is more suitable for the research and analysis of land-use conflicts in hilly areas.

Research on land-use conflicts in China started late. The relevant theories still need to be refined, the technical methods are not advanced, and there is a lack of a reliable scientific basis and objective evaluation criteria. In addition, the current research fields are mostly in large areas such as plain cities. Less attention has been given to hilly areas, and the existing studies do not reflect the influence of topographic features of hilly areas on land use conflicts.

Based on the theories of human–land relationships, landscape ecology and ecosystem balance, this study used spatial analysis with GIS, landscape ecological risk assessment and spatial autocorrelation analysis to explore land-use conflicts. Lin'an District, located in Hangzhou in the hilly areas of northern Zhejiang, was taken as a representative example. First, a model for land-use conflict measurement was constructed to identify the land-use conflicts between agricultural space and ecological space. Second, spatial autocorrelation and topographic gradient characteristics were analyzed. Finally, several trade-off mechanisms involved in land-use conflicts were proposed.

2. Study Area and Data Sources

2.1. Study Area

Lin'an District (Figure 1) in Hangzhou city is located in northwestern Zhejiang Province, bordering the main city of Hangzhou in the east and Huangshan in the west. In July 2017, Lin'an was established as a district, making it the youngest district in Hangzhou, with five streets and 13 towns (towns and streets are the smallest units in China's administrative divisions). The eastern part of Lin'an has been basically integrated into the main city of Hangzhou and is a highly urbanized area. In contrast, the western part is low hilly area with a good ecological environment and rich species diversity. It is approximately 100 km wide from east to the west and 50 km long from north to south. The hilly area accounts for 86% of this district, which is referred to as having 'nine mountains, half water and half farmland'. The elevation is high in the northwest and low in the southeast. The mountains in the northwestern part of the district are rugged, and the eastern part of the district is characterized by a staggered distribution of hilly and wide valley basins. Lin'an District is located on the southern edge of the subtropical monsoon climate zone and experiences a monsoon climate. Streams and ditches in the district show a crisscrossed pattern. The Changhua Stream, Tianmu Stream and Tiao River are part of the Qiantang River and Yangtze River systems. The agricultural industry in Lin'an District is dominated by *Carya cathayensis* (Chinese hickory) and *Phyllostachys praecox* (a cultivated species of bamboo). In 2018, the total household registered population of the whole region was 537,600, the GDP reached 53.963 billion yuan, the output value of the first, second and third largest industries accounted for 8.0%, 44.2% and 47.8%, respectively, and industry was developing well.

Figure 1. Geographic location of Lin'an District.

2.2. Data Sources and Processing

The data used in this study mainly included natural geographical data, socioeconomic statistical data and land-use data. The natural geographical data included Lin'an District DEM data with a resolution of 30 m × 30 m from the geospatial data cloud that was processed to provide slope data. Socioeconomic statistical data were obtained from the 'Bulletin of National Economic and Social Development Statistics of Lin'an District in 2018'. Land-use data included land-use change survey data from 2008, 2013 and 2018, cultivated land quality data, nature reserves and other relevant data. Three land-use categories were established based on land-use planning classification standards, namely, agricultural space (cultivated land, garden land and other agricultural land), ecological space (forestland, grassland, water areas and natural reserves) and urban space (construction land).

3. Methods

3.1. Land-Use Conflict Measurement Methods

The spatial conflict caused by human activities is in essence a game of spatial resource occupation by various land users. Spatial conflict is accompanied by the utilization of land resources, resulting in the change of regional spatial pattern and function, which further affects the original physical, chemical and ecological processes and affects the nature conditions [33,34].

Referring to previous studies [35,36], we established a measurement model for land-use conflict based on the conceptual model of ecological risk assessment [33,34,37]. In terms of describing the pressure from human disturbance and deterioration of natural conditions, the ability of the land resources themselves to withstand conflict pressure, and the stability of the land system, three indexes including the land complexity index, land fragility index and land stability index were selected to evaluate the land-use conflict [34,38,39]. Mathematical linear models were used to characterize a comprehensive index of land-use conflict (*CCI*):

$$CCI = LCI + LFI - LSI \tag{1}$$

In this formula, *CCI* refers to the comprehensive index of land use conflict; *LCI* refers to the land complexity index; *LFI* refers to the land fragility index; and *LSI* refers to the land stability index. Since land use spatial conflict is a complex scientific problem, it is difficult to determine the impact of complexity, vulnerability and stability on land-use spatial conflict. Therefore, this study uses equal weight to calculate the comprehensive index of land use spatial conflict without considering its possible nonlinear influence.

(1) Land complexity index (*LCI*) [34,39]:

The complexity of land-use systems mainly reflects the external pressures associated with human activities and land development intensity. For the measurement of external pressure, it is mainly considered from the perspective of the comprehensive influence of the space unit and its surrounding area. Fractals are an effective tool for describing the spatial pattern of nature and the spatial complexity of geographical phenomena. Therefore, the area-weighted mean patch fractal dimension (*AWMPFD*) was used as a spatial index to describe the complexity of land use patches, with a view to reflecting the influence of land-use processes in neighborhoods on current land-use types. In general, the fractal values of natural landscapes that are less disturbed by human activities are high, while the fractal values of man-made landscapes that are highly influenced by human activities are low [40]. The calculation method was as follows (Formula (2)):

$$LCI = AWMPFD = \sum_{i=1}^{m} \sum_{j=1}^{n} \frac{2\ln(0.25P_{ij})}{\ln a_{ij}} \times \frac{a_{ij}}{A} \tag{2}$$

In this formula, P_{ij} refers to the perimeter of the *j*th landscape patch of space type *i*; a_{ij} refers to the area of the *j*th landscape patch of space type *i*; *A* refers to the total area of the spatial unit; *m* refers to the number of space types; and *n* refers to the number of landscape patches. To simplify and standardize the calculations, the obtained index values were standardized to 0~1.

(2) Land fragility index (*LFI*) [39,41]:

The *LFI* represents the resistance of landscape patches to external pressures. The role of different landscape types in maintaining biodiversity, and improving overall structure and function varies. This difference is related to the stages in which different landscapes are in the natural succession process. In general, ecosystems in the primary succession stage, with simpler food chain structure and lower biodiversity are more fragile [42]. The fragility of each landscape in the space was used to calculate the *LFI*. The calculation method was as follows (Formula (3)):

$$LFI = \sum_{i=1}^{m} \sum_{s=1}^{r} fis \times \frac{ais}{A}$$

(3)

In this formula, f_{is} refers to the landscape fragility of land use types in space type i; a_{is} refers to the landscape patch area of land-use types in space type i; A refers to the total area of the spatial unit; m refers to the number of space types; and r refers to the number of land-use types.

Land-use types were assigned landscape fragility values [31,43], as shown in Table 1. To simplify and standardize the calculations, the obtained index values were standardized to 0~1.

Table 1. Assessment of the landscape fragility of agricultural spaces and ecological spaces in Lin'an District.

Space Type	Land-Use Type	Landscape Fragility Value
Agricultural space	Cultivated land	3
	Garden land	5
	Other agricultural land	1
Ecological space	Forestland	2
	Grassland	4
	Water area	6
	Nature Reserve	7

(3)　Land stability index [31,39] (*LSI*):

One of the most significant effects of land-use conflicts on regional spatial patterns is the fragmentation of landscape patches, which presents as a changing process from a continuously varying structure to a mosaic of patches that tend to be complex, heterogeneous and discontinuous. The conservation of many biological species requires large areas of natural habitat. With the fragmentation of the landscape and the shrinking area of patches, the environment suitable for living organisms is decreasing, which will directly affect the reproduction, dispersal, migration and conservation of species. Therefore, landscape fragmentation is one of the main reasons for the loss of biodiversity and decline of ecosystem stability.

Patch density (*PD*), a commonly used landscape index, was used to represent landscape fragmentation and reflect the degree of land stability. A higher *PD* value indicates a higher degree of landscape fragmentation, poorer land stability, and the more intense the conflict in a unit area [35,44].

Therefore, the *LSI* can be expressed as the reverse of the *PD* and was calculated as follows (Formula (4)):

$$LSI = 1 - PD$$

(4)

The *PD* was calculated as follows (Formula (5)):

$$PD = \frac{n_i}{A}$$

(5)

In this formula, n_i refers to the number of landscape patches of space type i, and A refers to the total area of the spatial unit. To simplify and standardize the calculations, the index values obtained were standardized to 0~1.

A 500 m × 500 m grid was selected as the measurement unit based on the research scale, the total amount of data, and the landscape patch situation. Based on the results of existing research [44], the conflict values were classified into four grades: stable and controlled [0,0.4], basically controlled (0.4,0.6], basically uncontrolled (0.6,0.8] and severely uncontrolled (0.8,1.0].

3.2. Land-Use Conflict Spatial Analysis Methods

3.2.1. Spatial Autocorrelation Analysis

Spatial autocorrelation analysis reflects the correlation between the values of a variable in a space and the surrounding space and can also be used to determine whether there is autocorrelation between the different spaces. To study the spatial distribution of spatial units with different degrees of land-use conflict, this research used GeoDa 1.14 software to establish spatial weights based on adjacency relationships and used the global spatial autocorrelation index *Moran's I* and the local spatial autocorrelation index LISA [45–47] to measure the spatial autocorrelation characteristics of land-use conflict in Lin'an District.

(1) Global spatial autocorrelation index. The global spatial autocorrelation index *Moran's I* was calculated as follows (Formula (6)):

$$Moran's\ I \ = \ \frac{n}{\sum\limits_{i=1}^{n}\sum\limits_{j=1}^{n}W_{ij}} \times \frac{\sum\limits_{i=1}^{n}\sum\limits_{j=1}^{n}W_{ij}(x_i - \overline{x})(x_j - \overline{x})}{\sum\limits_{i=1}^{n}(x_i - \overline{x})^2} \tag{6}$$

In this formula, n refers to the total number of sample points of variable x; x_i and x_j refer to the values of variable x at spatial locations i and j, respectively; $\overline{x}$ refers to the average values of n location attribute values; and W_{ij} refers to the elements of the binary spatial weight matrix W in general cross-product statistics, which reflect the location similarity of spatial units.

Moran's I reflects the degree of similarity in the comprehensive index of land use conflict in units around a space. Its value range is [–1,1]; values of (0,1] indicate positive spatial autocorrelation, 0 indicates no spatial autocorrelation, and values of [–1,0) indicate negative spatial autocorrelation [48]. Based on this analysis, the Monte Carlo simulation method was used to calculate the Z value and the P value for further testing.

(2) Local spatial autocorrelation index. The local spatial autocorrelation index *LISA* was calculated as follows (Formula (7)):

$$LISA \ = \ \frac{n(x_i - \overline{x})}{\sum\limits_{i}(x_i - \overline{x})^2}\sum\limits_{j}W_{ij}(x_i - \overline{x}) \tag{7}$$

When the *LISA* index value is positive, the similarity values around a unit are clustered in space; when the *LISA* index value is negative, non-similar values around the unit are clustered in space.

3.2.2. Terrain Gradient-Based Analysis

Topographic factors have a key influence on a land-use pattern. Lin'an District is a typical hilly area with large topographic undulations, scarce arable land resources, and tense human–land relations. The characteristics of land use conflicts vary significantly with changes in topographic features. Therefore, the slope factor was chosen to explore the topographic gradient characteristics of land-use conflicts in Lin'an District.

The reclassification module of ArcGIS was used to classify the slope index to quantitatively explore the land-use conflict characteristics in Lin'an District. With reference to the current classification standards and the actual surface morphology of the study area, the slope reclassification of Lin'an District was divided into three categories, namely, low slope [0°, 6°], medium slope (6°, 25°] and high slope (25°, 90°]. After that, the spatial distribution of the land-use conflict degree in 2008, 2013 and 2018 were overlaid with the slope classification to explore the characteristics of land-use conflicts under different slope conditions in Lin'an District.

4. Results

4.1. Changes in Potential Land-Use Conflicts

The quantitative results (Table 2) showed that the potential land-use conflicts between agricultural space and ecological space in Lin'an District showed an increasing trend during the study period. Specifically, the proportion of spatial units with controlled grades in 2008–2018 was 86.19~88.33%, which accounted for nearly 90% of the total units of agricultural and ecological spaces in Lin'an District. The proportion of spatial units with stable and controlled grades fluctuated, showing a trend of first decreasing and then increasing; the proportion increased by 0.47% during 2013 and 2018, but that increase was far lower than the decline during 2008 and 2013 (0.83%). The proportion of basically controlled spatial units continued to decline during 2008 and 2018. The proportion of basically uncontrolled spatial units increased over time, and the increase in 2013–2018 was 2.14 times that in 2008–2013. The proportion of severely uncontrolled spatial units also continuously increased, from 3.23% in 2008 to 4.24% in 2018.

Table 2. Changes in the degree of land-use conflict between agricultural and ecological spaces in Lin'an District, 2008–2018.

Conflict Grade	Conflict Value	Number of Spatial Units			Percentage of Spatial Units			Change Rate	
		2008	2013	2018	2008	2013	2018	2008–2013	2013–2018
Stable and controlled	0~0.4	368	268	322	3.09	2.26	2.73	−0.83	0.47
Basically controlled	0.4~0.6	10,163	10,060	9843	85.24	84.84	83.46	−0.4	−1.38
Basically uncontrolled	0.6~0.8	1006	1043	1129	8.44	8.80	9.57	0.36	0.77
Severely uncontrolled	0.8~1.0	385	487	500	3.23	4.10	4.24	0.87	0.14
Total		11,922	11,858	11,794	100.00	100.00	100.00	0.00	0.00

The spatial distribution (Figure 2) clearly shows the expansion trend of uncontrolled (basically uncontrolled and severely uncontrolled) spatial units, and the distribution area of controlled spatial units continuously decreased over the study period. The distribution of stable and controlled spatial units in 2008–2018 was scattered throughout the Towns and Streets. The basically controlled spatial units were the most widely distributed, and their distribution area decreased during 2008 and 2018. The land-use types inside these spatial units mainly included forestland. In 2008, Yuqian town was the center of the zonal distribution of basically uncontrolled spatial units, and cultivated land was the main land-use type in these spatial units. Between 2013 and 2018, these spatial units were increasingly widely distributed and scattered. In 2008, the severely uncontrolled spatial units were mainly distributed in Qingshanhu Street, Jinnan Street, Qianchuan town, Qingliangfeng town and so on. With the passage of time, these spatial units gradually spread to the surrounding areas. By 2018, some spatial units gradually changed from basically uncontrolled to severely uncontrolled. The main areas of this change included Tianmushan town, Yuqian town and Longgang town.

4.2. Spatial Pattern Analysis of Potential Land-Use Conflict

Table 3 shows that the *Moran's I* values for land use conflicts in Lin'an District in 2008, 2013 and 2018 were 0.238, 0.253 and 0.232, respectively, indicating that the comprehensive index of land-use conflict in Lin'an District showed significant and positive global spatial autocorrelation.

To reveal the spatial relationships among local spatial units in Lin'an District, this research further calculated the local spatial autocorrelation index *LISA* and obtained clustering and significance results for land-use conflict. In the quantitative clustering results (Figure 3), except for the non-significant spatial units, the spatial units were dominated by high-high aggregation, which accounted for 5.54%, 5.92% and 5.04% of the units in 2008, 2013 and 2018, respectively. The spatial units with low-low aggregation changed significantly, increasing from 4.59% in 2008 to 5.70% in 2013 and then decreasing to 4.65%

in 2018. Over the 10 years, the proportion of spatial units with low-high aggregation remained between 4.93% and 5.67%. The proportions of spatial units with the above three clustering grades were similar; all of them were approximately 5% and increased first and then decreased. The proportion of spatial units with high-low aggregation was the smallest, but it continued to increase, reaching its highest proportion of 0.81% by 2018. The proportion of nonsignificant spatial units was the highest; it remained higher than 80% during the period 2008–2018 and showed a trend of first decreasing and then increasing. The above results show that the spatial unit clustering of land-use conflict was the most significant in 2013.

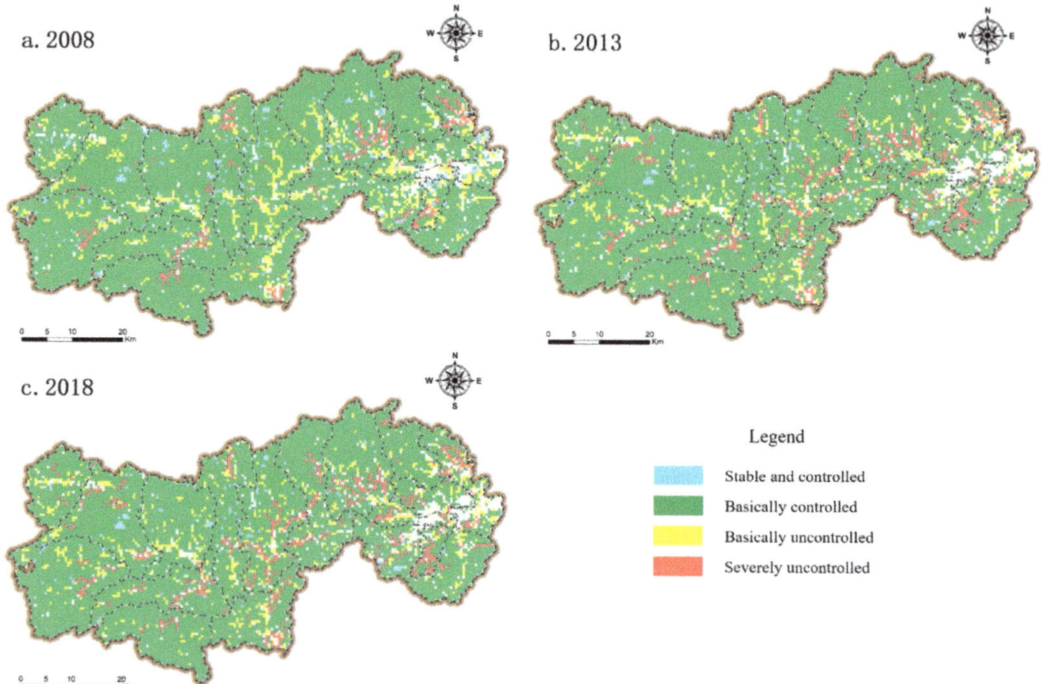

Figure 2. The degrees of land-use conflict between agricultural and ecological spaces in Lin'an District, 2008–2018. (**a**) 2008. (**b**) 2013. (**c**) 2018.

Table 3. Changes in the global spatial autocorrelation of land-use conflict in Lin'an District, 2008–2018.

	Value			Variance Value	
	2008	**2013**	**2018**	**2008–2013**	**2013–2018**
Moran's I	0.238	0.253	0.232	0.015	−0.021
Z	51.792	52.970	48.680	1.178	−4.290
P	0.001	0.001	0.001	0.000	0.000

The distribution areas of high-high aggregation, low-low aggregation and low-high aggregation spatial units in the clustering results were relatively similar (Figure 4). The high-high aggregation spatial units were mainly distributed zonally. In addition, there were obvious aggregation areas in the eastern region. The cultivated land, garden land and water areas were interlaced in these spatial units, and the land-use types were more complex. The low-low aggregation spatial units were scattered, and the streets of each town had low-low aggregation spatial units. Their distribution in Sun town was the most obvious; this area was mainly grassland, the surrounding land-use type was uniform, and the conflict level was low. The low-high and high-low aggregation spatial units were distributed on the periphery of the high-high and low-low aggregation spatial units, respectively, and the latter had a smaller area of distribution. The non-significant spatial units were the most widely distributed and had a continuous distribution; these units were mainly forests.

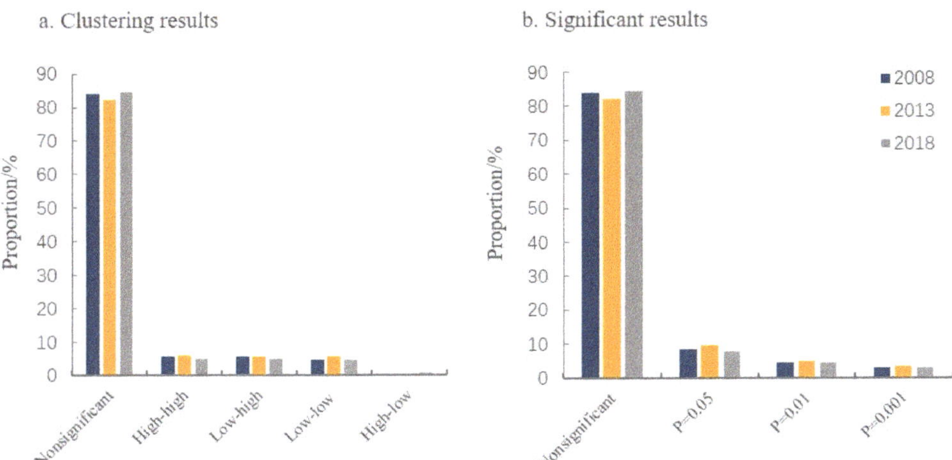

Figure 3. Changes in the local spatial autocorrelation of land-use conflict in Lin'an District, 2008–2018. (**a**) Clustering results. (**b**) Significant results.

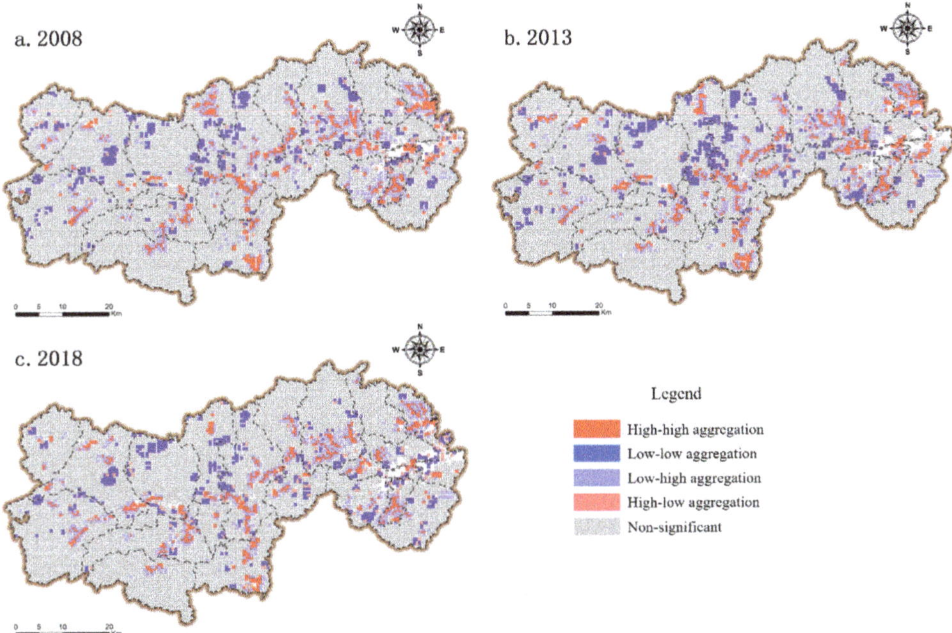

Figure 4. Local spatial autocorrelation of potential land-use conflict in Lin'an District, 2008–2018. (**a**) 2008. (**b**) 2013. (**c**) 2018.

The spatial units with $p = 0.001$ in the significance results (Figure 5) have obvious clustering areas, mainly in the eastern part, Qianchuan town, Heqiao town, Longgang town and Qingliangfeng town. The spatial units with $p = 0.01$ and $p = 0.05$ were distributed sequentially at their periphery, with the latter having the largest distribution area except for the non-significant spatial units. The distribution characteristics of the non-significant spatial units were consistent with the clustering results.

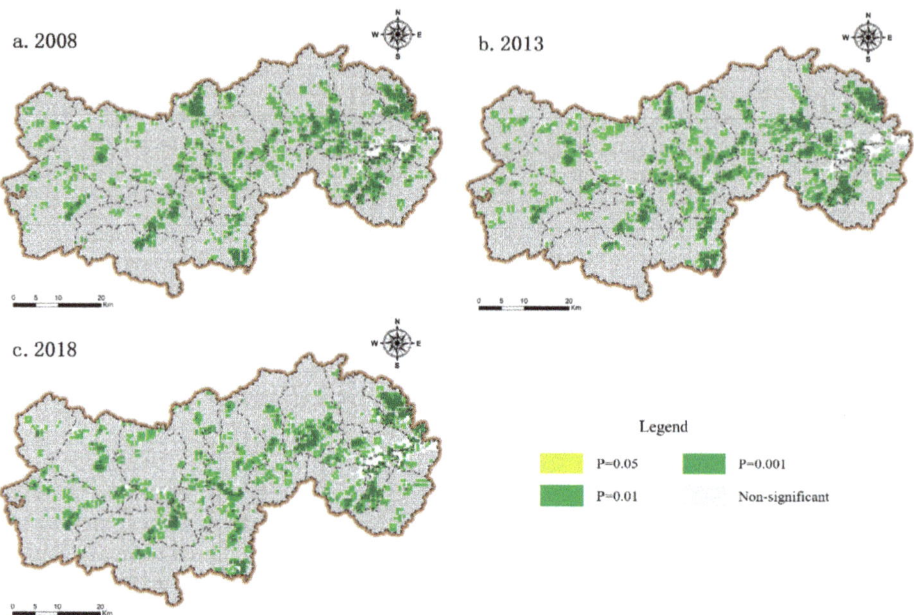

Figure 5. Significant results for local spatial autocorrelation of potential land-use conflict in Lin'an District. (**a**) 2008. (**b**) 2013. (**c**) 2018.

4.3. Topographic Gradient Feature of Potential Land-Use Conflict

To study the spatial-temporal differentiation of land-use conflict between agricultural space and ecological space under different topographic gradient conditions in Lin'an District, the slope factor was selected to analyze the topographic gradient characteristics of the areas with different conflict degrees.

The land-use conflict degree in the low-slope areas was the highest over the ten years (Figure 6). Controlled spatial units did not show an obvious advantage, and the proportion of uncontrolled spatial units increased over the 10 years. In 2018, more than half of the total low-slope area had an uncontrolled grade of land-use conflict. The degree of land-use conflict in the medium-slope area from 2008–2018 was low, but it showed an increasing trend. The proportion of basically controlled spatial units was maintained at higher than 80%, but it showed a downward trend because of the expansion of uncontrolled spatial units and was reduced by 3.12% in total. The degree of land-use conflict in the high-slope area was the lowest over the 10 years. The proportion of controlled spatial units remained between 93.85% and 94.88%, and the severely uncontrolled spatial units gradually became basically controlled spatial units, indicating that they became better controlled. The proportion of severely uncontrolled spatial units decreased continuously by 0.35% from 1.64% in 2008.

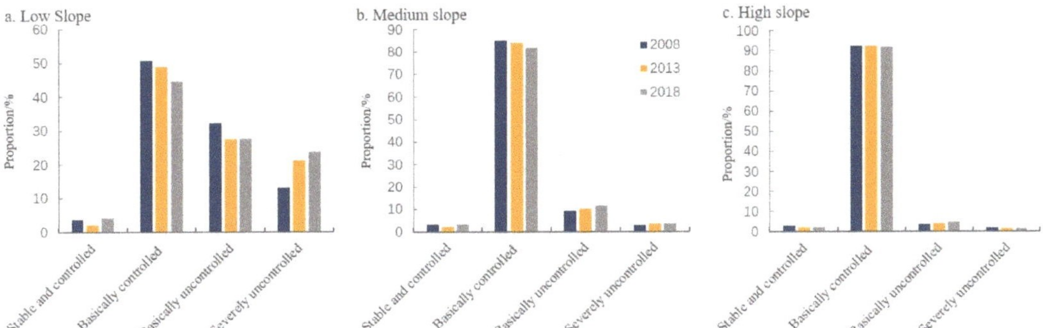

Figure 6. Changes in the degree of land-use conflict in areas with different slopes in Lin'an District, 2008–2018. (**a**) Low slope. (**b**) Medium slope. (**c**) High slope.

5. Discussion

5.1. Characteristics of Land-Use Conflict

It is important to carry out land-use conflict research to realize the coordinated development of regional agricultural space and ecological space, ease the pressure on land resources and human–land conflicts, establish a harmonious human–land relationship between production and ecology, and give full play to regional resource advantages for scientific development.

In this paper, a land-use conflict measurement model [35] was constructed to identify the land use conflict between agricultural space and ecological space taking Lin'an District in Hangzhou City, a hilly area in northern Zhejiang Province, as an example. The study found that although the conflict between agricultural space and ecological space in Lin'an District was generally controlled, there were also areas in which the conflict was uncontrolled due to the complex shape and high fragmentation of the patches [49]. The rapid development of the economy has resulted in the continuous expansion of agricultural land such as cultivated land and the occupation of ecological land such as forestland and grassland [50]. On the one hand, the increase in the output of agricultural production has led to the continuous expansion of agricultural land such as arable land and garden land, while large areas of ecological land such as water surface, woodland and pasture were occupied. On the other hand, as arable land with high slopes is not suitable for agricultural production, the arable land was abandoned and evolved into ecological land naturally, resulting in the phenomenon of the compound use of agricultural space and ecological space is more common in Lin'an. These changes have led to increasing degrees of conflict, the gradual expansion of uncontrolled areas, and the massive extrusion of controlled areas. Therefore, management measures should be implemented in time to prevent further deterioration and to ensure the overall balance and sustainable development of agriculture and ecology in Lin'an District.The main land types of agricultural space include arable land, garden land and other agricultural land, which are closely related to human activities and are, therefore, subject to more human activities, with increasing landscape fragmentation and consequent declines in land stability; therefore, their land-use conflicts are also higher [51]. Lin'an District has adopted strict ecological protection policies, delineating ecological red lines and focusing on protecting land with high ecological benefits such as water areas, and these protection policies have played an important role in promoting regional development. These conservation policies have played an important role in promoting the sustainable and healthy development of the region and alleviating the regional land-use conflicts.

The degree of land-use conflict (Figure 7) in Lin'an District is closely related to economic development, land use and other related factors [52]. As the process of urban–rural integration progressed, adjacent town streets showed high similarity in these aspects, and these similarities resulted in spatial autocorrelation in land-use conflict. The spatial aggregation effect of land use conflict in Lin'an District in 2013 was the most obvious. The overall spatial pattern was relatively stable over the 10 years, with no major changes. The spatial autocorrelation in the eastern region and in towns such as Qianchuan town and Heqiao town was significant due to the complexity of the land-use types.

5.2. Differences in Land-Use Conflict Gradients

To study the spatial and temporal land-use conflicts between agricultural and ecological spaces under different gradient conditions in Lin'an District, this paper selected the slope factor to analyze the topographic gradient characteristics of the conflicts.

The performance of uncontrolled conflicts was the most obvious in the low-slope area of Lin'an District, and the distribution of spatial units with basic controllable conflict levels in the areas with medium- and high- slopes had an absolute advantage.

The main land-use conflict categories in the low-slope area from 2008 to 2018 were arable land and water and were distributed in a band with Yu Qian town as the center. The uncontrolled conflict units in this area were more aggregated and gradually expanded to squeeze the controllable grade with the passage of time, and the degree of land-use conflict was increasingly intensified. This was due to the flat terrain, fertile soil and abundant water in the low- slope area of Lin'an District, which is suitable for human production activities. The increased intensity of land use makes the land structure ratio increasingly imbalanced, and the land use conflict is more serious [53].

The main land types of land-use conflicts in the medium-slope region from 2008 to 2018 were mainly forestland, and the spatial units of the basic controllable conflict levels were the most widely distributed and fragmented. The spatial units of the uncontrollable grade were also scattered, mainly distributed at the junction of the medium-slope and low-slope areas, and showed a gradual expansion trend. This is because the production conditions in the medium-slope areas are inferior to those in the low-slope areas, their land-use conflicts are more moderate, and the proportion of land-use structure is more reasonable [54].

High-slope areas had forestland and grassland as the main land types from 2008 to 2018, which were concentrated in the central and western areas. Spatial units with basic controllable conflict levels occupied the majority of the distribution and had a high degree of contiguity. The spatial units with severe uncontrollable conflict grades were scattered, and the distribution area gradually decreased; only in Jinan Street in 2018 was there a more obvious performance. This is due to the steep terrain and inconvenient transportation in the high-slope area, which is not suitable for production activities; in addition, the current land risk in the high-slope area is low, and the scale and pattern of the regional landscape are well protected [55].

5.3. Trade-off Mechanisms in Land-Use Conflict

Under the current territorial spatial planning system in China, the Lin'an District Government manages the territorial space mainly by preparing territorial spatial planning and delineating "three zones and three lines". The "three zones" refer to ecological, agricultural and urban function spaces, and the "three lines" refer to the "three control lines" of ecological protection red line, permanent basic agricultural land and urban development boundary. It provides for special protection of arable land and strict control of conversion of arable land to other agricultural land such as forest, grassland and garden land. Land with important ecological functions or ecologically sensitive land should be designated ecological protection red line in accordance with the law and implement strict protection. Under the planning constraints, spatial conflicts between ecology and agriculture often do not occur in strictly controlled areas, but in the transition zone between ecology and agriculture or in the multi-appropriate areas for ecological and agricultural use. Based on the identification and analysis of land-use conflicts between agricultural space and ecological space in Lin'an District, the trade-off mechanisms for land-use conflict were identified as follows:

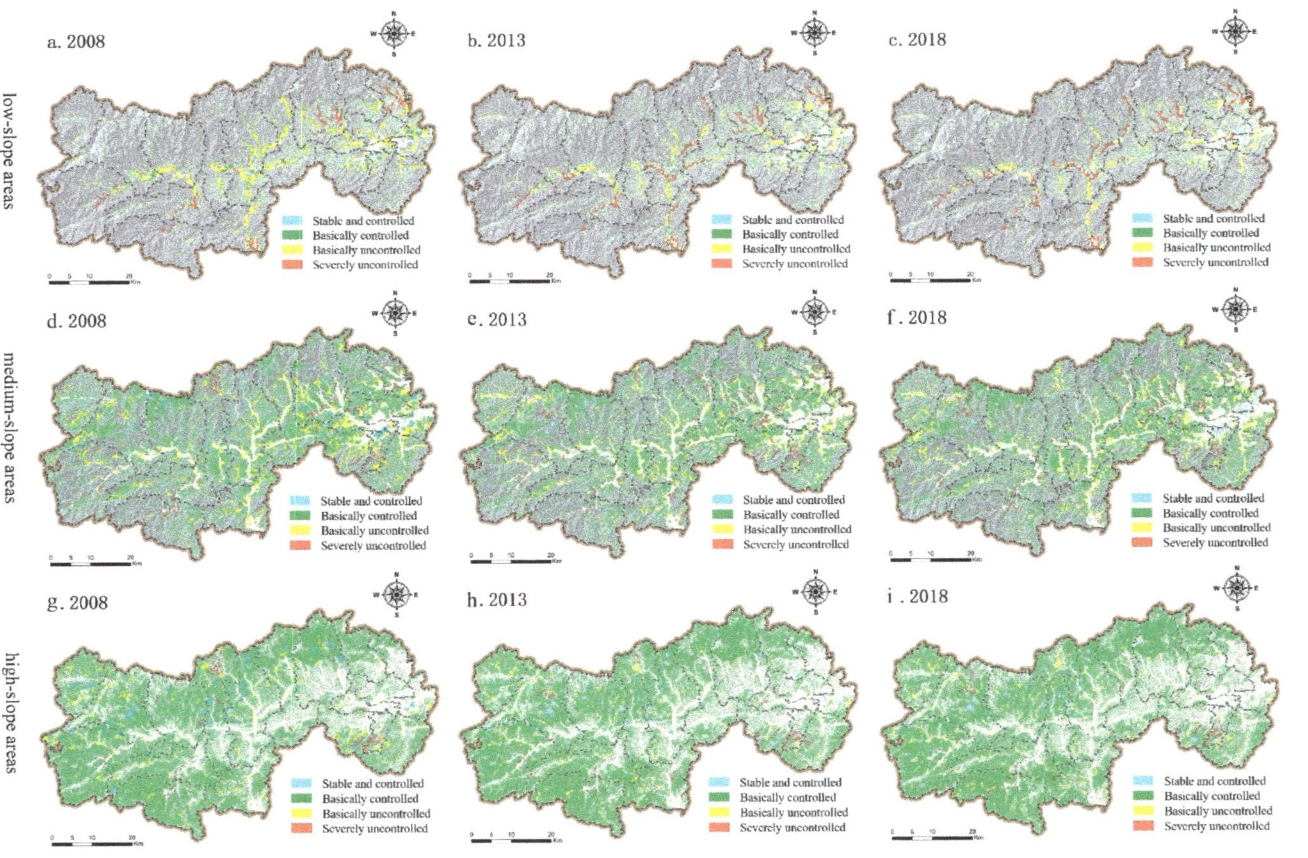

Figure 7. Spatial distribution of potential land-use conflict between agricultural and ecological spaces in low-, medium- and high-slope areas. (**a**) 2008 low-slope areas. (**b**) 2013 low-slope areas. (**c**) 2018 low-slope areas. (**d**) 2008 medium-slope areas. (**e**) 2013 medium-slope areas. (**f**) 2018 medium-slope areas. (**g**) 2008 high-slope areas. (**h**) 2013 high-slope areas. (**i**) 2018 high-slope areas.

(1) Optimizing land-use structure and improving land-use efficiency

To realize the sustainable development of its society, Lin'an District should strengthen environmental protections on the basis of ensuring the supply of production land. In addition, the district should continuously improve the efficiency of land resource utilization by adjusting the land resource utilization structure and implementing land recycling to realize a 'win-win' situation for production development and ecological protection.

Lin'an District should actively optimize its ecological structure on the basis of ensuring food security; return cultivated land with low production efficiency to forestland, grasslands and lakes; carry out afforestation and greening; increase landscape connectivity; reduce patch fragmentation; improve land stability; and enhance the quality of the natural environment and the suitability of the regional landscape ecology [56]. In areas with advantageous production conditions, Lin'an District should make effective use of the surrounding production resources. The government should actively promote regional agricultural restructuring and the reasonable cultivation of woodlands, grasslands and other ecological areas to ensure ecological security; cultivate agricultural lands appropriately for the actual local situation; build economic garden belts such as vegetable and fruit gardens; improve connectivity with existing agricultural lands; increase land-use efficiency; and slow the increase in land-use conflicts.

(2) Coordinating land resource allocation and promoting coordinated regional development

Lin'an District should coordinate its resource elements, rationally allocate its land resources, optimize the structural proportion and spatial layout of the land, construct a balanced and unified spatial planning system, persist in 'drawing a blueprint to the end', continuously promote the integration of land resources, strengthen the optimal allocation of land resources, and achieve an efficient supply of land resources [57]. The terrain in Lin'an District is mainly hilly, and land-use conflict varied obviously with the difference of topography and geomorphology. According to the different characteristics of land-use conflicts, allocating land resources according to local conditions is of great significance to rational land development and utilization and can have substantive effects on promoting coordinated development in the region.

(3) Clearing classification protection and implementing differential use control

Lin'an District should further refine and classify agricultural and ecological spaces based on the importance of their land functions and the policy protection level. This further classification could form a classification control system to determine the 'red line' for permanent basic farmland protection and the 'bottom line' for ecological protection, establish a classification-based spatial protection system, and improve the quality and stability of natural and agricultural ecosystems [58]. At the same time, strengthening the top-level design of controls on national territory use, refining the rules for the control of the differential uses of agricultural space and ecological space, and planning for and taking into account the functional attributes of different land resources as a whole are conducive to adapting to the existing management system and its daily regulatory needs as well as to promoting regional economic development.

(4) Realizing multiple land values and improving the structural layout of industry

Land has multiple functional values, such as agricultural production and ecological services [59]. In the process of land development and utilization, the one-sided value of a certain function of land should not be evaluated alone; rather, multiple values of land should be considered together to maximize the benefit of land resources. Lin'an District has abundant resources, such as natural ecosystems and technological industries. It should make rational use of its resource advantages and follow the patterns of social development to perfect the land structure layout of primary, secondary and tertiary industries. The district should also actively carry out scientific and technological innovations, reduce industry dependence on land, improve land-use efficiency, and increase the land output rate and income of enterprises. Continuing to promote scientific and technological innovation, strengthening the development of industrial science and technology as well as the environment, and increasing the yield per mu through the cultivation of new scientific and technological industries will reduce land-use conflicts.

(5) Adhering to people-oriented concepts and strengthening public messaging and guidance

The process of land resource allocation in Lin'an District should adhere to people-oriented concepts and comprehensively consider the production, ecological and other functions of land resources [60]. Through this process, the total interests of all land-use stakeholders can be maximized and the global optimum can be achieved. At the same time, policy makers should establish a benign interaction mechanism, strengthen information transmission and guidance, and strive to coordinate the interests of all parties. By allowing full participation of the majority of stakeholders and incorporating public feedback into the revision and improvement of the policy system in a timely manner, land-use conflicts can be effectively mitigated.

For its theoretical framework, this research applied the theory of conflict identification and evaluation to the process of land use management and constructed a theoretical framework for research on conflicts between agricultural space and ecological space. In practical applications, hilly areas are usually selected as research areas at the county scale. However, the identification of land-use conflict in agricultural space and ecological space involves many aspects of production and ecology, and there are some limitations to constructing conflict identification models by using the landscape ecological risk assessment method alone.

However, this paper does not take into account the theory of ecology-production-economy compound zones. This theory classifies human activities and land-use types by defining the types and restrictions of human activities and then defines two types of regional zones: urban tourism economic zones and urban production economic zones. Urban tourism economic zones are distributed in areas with convenient transportation and lifestyle amenities, and the tourist landscape is mainly a human-based landscape. Urban production economic zones are mainly paddy fields, which provide ecological value for cities and towns to some extent. The theory of ecology-production complex areas may provide a new interpretation of the current conflicts in the urban ecological agricultural environment. In the future, we should broaden the methods used in this study, further verify and refine the model, and generate results that are more scientific.

6. Conclusions

Taking Lin'an District in Hangzhou City in the hilly areas of northern Zhejiang Province as an example, this research constructed a measurement model for land-use conflict by using the landscape ecological risk assessment model method. The land-use conflicts between agricultural space and ecological space were identified and analyzed, and trade-off mechanisms for land-use conflict were proposed. The main conclusions are as follows:

Land-use conflicts are a direct driver of the changing land-use structure of agricultural and ecological spaces in Lin'an District. The degree of land-use conflict in Lin'an District from 2008 to 2018 was generally dominated by controlled conflict. However, with continuous urbanization, the degree of land-use conflict showed an increasing trend, the expansion trend of uncontrolled spatial units was clear, and the distribution area of controlled spatial units decreased continuously.

The comprehensive index of land-use conflict showed significant and positive global spatial autocorrelation and clear spatial aggregation effects at the 99.9% confidence level. Except for the non-significant spatial units, the other spatial units were mainly highly aggregated ($p = 0.05$), and the spatial unit clustering and significance for land-use conflict were the most prominent in 2013.

The characteristics of land-use conflict were obviously different among areas with different topographies. The conflict levels in the low-, medium- and high-slope areas were all dominated by basic controllability. The slope was negatively correlated with the degree of land-use conflict, with uncontrolled conflict being most evident on low slopes and the distribution of basically controllable spatial units being absolutely dominant on medium and high slopes.

To characterize land-use conflict in Lin'an District, this research proposed specific trade-off mechanisms of land-use conflict from five perspectives. These mechanisms provide a theoretical basis and practical support for the control of conflicts between agricultural space and ecological space.

Author Contributions: Conceptualization, J.Z. and M.G.; methodology, J.Z., and M.G.; software, Y.C.; formal analysis, J.Z.; writing-original draft preparation, J.Z.; writing-review and editing, Y.C., C.Z.; B.H.; and M.G.; writing-review and editing, Y.C., C.Z, and M.G. All authors have read and agreed to the published version of the manuscript.

Funding: This work was financially supported by the Project Supported by the Open Fund of Key Laboratory of Urban Land Resources Monitoring and Simulation, Ministry of Natural Resources (KF-2020-05-073), the Project Supported by the Open Fund of Technology Innovation Center for Land Spatial Eco-restoration a Metropolitan Area, Ministry of Natural Resources (CXZX202010), supported by the Fundamental Research Funds for the Central Universities (2021QN81014).

Conflicts of Interest: The authors declare no conflict of interest.

References

1. Wang, X.; He, J.; Wang, B.; Zhang, X. Thoughts and practice of ecological protection and restoration of mountains, rivers, forests, farmlands, lakes and grasslands. *J. Landsc. Res.* **2020**, *3*, 5–8.
2. Wen, D. A preliminary discussion on agroecosystems and agro-landscapes. *Rural. Eco-Environ.* **1990**, *3*, 52–55.
3. Yu, Z.; Hu, X. Research on the relation of food security and cultivated land's quantity and quality in China. *Geogr. Geo-Inf. Sci.* **2003**, *19*, 45–49.
4. Lin, Y.; Fan, J.; Wen, Q.; Liu, S.; Li, B. Primary exploration of ecological theories and technologies for delineation of ecological redline zones. *Acta Ecol. Sin.* **2016**, *36*, 1244–1252.
5. Yi, D.; Zhao, X.; Guo, X.; Han, Y.; Jiang, Y.; Lai, X.; Huang, X. Study on the spatial characteristics and intensity factors of "three-line conflict" in Jiangxi province. *J. Nat. Resour.* **2020**, *35*, 2428.
6. Gao, W. Research on Tourist Island Landscape Function Conflict Trade—Focus on Dongshan Island in Fujian Province. Master's Thesis, Huaqiao University, Quanzhou, China, 2016.
7. Brown, G.; Raymond, C.M. Methods for identifying land use conflict potential using participatory mapping. *Landsc. Urban Plan.* **2014**, *122*, 196–208. [CrossRef]
8. Henderson, S.R. Managing land-use conflict around urban centres: Australian poultry farmer attitudes towards relocation. *Appl. Geogr.* **2005**, *25*, 97–119. [CrossRef]
9. Sadomba, W.Z. Retrospective community mapping: A tool for community education. *PLA NOTES* **1996**, 9–13. Available online: https://pubs.iied.org/sites/default/files/pdfs/migrate/G01610.pdf (accessed on 23 September 2021).
10. Zhang, Q. Research on the Implementation Plan Model of Public-Private Partnership Project Based on Logical Framework Approach. Master's Thesis, Chongqing University, Chongqing, China, 2017.
11. Shao, X.; Hu, Y. Conflict and balance between farmland ownership and income rights: A theoretical analysis framework for realizing farmers' land property rights. *Macroecon. Res.* **2016**, *12*, 3–13.
12. Sekeris, P.; Luca, D.G. Land inequality and conflict intensity. *Public Choice* **2011**, *150*, 119–135.
13. Becker, N.; Easter, K.W. Water diversion from the Great Lakes: Is a cooperative approach possible? *Int. J. Water Resour. Dev.* **1997**, *13*, 53–66. [CrossRef]
14. Dunk, A.; Grêt-Regamey, A.; Dalang, T.; Hersperger, A.M. Defining a typology of peri-urban land-use conflicts–A case study from Switzerland. *Landsc. Urban Plan.* **2011**, *101*, 149–156. [CrossRef]
15. Orr, A.; Mwale, B. Adapting to adjustment: Smallholder livelihood strategies in Southern Malawi. *World Dev.* **2001**, *29*, 1325–1343. [CrossRef]
16. Jiang, C.; Li, H. Study on post-evaluation of engineering supervision based on logical framework approach. *J. Hefei Univ. Technol. Nat. Sci.* **2008**, *31*, 248–252.
17. Yu, B. Land Use Conflicts in the Urban Fringe: A Theoretical Framework and Case Studies. Ph.D. Thesis, Chinese Academy of Sciences, Beijing, China, 2006.
18. Jiang, S.; Meng, J.; Zhu, L. Spatial and temporal analyses of potential land use conflict under the constraints of water resources in the middle reaches of the Heihe River. *Land Use Policy* **2020**, *97*, 104773. [CrossRef]

19. Amman, H.M.; Duraiappah, A.K. Modeling instrumental rationality, land tenure and conflict resolution. *Comput. Econ.* **2001**, *18*, 251–257. [CrossRef]
20. Duraiappah, A.K. *Land Tenure, Land Use, Environment Degradation and Conflict Resolution: A PASIR Analysis for the Narok District, Kenya*; IIED: London, UK, 2000.
21. Deininger, K.; Castagnini, R. Incidence and impact of land conflict in Uganda. *J. Econ. Behav. Organ.* **2006**, *60*, 321–345. [CrossRef]
22. Zeng, F.; Wei, Y. Study on the influencing factors of ecological conflict of urban land use in China—Take Guiyang as example. *Reform. Strategy* **2016**, *9*, 107–113.
23. Lu, C.; Ittersum, M.; Rabbinge, R. A scenario exploration of strategic land use options for the Loess Plateau in northern China. *Agric. Syst.* **2004**, *79*, 145–170. [CrossRef]
24. Kächele, H.; Dabbert, S. An economic approach for a better understanding of conflicts between farmers and nature conservationists—An application of the decision support system MODAM to the Lower Odra Valley National Park. *Agric. Syst.* **2002**, *74*, 241–255. [CrossRef]
25. Torre, A.; Melot, R.; Magsi, H.; Bossuet, L.; Cadoret, A.; Caron, A.; Kolokouris, O. Identifying and measuring land-use and proximity conflicts: Methods and identification. *SpringerPlus* **2014**, *3*, 1–26. [CrossRef]
26. Yang, Y.; Zhu, L. Theories and diagnostic methods of land use conflicts. *Asian Agric. Res.* **2013**, *5*, 63–70.
27. Guan, D.; Chen, T.; He, X.; Luo, X.; Luo, L.; Deng, H. Spatial conflict type identification and its driving mechanism of land use in the three gorges reservoir area(Chongqing section). *J. Chongqing Jiaotong Univ. (Nat. Sci.)* **2019**, *38*, 65–71.
28. Zheng, L.; Song, Z.; Yang, J. Study on the relief mechanism of land use conflict in mine-grain mixed zone from the perspective of social-ecological system. *China Min. Mag.* **2018**, *27*, 13–18, 32.
29. Feng, C.; Cao, M.; Xie, T. Optimization of land use structure in Tongling city based on different ecological conservation scales. *Geogr. Res.* **2014**, *33*, 2217–2227.
30. Yang, X. The Strategy Analysis of Profit Conflict and Planning Regulation in the Progress of Redevelopment of Urban Land. Master's Thesis, Suzhou University of Science and Technology, Suzhou, China, 2014.
31. Zhou, D.; Xu, J.; Wang, L. Land use spatial conflicts and complexity: A case study of the urban agglomeration around HangZhou Bay, China. *Geogr. Res.* **2015**, *34*, 1630–1642.
32. Ran, N.; Jin, X.; Fan, Y.; Xiang, X.; Liu, J.; Zhou, Y. Three Lines delineation based on land use conflict identification and coordination in Jintan district, Changzhou. *Resour. Sci.* **2018**, *40*, 284–298.
33. Lin, G.; Jiang, D.; Fu, J.; Cao, C.; Zhang, D. Spatial conflict of production-living-ecological space and sustainable-development scenario simulation in Yangtze River delta agglomerations. *Sustainability* **2020**, *12*, 2175. [CrossRef]
34. Zhou, D.; Lin, Z.; Lim, S.H. Spatial characteristics and risk factor identification for land use spatial conflicts in a rapid urbanization region in China. *Environ. Monit. Assess.* **2019**, *191*, 1–22. [CrossRef] [PubMed]
35. Pei, B.; Pan, T. Land use system dynamic modeling: Literature review and future research direction in China. *Prog. Geogr.* **2010**, *29*, 1060–1066.
36. Liao, W.; Dai, W.; Chen, J.; Huang, W.; Jiang, F.; Hu, Q. Spatial conflict between ecological-production-living spaces on Pingtan Island during rapid urbanization. *Resour. Sci.* **2017**, *39*, 1823.
37. Peng, J.; Zhou, G.; Tang, C.; He, Y. The analysis of spatial conflict measurement in fast urbanization region based on ecological security: A case study of Changsha-Zhuzhou-Xiangtan urban agglomeration. *J. Nat. Resour.* **2012**, *27*, 1507–1517.
38. Ma, W.; Jiang, G.; Chen, Y.; Qu, Y.; Zhou, T.; Li, W. How feasible is regional integration for reconciling land use conflicts across the urban–rural interface? Evidence from Beijing-Tianjin-Hebei metropolitan region in China. *Land Use Policy* **2020**, *92*, 104433. [CrossRef]
39. Jiang, S.; Meng, J.; Zhu, L.; Cheng, H. Spatial-temporal pattern of land use conflict in China and its multilevel driving mechanisms. *Sci. Total. Environ.* **2021**, *801*, 149697. [CrossRef]
40. Song, Z.; Yu, L.; Li, T. A study on the generalised space of urban–rural integration in Beijing suburbs during the present day. *Urban Stud.* **2015**, *52*, 2581–2598. [CrossRef]
41. Sun, P.; Xu, Y.; Wang, S. Terrain gradient effect analysis of land use change in poverty area around Beijing and Tianjin. *Trans. Chin. Soc. Agric. Eng.* **2014**, *30*, 277–288.
42. Pimm, S.L. The complexity and stability of ecosystems. *Nature* **1984**, *307*, 321–326. [CrossRef]
43. Fan, J.; Wang, Y.; Zhou, Z.; You, N.; Meng, J. Dynamic ecological risk assessment and management of land use in the middle reaches of the Heihe River based on landscape patterns and spatial statistics. *Sustainability* **2016**, *8*, 536. [CrossRef]
44. Zhou, G.; Peng, J. The evolution characteristics and influence effect of spatial conflict: A case study of Changsha-Zhuzhou-Xiangtan urban agglomeration. *Prog. Geogr.* **2012**, *31*, 717–723.
45. Li, W.; Zhu, C.; Wang, H.; Xu, B. Multi-scale spatial autocorrelation analysis of cultivated land quality in Zhejiang province. *Trans. Chin. Soc. Agric. Eng.* **2016**, *32*, 239–245.
46. Liu, Y.; Li, Y.; Yi, X.; Cheng, X. Spatial evolution of land use intensity and landscape pattern response of the typical basins in Guizhou Province, China. *Ying Yong Sheng Tai Xue Bao J. Appl. Ecol.* **2017**, *28*, 3691–3702.
47. Cressie, N.; Kang, E.L. Hot enough for you? A spatial exploratory and inferential analysis of North American climate-change projections. *Math. Geosci.* **2016**, *48*, 107–121. [CrossRef]
48. Xie, H. Spatial characteristic analysis of land use eco-risk based on landscape structure: A case study in the Xingguo County, Jiangxi Province. *China Environ. Sci.* **2011**, *31*, 688–695.

49. Dai, Z. Land Use Characteristics and Conflict Evaluation of Ecological-Production-Living Space in Hilly and Mountainous Areas. Master's Thesis, Southwest University, Chongqing, China, 2019.
50. Liu, J. Study on the Evolution and Formation Mechanism ofLand Use Conflict in Institutional Ecological Space—Take the Southern Mountain Area of Jinan City as an Example. Master's Thesis, Shandong Jianzhu University, Jinan, China, 2018.
51. Chen, L.; Fu, B. Analysis of impact of Human activity on landscape structiure in yellow river delta-a case study of dongying region. *Acta Ecol. Sin.* **1996**, *16*, 337–344.
52. Yang, Y.; Zhu, L. The theory and diagnostic methods of land use conflicts. *Resour. Sci.* **2012**, *34*, 1134–1141.
53. Li, J.; Lü, Z.; Shi, X.; Li, Z. Spatiotemporal variations analysis for land use in Fen River Basin based on terrain gradient. *Trans. Chin. Soc. Agric. Eng.* **2016**, *32*, 230–236.
54. Zhao, Y.; Cao, J.; Zhang, X.; He, G. Topographic gradient effect and spatial pattern of land use in Baota District. *Arid. Land Geogr.* **2020**, *43*, 1307–1315.
55. Yang, B.; Wang, Z.; Yao, X.; Zhang, L. Terrain gradient effect and spatial structure characteristics of land use in mountain areas of northwestern hubei province. *Resour. Environ. Yangtze Basin* **2019**, *28*, 313–321.
56. Cui, S.; Liu, Q.; Wang, J. Scale effect of landscape pattern index and its response to land use change in the coastal development zone: A case study of Dafeng city in Jiangsu province. *Geogr. Geo-Inf. Sci.* **2016**, *32*, 87–93.
57. The Communist Party of China Central Committee and the State Council. *A Guideline on Establishing a National Territory Spatial Planning System and Supervision of Its Implementation*; People's Publishing House: Beijing, China, 2019.
58. Wei, X.; Kai, X.; Wang, Y.; Yu, H. Discussions on the methods of "three zones and three lines" implementation at the spatial levels of city and county based on "double evaluations". *City Plann. Rev.* **2019**, *43*, 10–20.
59. Zhen, L.; Cao, S.; Wei, Y.; Xie, G.; Li, F.; Yang, L. Land use functions: Conceptual framework and application for China. *Resour. Sci.* **2009**, *31*, 544–551.
60. Yang, Y.; Liu, Y.; Zhu, L. Theories and methods on trade-offs of land-use conflicts. *Areal Res. Dev.* **2013**, *31*, 171–176.

 *land*

Article

Land Use Multi-Suitability, Land Resource Scarcity and Diversity of Human Needs: A New Framework for Land Use Conflict Identification

Guanglong Dong [1], Yibing Ge [1], Haiwei Jia [2,*], Chuanzhun Sun [2] and Senyuan Pan [2]

[1] School of Management Engineering, Shandong Jianzhu University, Jinan 250101, China; dongguanglong18@sdjzu.edu.cn (G.D.); geyibinging@163.com (Y.G.)
[2] School of Public Management, South China Agriculture University, Guangzhou 510642, China; suncz.11b@igsnrr.ac.cn (C.S.); pansy09210804@163.com (S.P.)
* Correspondence: hwjia@scau.edu.cn

Abstract: Land use conflicts are intensifying due to the rapid urbanization and accelerated transformation of social and economic development. Accurate identification of land use conflicts is an important prerequisite for resolving land use conflicts and optimizing the spatial pattern of land use. Previous studies on land use conflict using multi-objective evaluation methods mainly focused on the suitability or competitiveness of land use, ignoring land resource scarcity and the diversity of human needs, hence reducing the accuracy of land use conflict identification. This paper proposes a new framework for land use conflict identification. Considering land use multi-suitability, land resource scarcity and the diversity of human needs, the corresponding evaluation index system was constructed, respectively, and the linear weighted sum model was used to calculate the land use conflict index. Taking Jinan as the study area, the spatial distribution characteristics of land use conflicts are accurately identified and analyzed. The results show that: (1) Land use multi-suitability in Shanghe county and Jiyang district is high, but the intensity of land use conflict is not. This indicates that land use multi-suitability is the premise and basis of land use conflict, but it is not the only determinant, which is consistent with our hypothesis. (2) Land use conflicts in Jinan were dominant by medium conflict, accounting for 43.89% of the conflicts, while strong and weak land use conflicts accounted for 25.21% and 30.90% of the conflicts, respectively. The spatial distribution of land use conflicts is obviously different, with high conflicts in the north and low conflicts in the south. Strong land use conflicts are concentrated in the urban and rural transition zones of Tianqiao, Huaiyin and Shizhong districts and in the northern parts of Licheng and Zhangqiu districts. (3) Inefficient land use and land resource waste aggravated regional land use conflicts in Licheng and Zhangqiu districts. (4) The new framework for land use conflict identification proposed in this study can accurately identify land use conflicts, providing a scientific reference and new ideas for accurate identification of land use conflicts.

Keywords: land use conflict; conflict identification; suitability evaluation; multi-objective evaluation; multifunction

Citation: Dong, G.; Ge, Y.; Jia, H.; Sun, C.; Pan, S. Land Use Multi-Suitability, Land Resource Scarcity and Diversity of Human Needs: A New Framework for Land Use Conflict Identification. *Land* **2021**, *10*, 1003. https://doi.org/10.3390/land10101003

Academic Editors: Dong Jiang, Jinwei Dong and Gang Lin

Received: 28 August 2021
Accepted: 16 September 2021
Published: 23 September 2021

1. Introduction

According to the World Population Change Report published by the World Bank, the global population has increased by approximately 1 billion every 12 years since 1975. At the same time, the world has experienced rapid urbanization and a large influx of people into cities, with more than half of the population now living in urban areas, and this proportion is predicted to rise to two-thirds by 2050 [1]. Transformation of the social economy increases the pressure on land use, and various types and forms of land use conflicts are widespread across the world [2–6]. As the most populous country in the world, China has experienced unprecedented rapid urbanization and industrialization, with the urbanization level rising from 17.9% before the reform and opening-up to 63.89% in 2020. This extensive development model has made China face increasingly severe land use conflicts, which are mainly manifested in the disorderly expansion of urban and rural settlements occupying high-quality farmland and ecological spaces, leading to the fragmentation of cultivated land and a decline in the quality of the ecological environment, further threatening food security, ecological security and the sustainable development of urban and rural areas [7–10].

Land use conflict identification is the premise of land use conflict prevention, mediation and settlement and is the core focus of land-use conflict research. Land use conflict identification methods mainly includes participatory assessment [11,12], game theory [13], landscape ecological risk assessment [14,15] and multi-objective comprehensive assessment [8,16,17]. Participatory evaluation has strong subjectivity and is suitable for small-scale research. Game theory focuses on strategic analysis, which is difficult to achieve quantitative evaluation of land use conflicts [16]. The landscape ecological risk assessment method only uses land use data and pays attention to landscape pattern characteristics of land use, but does not take social and economic factors into account [15,16]. The multi-objective comprehensive evaluation method has obvious advantages compared with other methods [16,18]. It can quantitatively evaluate land use conflicts at multiple scales, such as administrative divisions and grids, and can take socioeconomic and natural attributes into account.

Iojă, Niţă, Vânău, Onose and Gavrilidis [19] selected 10 indicators from two variables: space and urban development, to evaluate the spatial distribution of land use conflicts at administrative units in Bucharest Metropolitan Area. Jiang, Meng and Zhu [16] applied the multi-objective evaluation method and constructed a competitive evaluation index system for cultivated, construction and ecological land using two dimensions: land suitability and driving force of land use conversion and analyzed the temporal and spatial characteristics of land use conflicts in the middle reaches of the Heihe River. Zou, Liu, Wang, Yang and Wang [8] evaluated the suitability of construction, agricultural and ecological land by selecting evaluation indices from four factors: nature, location, society and policy and further identified the types and intensities of land use conflicts. Jing et al. [20] constructed a suitability evaluation index system for production, living and ecological land and identified potential land use conflicts. Kim and Arnhold [21] measured land use conflicts in agricultural basins using two dimensions: land use preference and location importance.

These studies have provided many reference methods and a large number of important conclusions for the identification of land use conflicts and form an important starting point for this study. Although there is still a lack of unified standards and rules for the construction of the evaluation index system of land use conflicts, consensus has been reached on the causes of land use conflicts, and it is believed that land use multi-suitability, land resource scarcity and diversity of human needs are the fundamental causes of land use conflicts [16,22–25]. They are also a necessary condition for land use conflicts. All three are indispensable. Unfortunately, there are a few quantitative assessments of land resource scarcity and diversity of human needs, although some socio-economic and policy indicators, such as population density [16,20], impact of central cities [20,23] and land use planning [8,16,23], have been introduced. This will reduce the accuracy of land use conflict identification and the feasibility of land use conflict mediation.

In view of this, this study attempts to build a new framework for land use conflict identification based on land use multi-suitability, land resource scarcity and diversity of human needs and uses Jinan city as an example to conduct empirical research. This study expands the identification method of land use conflict, which is helpful to land use conflict mediation, promote rational use of land resources and optimize land spatial pattern.

2. Conceptual Framework

The essence of land use conflict is the contradiction between humans and land. Specifically, land use conflict is the result of the comprehensive effects of land use multi-suitability [8,17], land resource scarcity [22,24] and diversity of human needs [24,26] (Figure 1).

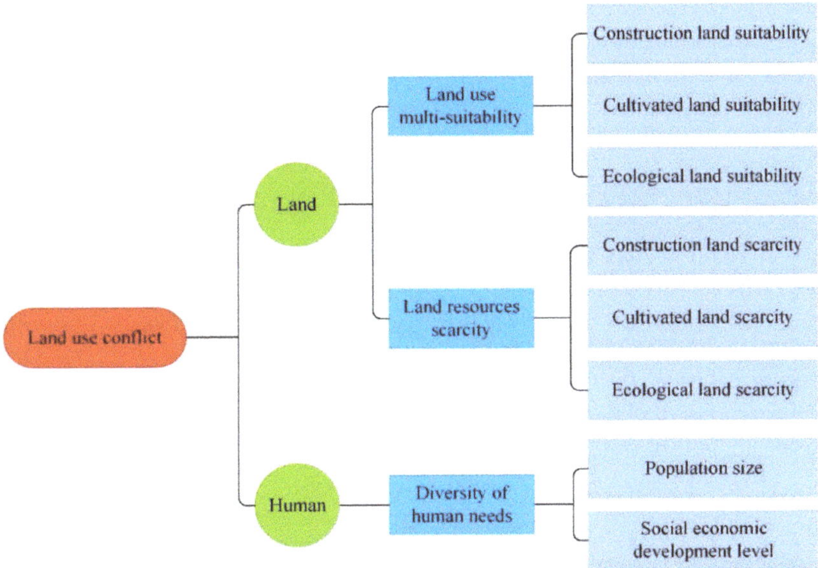

Figure 1. Conceptual framework for land use conflict identification.

Land use multi-suitability is an inherent attribute of land resources. The better the natural endowment and location conditions, the higher the land use multi-suitability. Land use suitability is divided into construction, cultivated and ecological land suitability. Land use multi-suitability can be determined by overlaying the evaluation results of construction, cultivation and ecological land suitability. Land accounts for only 29.2% of the Earth's surface area, and the total amount of land resources is limited and scarce. However, scarcity is a relative concept, and the scarcity of land resources is relative to human needs. Therefore, land resource scarcity should be measured by the index of per capita land use, such as per capita construction land area, per capita cultivated land area and per capita ecological land area. In addition, it is worth noting that even if land resources are suitable for multiple uses, there will be no land use conflict if there is no human demand for them. Therefore, the diversity of human needs is also a key factor in land use conflicts [8,26]. The diversity of human needs is affected by the size of the population and the level of social and economic development. The larger the population and the higher the level of social and economic development, the greater the diversity of human needs.

3. Materials and Methods

3.1. Study Area

Jinan is located at the central and western parts of Shandong province, the southeastern edge of the North China Plain, the intersection of the low mountains and hills in the south of Shandong province and the alluvial plain in the northwest of Shandong province (Figure 2). It is close to Mount Tai in the south and straddles the Yellow River to the north. The terrain is high in the south and low in the north. The city has jurisdiction over 12 county-level administrative regions, including 10 municipal districts and 2 counties, with a total area of 10,244.45 km^2. In 2020, the permanent resident population in Jinan was 9,202,432, the urbanization rate was 73.46%, and the GDP was 944.337 billion yuan. Due to the rapid advancement of urbanization and industrialization, as well as the accelerated transformation and development of the social economy, the disparity between population size and land use in Jinan has increased, causing a series of problems including urban sprawl, declining quality of the ecological environment and air pollution. In April 2021, the State Council agreed in principle to the Implementation Plan for the Construction of the Starting Area for the Conversion of Old and New Growth Drivers in Jinan, which provided new development opportunities for Jinan and put forward higher requirements for rational land use and the optimization of territorial spatial patterns.

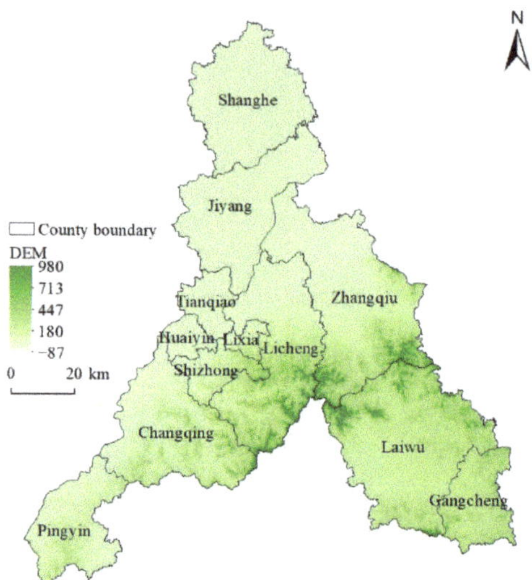

Figure 2. Map of study area.

3.2. Data Sources

The data required for land use conflict evaluation include a digital elevation model (DEM), geological hazard susceptibility rate, land use status map, grade of cultivated land use, normalized difference vegetation index (NDVI) and total population and gross domestic product (GDP) at county level. DEM (30 m spatial resolution) and NDVI (1 km spatial resolution) data were obtained from the Data Center of Resources and Environmental Science, Institute of Geographic Resources and Resources (https://www.resdc.cn, accessed on 22 August 2021). The geological hazard distribution map in JPG format was collected by our research group from geological disaster department of Shandong Province in the process of project research and was registered and vectorized. The 2018 land use status map was provided by Shandong Land Survey and Planning Institute, and the cultivated land use grade data were extracted from the agricultural land grading database. The total population and GDP data at county level were from Jinan Statistical Yearbook in 2019.

3.3. Construction of Land Use Conflict Evaluation Index System

3.3.1. Construction of Land Use Multi-Suitability Evaluation Index System

Land use multi-suitability refers to the suitability of using land resources for different functions; for example, fertile and flat land can be used as agricultural, construction or ecological land. However, different uses require different conditions [27]. Therefore, a land suitability evaluation index system should be constructed depending on the use. Based on relevant studies [16,17,23,27], according to the principles of (1) reflecting the connotation of suitability, (2) high frequency of use and (3) strong representativeness and combined with the accessibility of data [28], evaluation indexes were selected to construct the evaluation index system of construction land, cultivated land and ecological land suitability.

(1) Evaluation index system for suitability of construction land.

The suitability of construction land is mainly affected by the natural and location conditions (Table 1). The natural conditions include elevation, slope and geological hazard susceptibility. The higher the elevation and the steeper the slope, the more difficult the construction, the higher the cost and the lower the suitability of the land for construction. Geological disasters can damage buildings and even endanger residents' lives and property. The higher the rate of occurrence of geological disasters, the lower the suitability of the land for construction. The location condition is characterized by the distance from the city and the distance from the road. The city is the center of social and economic activities, and the road is the reflection of the degree of transportation convenience. The closer the distance to the city and the road, the higher the suitability of the land for construction.

Table 1. Evaluation index system for construction land suitability.

Factor (Weight)	Indicator	Weight	Indicator Classification and Score				
			100	80	60	40	20
Natural conditions(0.6667)	DEM (m)	0.1634	<95	95–221	221–347	347–502	>502
	Slope (°)	0.2970	<2	2–6	6–15	15–25	>25
	Geological hazard susceptibility	0.5396	None	—	Low	Medium	High
Location conditions(0.3333)	Distance to city (m)	0.75	<1279	1279–2934	2934–4806	4806–7358	>7358
	Distance to main road (m)	0.25	<1295	1295–2974	2974–5147	5147–8212	>8212

(2) Evaluation index system for cultivated land suitability.

Cultivated land suitability was evaluated using two variables: natural endowment and farming convenience (Table 2). Low elevation, flat terrain and fertile soil are ideal for arable land, where soil organic matter content, available soil thickness, soil texture, soil pH, irrigation conditions and other factors can affect soil fertility. Based on the above indicators, China's agricultural authorities have implemented a classification and grading system for agricultural land; therefore, the grade was selected to represent the fertility of cultivated land. Farming convenience is also an important aspect of the suitability of cultivated land. Remote and inconveniently located land often has poor natural endowment and is more likely to be abandoned; thus, its cultivated land suitability is low.

Table 2. Evaluation index system for cultivated land suitability.

Factor (Weight)	Indicator	Weight	Indicator Classification and Score				
			100	80	60	40	20
Natural endowments(0.6)	DEM (m)	0.1311	<95	95–221	221–347	347–502	>502
	Slope (°)	0.2081	<2	2–6	6–15	15–25	>25
	Grade of cultivated land	0.6608	6/7	8/9	10	11	Uncultivated area
Farming convenience(0.4)	Distance to rural settlement (m)	0.75	<301	301–730	730–1574	1574–3236	>3236
	Distance to rural road (m)	0.25	<500	500–1500	1500–3000	3000–5000	>5000

(3) Evaluation index system for ecological land suitability.

There are relatively few limiting factors for ecological land suitability. The suitability of ecological land was evaluated using two variables: natural endowment and human disturbance (Table 3). The better the natural endowment, the less human disturbance and the more suitable the land is for ecological activity. NDVI and land use type are used to represent natural endowment and human disturbance, respectively. The higher the vegetation index, the better the natural endowment and the higher the ecological suitability grade. According to the intensity of human activities on different land use types, land use types are graded and assigned. The higher the intensity of human activities, the lower the ecological suitability. For example, the construction land has the highest intensity of human activities and lowest ecological suitability, and the value is 20. The intensity of human activities in forestland is weak and the ecological suitability is high, and the value is 100.

Table 3. Evaluation index system for ecological land suitability.

Factor	Indicator	Weight	Indicator Classification and Score				
			100	80	60	40	20
Natural endowment	NDVI	0.55	>0.792	0.704–0.792	0.596–0.704	0.448–0.596	<0.448
Human disturbance	Land use type	0.45	Forest	Grassland	Water area, unused land, inland tidal flats	Cultivated land, garden land, ditches, agricultural facilities	Construction land

3.3.2. Construction of Land Resource Scarcity Evaluation Index System

The per capita construction land area, per capita cultivated land area and per capita ecological land area were selected to represent the scarcity of construction land, cultivated land and ecological land (Table 4). The larger the per capita land area, the lower the scarcity.

Table 4. Evaluation index system for land resource scarcity.

Factor	Indicator	Indicator Classification and Score				
		100	80	60	40	20
Construction land scarcity	Per capita construction land area	<165.52	165.52–219.44	219.44–307.98	307.98–321.21	>321.21
Cultivated land scarcity	Per capita cultivated land area	<180.18	180.18–443.57	443.57–606.61	606.61–886.07	>886.07
Ecological land scarcity	Per capita ecological land area	<91.40	91.40–128.54	128.54–176.57	176.57–317.60	>317.60

3.3.3. Constructing of Human Needs Diversity Evaluation Index System

Population size and the level of economic development are two key factors that affect the diversity of human needs (Table 5). The larger the population and the higher the level of economic development, the richer the diversity of human needs; therefore, total population and GDP were selected to represent the population and the economic development level, respectively, to evaluate the diversity of human needs.

Table 5. Human needs diversity evaluation index system.

Factor	Indicator	Indicator Classification and Score				
		100	80	60	40	20
Population	Total population	>68.37	59.22–68.37	43.60–59.22	35–43.60	<43.60
Economic development level	GDP	>864.5	593.51–864.5	366.9–593.51	200–366.9	<200

3.4. Index Quantification, Grading Assignment and Weight Determination

According to the characteristics of the data, we adopt a variety of methods to quantify, grade and assign the evaluation index, including the classification assignment method and natural breakpoint method. For classified data, such as geological disaster-prone areas and land use types, the corresponding relationship between types and grades is established by using a classification assignment method based on relevant studies, and corresponding scores are assigned. Specifically, geological hazard susceptibility was divided into four grades—none, low, medium and high—and assigned scores of 100, 60, 40, and 20, respectively. The grade of the cultivated land in Jinan city is 6–11, and 6 and 7 are assigned to 100, 8 and 9 to 80, 10 to 60, 11 to 40 and non-cultivated land to 20. Land use types were divided into five grades based on the intensity of human activities in each land use type and assigned values. For numerical data, such as DEM and distance index, the natural breakpoint method is used to divide them into 5 grades with the help of ArcGIS software, and the corresponding scores were assigned. Besides, based on the grading standards of the slope in the grading regulations of agricultural land, slopes were divided into five grades: <2°, 2°–6°, 6°–15°, 15°–25° and >25°.

Common methods for determining the weights include expert scoring, the analytic hierarchy process, the entropy value method and principal component analysis. Of these, the combination of expert scoring and analytic hierarchy process has the advantages of making full use of expert experience and being simple and easy to operate. Therefore, this study adopted the expert scoring method and the analytic hierarchy process to comprehensively determine the weight of evaluation indicators.

3.5. Land Use Conflict Index Calculation

According to the conceptual framework of land use conflict identification, the following linear weighted sum model was used to calculate the land use conflict index:

$$LUC = w_1 * MS + w_2 * Sca + w_3 * DHN \tag{1}$$

$$MS = \sqrt[3]{ConLSu \times CulLSu \times EcoLSu} \tag{2}$$

$$Sca = \sqrt[3]{ConLSca \times CulLSca \times EcoLSca} \tag{3}$$

$$DHN = \sqrt{Pop \times GDP} \tag{4}$$

Where *LUC* represents the land use conflict index, *MS* represents land use multi-suitability index, *Sca* represents land resource scarcity index, and *DHN* represents diversity of human needs. w_1, w_2 and w_3 are the weights of *MS*, *Sca*, and *DHN*, respectively. *ConLSu*, *CulLSu* and *EcoLSu* represent the suitability of construction land, cultivated land and ecological land, respectively. *ConLSca*, *CulLSca* and *EcoLSca* represent the scarcity of construction land, cultivated land and ecological land, respectively. *Pop* stands for the total population.

The natural breakpoint method was used in ArcGIS to divide the results of the evaluation of construction land suitability, cultivated land suitability, ecological land suitability, land use multi-suitability, land resource scarcity, human needs diversity and land use conflict into three levels: high, medium and low.

4. Results

4.1. Land Use Suitability

4.1.1. Construction Land Suitability

Areas with high suitability for construction accounted for the highest proportion (58.39%) and were concentrated in Shanghe county and Jiyang, Tianqiao, Huaiyin, Shizhong and Lixia districts, as well as the northern Licheng district, northern Zhangqiu city, western Changqing district and Pingyin county (Figure 3). Of these, Tianqiao, Huaiyin, Shizhong and Lixia districts are the central districts of Jinan with developed transportation and obvious regional advantages. Shanghe county and Jiyang district are in the North China

Plain where the terrain is flat and geological disasters do not occur easily. Areas with low construction land suitability accounted for the smallest proportion (17.23%) and were mainly distributed in the east of Changqing district, the south of Licheng district, the south of Zhangqiu city and the north of Laiwu district. These areas are in the southern mountainous region and have high terrain and steep slopes, are prone to geological disasters, are far away from towns, have inconvenient transportation and have low suitability for use as construction land. The proportion of areas with medium suitability for construction was 24.38% and were relatively spatially scattered.

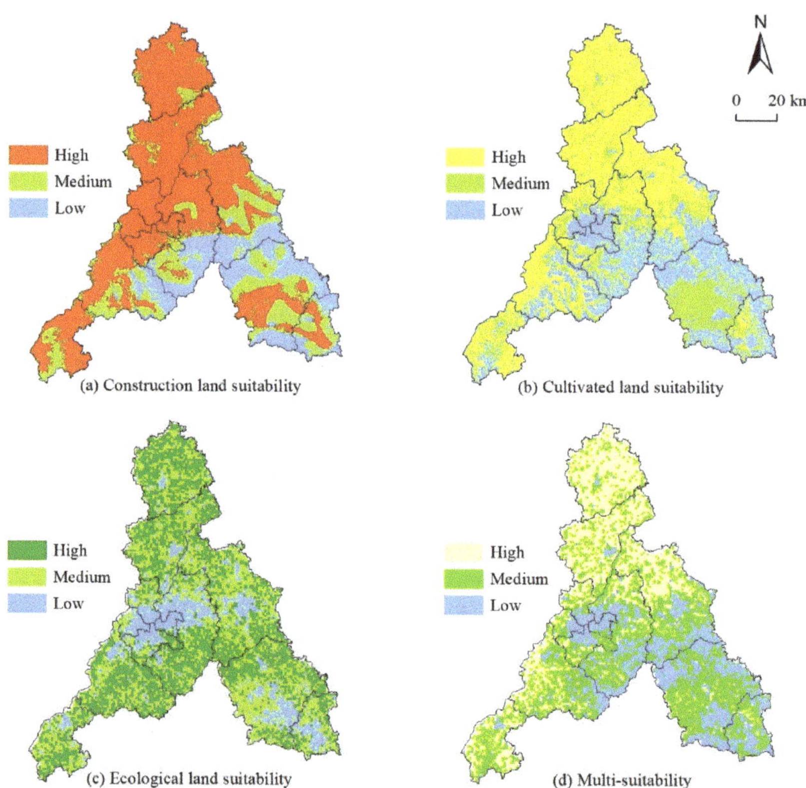

Figure 3. Spatial distribution of land use suitability.

4.1.2. Cultivated Land Suitability

The suitability of land for cultivation was mainly medium, accounting for 39.28% of the area studied. The proportion of areas with high suitability for cultivation was 34.00%. The area with low cultivated land suitability accounted for the lowest proportion (26.72%). Areas with high suitability for cultivation are mainly distributed in the north of Jinan city, particularly in Shanghe county and Jiyang district. They are rich in cultivated land resources, have flat terrain, fertile soil, abundant water resources and good irrigation and drainage conditions. They are important agricultural production areas and grain production bases. The suitability of land for cultivation in the southern mountainous area is low due to the influence of topography and farming convenience. The areas with medium suitability for cultivation were relatively scattered, but concentrated in the middle of Laiwu district.

4.1.3. Ecological Land Suitability

The proportions of areas with high and medium ecological land suitability were similar (42.24% and 42.05%), while the proportion of areas with low ecological land suitability was small (15.71%). The areas with low ecological land suitability are mainly urban land and rural residential land, which have low vegetation coverage and high intensity of human activities and are not suitable for use for ecological activities. The areas with high and middle ecological land suitability are adjacent and have a wide distribution range. Of these, the southern mountainous area has higher forest coverage and less human disturbance. Therefore, land with high ecological suitability is widely distributed here.

4.1.4. Land Use Multi-Suitability

Land use multi-suitability of in Jinan is relatively high and is dominated by medium to high suitability, accounting for 74.23% of the total land area, while the proportion of low multi-suitability is relatively small (25.77%). The spatial distribution of land use multi-suitability is affected by topography and is higher in the north and lower in the south (Figure 3). The areas with high land use multi-suitability mainly distributed in Shanghe county, Jiyang district, northern Tianqiao district, northern Licheng district, northern Zhangqiu district, western Changqing district and some parts of Pingyin county. In these regions, the terrain is flat, the incidence of geological disasters is low, urban and rural settlements are densely distributed, transportation is convenient, and idle land use is relatively low; thus, land use multi-suitability is high. Medium land use multi-suitability is concentrated in the southern parts of Licheng and Zhangqiu districts, the western part of Changqing district, Pingyin county, the central part of Laiwu district and some parts of Gangcheng district. The multi-suitability of land in urban built-up areas and southern mountainous areas is low.

4.2. Scarcity of Land Resources

The evaluation indicators of land resource scarcity and human needs diversity include socioeconomic data such as population size and GDP. In China, the smallest statistical unit of the above socioeconomic data is the county-level administrative division. Therefore, the county-level administrative division was taken as the scale for the analysis of land resource scarcity and human needs diversity.

The spatial distribution of land resource scarcity is obviously different (Figure 4). Tianqiao, Huaiyin, Lixia and Shizhong districts in the central part of Jinan have the highest land resource scarcity. These areas are located in the center of Jinan, with high urbanization levels, large populations, limited arable and ecological land and small per capita construction land scales. Land resource scarcity in Jiyang, Licheng, Changqing and Laiwu is low. Among them, Jiyang is located at the edge of north China Plain, rich in cultivated land resources; Changqing and Laiwu belong to mountainous areas and are rich in ecological resources. In addition, these areas are still in the stage of rapid urbanization, with extensive land use and a large per capita construction land scale. Shanghe, Zhangqiu, Pingyin and Gangcheng have medium land resource scarcity.

4.3. Diversity of Human Needs

The diversity of human needs in Jinan is greatly affected by the level of social and economic development, resulting in high diversity in the middle and low diversity in the north and south (Figure 4). Lixia, Shizhong, Licheng, Zhangqiu and Laiwu districts have the highest diversities of human needs. These areas have a high level of social and economic development, with a total population of 6484–10518 million and GDP of 593.51–149.48 billion yuan. The diversity of human needs in Jiyang, Tianqiao, Huaiyin and Changqing districts are medium. The population size and GDP are lowest in Shanghe and Pingyin counties and Gangcheng district, resulting in the lowest diversity of human needs.

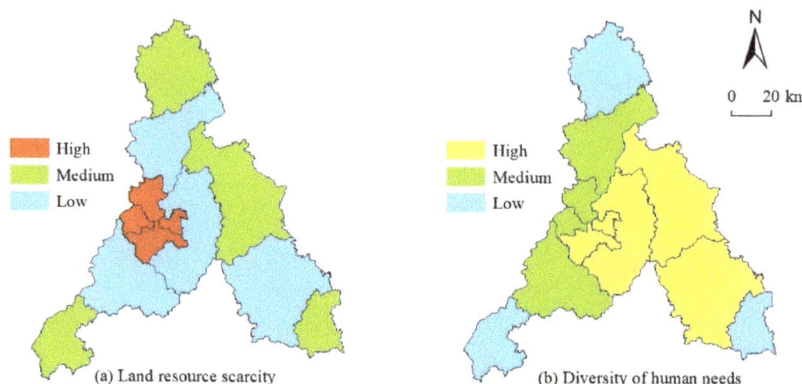

Figure 4. Scarcity of land resources and diversity of human needs.

4.4. Land Use Conflict

Land use conflicts in Jinan were dominated by medium conflicts (43.89%). Strong and weak land use conflicts accounted for 25.21% and 30.90%, respectively. The spatial distribution of land use conflicts differed, with high conflicts in the north and low conflicts in the south (Figure 5). Strong land use conflicts are concentrated in the urban and rural transition zones of Tianqiao, Huaiyin and Shizhong districts, as well as in the northern parts of Licheng and Zhangqiu districts. Due to proximity to the urban area, land resource scarcity is higher in the urban–rural transition zone. The rapid urbanization process leads to a massive influx of people into cities and towns, sharply increasing the demand for construction land. Urban expansion and sprawl will inevitably give priority to the occupation of land resources in the urban–rural transition zone. Therefore, the contradiction and conflicts over land use in this region will intensify. This has been confirmed by previous research on urban expansion and sprawl [29,30].

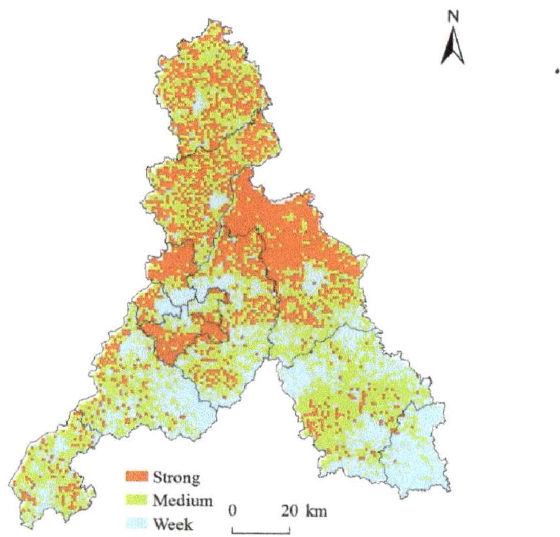

Figure 5. Spatial distribution of land use conflict.

Medium land use conflicts were mainly concentrated in Shanghe, Jiyang and Laiwu. Shanghe county and Jiyang district are in the North China Plain, which has flat terrain and fertile soils and is an important grain producing area. In summary, there are fewer factors restricting land use; thus, land use suitability is high. However, its urbanization level is relatively low and lacks cohesion and attraction to the population. In addition, these areas are dominated by migrant workers, and there is a certain degree of population loss, resulting in a large per capita cultivated land area, ranking them first and second in the per capita cultivated land area of Jinan. There is also a large per capita construction land scale, particularly in Jiyang district. Therefore, the diversity of human needs in Shanghe county and Jiyang district is not high, and land resource scarcity is low. Consequently, although Shanghe county and Jiyang district have high land use multi-suitability, the land resource scarcity is low and the diversity of human needs is not high, leading to a medium level of land use conflict. Laiwu district is surrounded by mountains to the north, east and south, with the Taishan mountains in the north and the Sorai Mountains in the south. It is gentle in the south and steep in the north, forming a semicircle basin protruding to the north. Due to limitations of topological factors, land use multi-suitability is medium in the basin and low in other areas. Laiwu district had a total population of 956,600 and a GDP of 59.351 billion yuan in 2018, with high diversity of human needs and medium land resource scarcity. The low land use multi-suitability, medium land resource scarcity and high diversity of human needs resulted in medium land use conflict in the district.

Weak land use conflicts are mainly distributed in the built-up areas of major towns, in the southern mountainous areas and in Gangcheng districts. The built-up areas are mostly made up of impervious surfaces and the possibility of converting them into cultivated or ecological land is very small. Therefore, the multi-suitability of the land is low. The main reasons for the weak land use conflicts in the southern mountainous area are the large topographic undulated area, frequent occurrence of geological disasters and inconvenient transportation. Gangcheng district is surrounded by mountains to the east, south and west; thus, the land use suitability is not high in most areas. The total population here is only 325,800, the GDP is 26.087 billion yuan, and the diversity of human needs and land resource scarcity is low. Therefore, land use conflict is weak.

5. Discussion

We designed a land use conflict identification framework based on land use multi-suitability, land resource scarcity and diversity of human needs, which can effectively and accurately identify land use conflicts. The areas with strong land use conflict are mainly distributed in the urban–rural transition zones, which is consistent with the results of previous studies [31,32]. However, we also find that the suburbs of cities and towns in the north of Licheng and Zhangqiu districts are also the areas where strong land use conflicts are concentrated. This is an interesting finding, which differs from previous studies. On the one hand, Jinan has implemented the "East Expansion" development strategy in recent years, and the urbanization process in eastern Jinan, including in Licheng and Zhangqiu districts, has advanced rapidly, and land use conflicts have intensified. On the other hand, the north of Licheng and Zhangqiu districts have flat terrains, low incidence of geological disasters, fewer land resource limiting factors, high land use suitability, large regional populations, large GDP and high diversity of land resource demand, resulting in a high intensity of land use conflict. However, it is worth noting that the per capita construction land area and per capita cultivated land area in these regions are both large, and land resource scarcity is low. There is a certain level of inefficient land use and even land resource waste, which aggravate regional land use conflicts.

The land use multi-suitability in Shanghe county and Jiyang district is high, but the intensity of land use conflict in most areas is not. This indicates that land use multi-suitability is the premise and basis of land use conflict, but it is not the only determinant. Even if land is suitable for multiple uses, it does not necessarily produce strong land use conflicts. Land use multi-suitability, land resource scarcity and diversity of human needs are three important determinants of land use conflicts, which together determine the occurrence and intensity of land use conflicts. However, previous studies focused only on the multi-suitability of, or competition over, land, and few studies have considered land use multi-suitability, land resource scarcity and human demand diversity simultaneously. In conclusion, compared with previous studies, the land use conflict identification framework constructed in this study has obvious advantages in the effective and accurate identification of land use conflicts.

The land use conflict identification and evaluation index system we have constructed here has the following advantages. First, the theoretical basis is sufficient. Using relevant studies and field research experience, we analyzed the root causes of land use conflicts and designed a conceptual framework for land use conflict identification. An evaluation index system for land use conflict identification was constructed by selecting evaluation indices from three variables: land use multi-suitability, land resource scarcity and diversity of human needs. Second, the land use conflict evaluation framework can identify the intensity of land use conflict on a grid scale, providing more accurate location information for land use decision making, hence contributing to the accurate land management.

Land use conflict arises from competing land uses. The theory of multifunctional land use can be used as a reference [16,33] and is based on the premise that the same parcel of land can have a dominant function while also having a variety of other functions. Based on the ecological service value trade-off/synergy theory and coupling coordination theory [28,34], the leading function should be rationally selected based on local conditions, and the tradeoff and coordination among different functions should be carried out to promote the coordinated development of multiple functions.

This study had some limitations. First, due to the limitation in the statistical unit of social and economic data, the evaluation unit of land resource scarcity and diversity of human needs was set at the county administrative division level and was not accurate to specific plots, reducing the accuracy of land use conflict identification. Second, this study evaluated the land use conflict in Jinan in 2018, but did not analyze the spatial–temporal evolution of the characteristics of land use conflicts. Finally, the ultimate purpose of land use conflict identification is to mediate land use conflicts. There are many more parameters (social, political, communicational, psychological, etc.) determining conflicts. Such a factor, e.g., is the ambiguity/unclearness of laws and policy contents [35]. Another one could be redistributive policies, uncertainty, etc. This study only presented a superficial discussion on the topic, and further research is needed. Therefore, future research should focus on using grid scale social and economic data to quantitatively evaluate the spatiotemporal dynamic characteristics of land use conflicts and to explore effective measures to resolve land use conflicts.

6. Conclusions

Considering land use multi-suitability, land resource scarcity and diversity of human needs, this paper proposes a new framework to identify land use conflicts and evaluates land use conflicts in Jinan. Land use conflicts in Jinan were dominated by medium conflicts (43.89%). Strong and weak land use conflicts accounted for 25.21% and 30.90%, respectively. Spatially, land use conflicts were high in the north and low in the south. Strong land use conflicts are concentrated in the urban–rural transition zones of Tianqiao, Huaiyin and Shizhong districts and in the northern parts of Licheng and Zhangqiu districts. Land resource scarcity and diversity of human needs have an important impact on land use conflicts. The new land use identification framework proposed in this paper can effectively identify land use conflicts. This method can be applied to other areas and is of great significance to promote the rational utilization of regional land resources. However, we only evaluated land resource scarcity and diversity of human needs at the county level using socio-economic data. How to accurately depict land resources scarcity and the diversity of human needs in order to improve the accuracy of land use conflict identification is the focus of future research.

Author Contributions: G.D. contributed to all aspects of this work; Y.G. and S.P. conducted data processing and analysis; H.J. and C.S. wrote the draft and obtained funding support. All authors have read and agreed to the published version of the manuscript.

Funding: This research was funded by National Key Research and Development Program of China, grant number 2019YFC0507802, National Natural Sciences Foundation of China, grant number 71673091, The Natural Sciences Foundation of China, grant number M-0342, National Natural Science Foundation of China, grant number 41801173, Doctoral Fund of Shandong Jianzhu University, grant number XNBS1803.

Institutional Review Board Statement: Not applicable.

Informed Consent Statement: Not applicable.

Data Availability Statement: The data presented in this study are available on request from the author.

Acknowledgments: Thanks to anonymous experts for their suggestions.

Conflicts of Interest: The authors declare no conflict of interest.

References

1. Malakoff, D.; Wigginton, N.S.; Fahrenkamp-Uppenbrink, J.; Wible, B. Use our infographics to explore the rise of the urban planet. *Science.* **2016**. Available online: http://www.sciencemag.org/news/2016/05/use-our-infographics-explore-rise-urban-planet (accessed on 21 September 2021). [CrossRef]
2. Hjalager, A.-M. Land-use conflicts in coastal tourism and the quest for governance innovations. *Land Use Policy* **2020**, *94*, 104566. [CrossRef]
3. Milczarek-Andrzejewska, D.; Zawalińska, K.; Czarnecki, A. Land-use conflicts and the Common Agricultural Policy: Evidence from Poland. *Land Use Policy* **2018**, *73*, 423–433. [CrossRef]
4. Bax, V.; Francesconi, W.; Delgado, A. Land-use conflicts between biodiversity conservation and extractive industries in the Peruvian Andes. *J. Environ. Manag.* **2019**, *232*, 1028–1036. [CrossRef]
5. Petrescu-Mag, R.M.; Petrescu, D.C.; Azadi, H.; Petrescu-Mag, I.V. Agricultural land use conflict management—Vulnerabilities, law restrictions and negotiation frames. A wake-up call. *Land Use Policy* **2018**, *76*, 600–610. [CrossRef]
6. Mugizi, F.M.P.; Matsumoto, T. From conflict to conflicts: War-induced displacement, land conflicts, and agricultural productivity in post-war Northern Uganda. *Land Use Policy* **2021**, *101*, 105149. [CrossRef]
7. Li, H.H.; Song, W. Expansion of Rural Settlements on High-Quality Arable Land in Tongzhou District in Beijing, China. *Sustainability* **2019**, *11*, 5153. [CrossRef]
8. Zou, L.; Liu, Y.; Wang, J.; Yang, Y.; Wang, Y. Land use conflict identification and sustainable development scenario simulation on China's southeast coast. *J. Clean. Prod.* **2019**, *238*, 117899. [CrossRef]
9. Zhou, Y.; Dong, J.; Xiao, X.; Liu, R.; Ge, Q. Continuous monitoring of lake dynamics on the Mongolian Plateau using all available Landsat imagery and Google Earth Engine. *Sci. Total Environ.* **2019**, *689*. [CrossRef] [PubMed]
10. Qin, Y.; Xiao, X.; Dong, J.; Zhang, Y.; Wu, X.; Shimabukuro, Y.; Arai, E.; Biradar, C.; Wang, J.; Zou, Z.; et al. Improved estimates of forest cover and loss in the Brazilian Amazon in 2000–2017. *Nat. Sustain.* **2019**, *2*, 764–772. [CrossRef]
11. Karimi, A.; Brown, G. Assessing multiple approaches for modelling land-use conflict potential from participatory mapping data. *Land Use Policy* **2017**, *67*, 253–267. [CrossRef]
12. Brown, G.; Raymond, C.M. Methods for identifying land use conflict potential using participatory mapping. *Landsc. Urban Plan.* **2014**, *122*, 196–208. [CrossRef]
13. Reuveny, R.; Maxwell, J.W.; Davis, J. On conflict over natural resources. *Ecol. Econ.* **2011**, *70*, 698–712. [CrossRef]
14. Zhou, D.; Lin, Z.; Lim, S.H. Spatial characteristics and risk factor identification for land use spatial conflicts in a rapid urbanization region in China. *Environ. Monit. Assess.* **2019**, *191*, 1–22. [CrossRef] [PubMed]
15. Gao, Y.; Wang, J.; Zhang, M.; Li, S. Measurement and prediction of land use conflict in an opencast mining area. *Resour. Policy* **2021**, *71*, 101999. [CrossRef]
16. Jiang, S.; Meng, J.; Zhu, L. Spatial and temporal analyses of potential land use conflict under the constraints of water resources in the middle reaches of the Heihe River. *Land Use Policy* **2020**, *97*, 104773. [CrossRef]
17. Lin, G.; Fu, J.; Jiang, D. Production–Living–Ecological Conflict Identification Using a Multiscale Integration Model Based on Spatial Suitability Analysis and Sustainable Development Evaluation: A Case Study of Ningbo, China. *Land* **2021**, *10*, 383. [CrossRef]
18. Cui, J.; Kong, X.; Chen, J.; Sun, J.; Zhu, Y. Spatially explicit evaluation and driving factor identification of land use conflict in yangtze river economic belt. *Land* **2021**, *10*, 43. [CrossRef]
19. Iojă, C.I.; Niţă, M.R.; Vânău, G.O.; Onose, D.A.; Gavrilidis, A.A. Using multi-criteria analysis for the identification of spatial land-use conflicts in the Bucharest Metropolitan Area. *Ecol. Indic.* **2014**, *42*, 112–121. [CrossRef]
20. Jing, W.; Yu, K.; Wu, L.; Luo, P. Potential Land Use Conflict Identification Based on Improved Multi-Objective Suitability Evaluation. *Remote Sens.* **2021**, *13*, 2416. [CrossRef]

21. Kim, I.; Arnhold, S. Mapping environmental land use conflict potentials and ecosystem services in agricultural watersheds. *Sci. Total Environ.* **2018**, *630*, 827–838. [CrossRef] [PubMed]

22. Zong, S.; Hu, Y.; Zhang, Y.; Wang, W. Identification of land use conflicts in China's coastal zones: From the perspective of ecological security. *Ocean Coast. Manag.* **2021**, *213*, 105841. [CrossRef]

23. Li, S.; Zhu, C.; Lin, Y.; Dong, B.; Chen, B.; Si, B.; Li, Y.; Deng, X.; Gan, M.; Zhang, J.; et al. Conflicts between agricultural and ecological functions and their driving mechanisms in agroforestry ecotone areas from the perspective of land use functions. *J. Clean. Prod.* **2021**, *317*, 128453. [CrossRef]

24. Jiang, S.; Meng, J.; Zhu, L.; Cheng, H. Spatial-temporal pattern of land use conflict in China and its multilevel driving mechanisms. *Sci. Total Environ.* **2021**, *801*, 149697. [CrossRef] [PubMed]

25. Zou, L.; Liu, Y.; Wang, Y. Research progress and prospect of land-use conflicts in China. *Progress in Geography.* **2020**, *39*, 298–309. [CrossRef]

26. Tudor, C.A.; Iojă, I.C.; Pătru-Stupariu, I.; Niță, M.R.; Hersperger, A.M. How successful is the resolution of land-use conflicts? A comparison of cases from Switzerland and Romania. *Appl. Geogr.* **2014**, *47*, 125–136. [CrossRef]

27. Zou, L.; Liu, Y.; Wang, J.; Yang, Y. An analysis of land use conflict potentials based on ecological-production-living function in the southeast coastal area of China. *Ecol. Indic.* **2021**, *122*, 107297. [CrossRef]

28. Dong, G.; Ge, Y.; Zhu, W.; Qu, Y.; Zhang, W. Coupling Coordination and Spatiotemporal Dynamic Evolution Between Green Urbanization and Green Finance: A Case Study in China. *Front. Environ. Sci.* **2021**, *8*, 621846. [CrossRef]

29. Deng, Y.; Qi, W.; Fu, B.J.; Wang, K. Geographical transformations of urban sprawl: Exploring the spatial heterogeneity across cities in China 1992-2015. *Cities* **2020**, *105*. [CrossRef]

30. Gao, B.; Huang, Q.X.; He, C.Y.; Sun, Z.X.; Zhang, D. How does sprawl differ across cities in China? A multi-scale investigation using nighttime light and census data. *Landsc. Urban Plan.* **2016**, *148*, 89–98. [CrossRef]

31. Li, X.; Zhou, W.; Ouyang, Z. Forty years of urban expansion in Beijing: What is the relative importance of physical, socioeconomic, and neighborhood factors? *Appl. Geogr.* **2013**, *38*, 1–10. [CrossRef]

32. Zhang, Q.W.; Su, S.L. Determinants of urban expansion and their relative importance: A comparative analysis of 30 major metropolitans in China. *Habitat Int.* **2016**, *58*, 89–107. [CrossRef]

33. Zhou, D.; Xu, J.; Lin, Z. Conflict or coordination? Assessing land use multi-functionalization using production-living-ecology analysis. *Sci. Total Environ.* **2017**, *577*, 136–147. [CrossRef] [PubMed]

34. Zhang, Z.; Li, Y. Coupling coordination and spatiotemporal dynamic evolution between urbanization and geological hazards–A case study from China. *Sci. Total Environ.* **2020**, *728*, 138825. [CrossRef]

35. Hasanagas, N.D. Network Analysis Functionality in Environmental Policy:Combining Abstract Software Engineering with Field Empiricism. *Int. J. Comput. Commun. Control.* **2011**, *6*, 622–635. [CrossRef]

land

MDPI

Article

Optimizing the Production-Living-Ecological Space for Reducing the Ecosystem Services Deficit

Xinxin Fu [1], Xiaofeng Wang [2,3,*], Jitao Zhou [1] and Jiahao Ma [1]

[1] School of Earth Science and Resources, Chang'an University, Xi'an 710054, China; 2019127009@chd.edu.cn (X.F.); 2020127005@chd.edu.cn (J.Z.); 2020127004@chd.edu.cn (J.M.)
[2] School of Land Engineering, Chang'an University, Xi'an 710054, China
[3] The Key Laboratory of Shaanxi Land Consolidation Project, Chang'an University, Xi'an 710054, China
* Correspondence: wangxf@chd.edu.cn

Abstract: With rapid urbanization and industrialization, China's metropolises have undergone a huge shift in land use, which has had a profound impact on the ecological environment. Accordingly, the contradictions between regional production, living, and ecological spaces have intensified. The study of the optimization of production-living-ecological space (PLES) is crucial for the sustainable use of land resources and regional socio-economic development. However, research on the optimization of land patterns based on PLES is still being explored, and a unified technical framework for integrated optimization has yet to be developed. Ecosystem services (ES), as a bridge between people and nature, provide a vehicle for the interlinking of elements of the human-land system coupling. The integration of ES supply and demand into ecosystem assessments can enhance the policy relevance and practical application of the ES concept in land management and is also conducive to achieving ecological security and safeguarding human well-being. In this study, an integrated framework comprising four core steps was developed to optimize the PLES in such a way that all ecosystem services are in surplus as far as possible. It was also applied to a case study in the middle and lower reaches of the Yellow River Basin. A regression analysis between ES and PLES was used to derive equilibrium thresholds for the supply and demand of ES. The ternary phase diagram method was used to determine the direction and magnitude of the optimization of the PLES, and finally, the corresponding optimization recommendations were made at different scales.

Keywords: land-use transition; production-living-ecological space; spatial mismatch; balance threshold; ES management strategies

Citation: Fu, X.; Wang, X.; Zhou, J.; Ma, J. Optimizing the Production-Living-Ecological Space for Reducing the Ecosystem Services Deficit. *Land* **2021**, *10*, 1001. https://doi.org/10.3390/land10101001

Academic Editors: Dong Jiang, Jinwei Dong and Gang Lin

Received: 18 August 2021
Accepted: 17 September 2021
Published: 23 September 2021

1. Introduction

Since the 20th century, along with the acceleration of global urbanization and industrialization, the continued large-scale exploitation of land resources has been accompanied by environmental problems, such as the crowding of ecological space by urban construction land, atmospheric pollution, water pollution, and ecological imbalance [1–3]. Since the reform and opening up of China in 1978, urbanization and industrialization have advanced rapidly. At the end of 2018, 59.6% of China's land was urbanized, and China has entered a period of steady urbanization [4,5]. In this context, structural imbalances in land use have come to the fore, the contradiction between production-living-ecological space (PLES) has become increasingly prominent, and land use is facing enormous pressure and challenges [6,7]. Therefore, to promote regional sustainable development and the effective and efficient application of land space, it is necessary to reasonably allocate limited spatial resources [8–10]. Integrating the spatial functions and land use structure under the PLES linkage and promoting the coordinated development of the quantitative structure and spatial layout of the PLES has become an urgent issue to be addressed [11,12].

Ecosystem services (ES), as a bridge between natural ecosystems and human well-being, are the various benefits that humans derive directly or indirectly from ecosystems [13,14].

Ecosystem services depend on the interactions and feedback between ecological and socio-economic factors [15,16]. Humans manage ecosystem processes by modifying ecosystem components and structures to provide ES that better meet their needs [17,18]. A common way of doing this is improving human well-being by changing land use/cover, e.g., converting other lands to cropland can improve food production service. The conversion of arable land and grassland into forest land can improve water yield services, soil conservation services, and carbon sequestration services, etc. [19]. The gradual deepening and formation of the concept of ES provides a vehicle for interlinking the elements of human-land system coupling [20,21]. As a result, the concept of ES is now becoming increasingly important at the land management level [22,23].

With the deepening of ES research, a large number of researchers have begun to focus on both the ES supply (i.e., the capacity of ecosystems to provide ecosystem goods and services to humans) and the ES demand (i.e., the sum of ecosystem goods and services used or consumed by humans) [24,25]. The gradual intensification of global climate change, environmental pollution, and human-land conflicts have led to changes in ecosystem structure and function, affecting the supply capacity of ES [26,27]. Meanwhile, the increasing level of urbanization and industrialization has led to the emergence of a large number of ES demand aggregation centers [28,29], further exacerbating the mismatch between ES supply and demand. Incorporating ES supply and demand into ecosystem assessments can improve the policy relevance and practical application of the ES concept in land management. It is also conducive to achieving ecological security and safeguarding human well-being [30,31].

There is often a desire to maximize the ES supply through land management to reduce mismatches and shortages, but a major challenge is to integrate analysis to avoid unnecessary trade-offs in ES [32,33]. In this context, exploring spatial mismatches between ES supply and demand associated with urbanization-related land use is crucial for the proper integration of ES into land management strategies [34]. Many studies have considered both ES supply and demand and have identified potential mismatches between ES supply and demand at multiple scales [35]. The challenge is that most ES assessments have not yet been effective in influencing land management decisions and, in particular, lack holistic considerations [36,37]. The PLES covers the spatial range of activities of human social life and is the basic vehicle for human economic and social development [38]. The three are both independent and interrelated, with symbiotic integration and constraining effects, and the collaboration of PLES functions can produce a synergistic effect in which the overall function is greater than the sum of the partial functions [39]. The PLES optimization belongs to the problem of optimizing the allocation of national land resources. Based on land characteristics and land-use system principles, the structure and direction of land resource use are arranged, designed, combined, and laid out at a hierarchical level on a spatial and temporal scale to improve the efficiency and effectiveness of land use, maintain the relative balance of land ecosystems, and realize the sustainable use of land resources [40]. The main theoretical support for current PLES optimization comes from the theory of regional resource and environmental carrying capacity and the theory of coupling urbanization and ecological environment [41,42]. This study focuses on the consideration from the perspective of ecosystem services, and it is a new attempt to apply the assessment of ecosystem service supply and demand to the optimization of PLES.

The Yellow River Basin, an important ecological barrier, straddles three regions in the east, central, and west of China and is an ecological corridor connecting the Qinghai-Tibet Plateau, the Loess Plateau, and the North China Plain. Although breakthroughs in ecological construction and environmental management have been made in the Yellow River Basin in recent years, the fragile ecological environment, water scarcity, and water environment problems are outstanding. especially in the middle and lower reaches of the Yellow River, which have undergone rapid urbanization over the past decades, leading to huge changes in the spatial pattern of land use, accompanied by huge landscape changes and related degradation of ES. Therefore, this study takes the Yellow River basin as an example to explore the spatial patterns of ES supply and demand and the response to PLES changes and to identify optimization areas through response thresholds to provide optimization strategies for land use at multiple scales.

Land use optimization is a complex concept [43]. This study has envisaged an ideal area where optimal PLES management reduces ES deficits and mismatches, to which the land use pattern of the remaining areas should be as close as possible. The basic optimization four steps included: (1) classifying production-living-ecological spaces based on land use types; (2) choosing key ES, quantifying ES supply and demand, and identifying spatial mismatches; (3) identifying the impact of PLES on the spatial mismatch of ES and thresholds; (4) determining the direction of optimization and proposing optimization solutions for different spatial scales.

The selection of key ES was based on the following principles: (1) spatially quantifiable mapping; (2) consistent with the focus of regional governments and residents; (3) better representation of the coupling mechanisms between different ES; (4) availability of measurement data. In this study, the carbon sequestration service, water yield service, soil conservation service, and grain production service were selected as indicators for measuring the ES supply and demand in the Yellow River Basin to minimize the deficit and mismatch of these four ES and carry out corresponding PLES optimization.

2. Materials and Methods

2.1. Study Area

The study area was in the middle and lower reaches of the Yellow River basin ($103°36'$–$119°55'$ E, $41°03'$–$32°46'$ N), at an altitude of about 0–4082 m. The area was located in the central-eastern part of China (Figure 1). The main provinces involved included Inner Mongolia Autonomous Region, Henan Province, Shaanxi Province, Shanxi Province, Qinghai Province, and Gansu Province. The middle and lower reaches of the Yellow River Basin are dominated by plains and hills. The region has a temperate continental climate and a temperate monsoon climate. The middle and lower reaches of the Yellow River Basin have abundant light, high temperature, abundant precipitation, are suitable for crop growth, and are the main production areas for agricultural products. The Yellow River Basin is an important economic zone and an important base for energy, chemicals, raw materials, and basic industries in China.

2.2. Data Sources

In this study, we used data from five different sources. (1) Meteorological elements and daily precipitation for 2000, 2010, and 2018, supplied by the China Meteorological Data Sharing Network (http://data.cma.cn/ (accessed on 1 July 2021)), were batch interpolated using the professional meteorological interpolation software ANUSPLIN. (2) Monthly NDVI data for 2000, 2010, and 2018 at a spatial resolution of 1 km × 1 km, supplied by the Geospatial Data Cloud (http://www.gscloud.cn/ (accessed on 1 July 2021)), and annual NDVI data were obtained using the maximum synthesis method. (3) Population data for 2000, 2010, and 2018 at a spatial resolution of 1 km × 1 km was supplied by WorldPop (https://www.worldpop.org/ (accessed on 1 July 2021)). (4) Land use data for 2000, 2010, and 2018 at a spatial resolution of 1 km × 1 km was supplied by the Resource and Environmental Science and Data Centre (http://www.resdc.cn/ (accessed on 1 July 2021)). (5) Grain production, energy consumption, and water consumption by the municipality for 2000, 2010, and 2018 was obtained from the statistical yearbooks and water resources bulletins of each province.

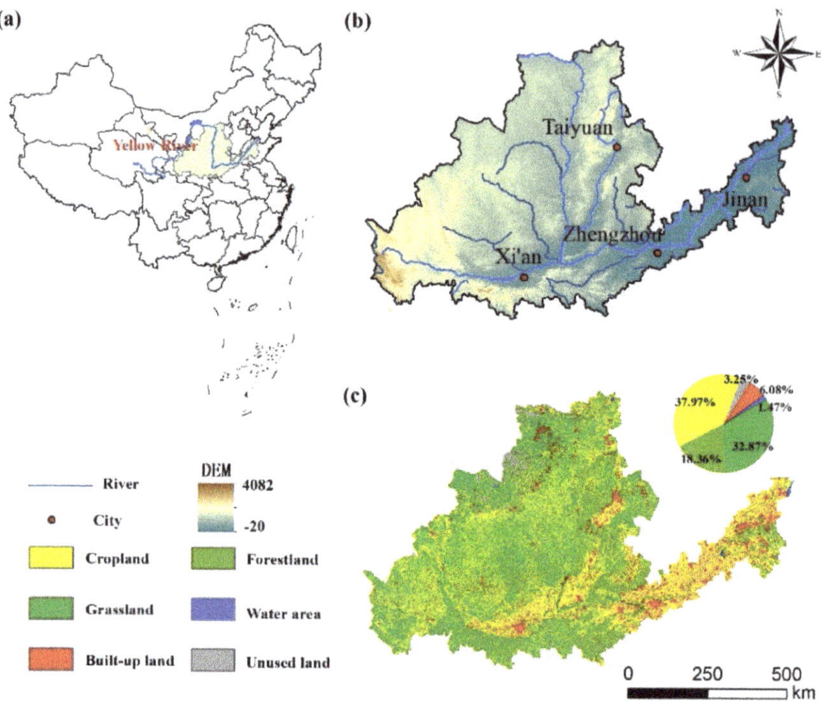

Figure 1. (a–c) Location, elevation, and 2018 land use of the study area.

2.3. Framework for Optimizing Production-Living-Ecological Space

Based on previous approaches and frameworks for land use optimization [23,44], this study identified ideal land-use patterns and optimized PLES at different scales by quantifying the mismatch between the supply and demand of ES associated with PLES. This was achieved through the following four core steps (Figure 2): In the first step, the composition, configuration, and spatial transition of PLES were analyzed based on land use data in the Yellow River Basin during 2000, 2010, and 2018. This step aims to examine the spatial changes in PLES in the Yellow River Basin during 2000, 2010, and 2018, and establish a basis for subsequent research. In the second step, those ES suitable for the Yellow River Basin were selected based on the basic principles for the selection of ES proposed above to assess the ES supply and demand. Mismatches and shortages between ES supply and demand were also identified. In the third step, based on the correlation between the ratio of production/living/ecological space and the supply and demand of ES, the thresholds were identified when ES supply and demand were imbalanced. This step aims to analyze the links that exist between the two, the most central part of which is the identification of thresholds. In the fourth step, the thresholds identified in the previous step were used to identify optimization areas using ternary phase diagrams, which were then optimized for PLES at different scales.

Step 1: The classification of production-living-ecological space based on land use types.

Production space is mainly the area that provides various products or services for people. Living space refers to the area that provides the function of carrying and guaranteeing human habitation and provides the function of residence, consumption, leisure, and recreation in the country. Ecological space refers to the area that can provide an ecological barrier and has the function of regulating the atmosphere, concealing water, and maintaining soil and water [11,12]. In this study, a classification system for PLES in the Yellow River

Basin was constructed based on geographical features and previous research results [45] (Table 1).

Table 1. Production-living-ecological space classification in the Yellow River Basin.

	LUCC Classification System
Production space	Paddy land (11), dry land (12), transport, industrial, and mining construction land (53)
Living space	Urban sites (51), rural settlements (52)
Ecological space	Wooded land (21), shrubland (22), high cover grassland (31), medium cover grassland (32), rivers and canals (41), lakes (42), reservoir ponds (43), open woodland (23), other woodlands (24), permanent glacial snow (44), mudflats (46) cover grassland (33), sandy land (61), gobi 62), saline land (63), marshland (64), bare land (65), bare rocky ground (66)

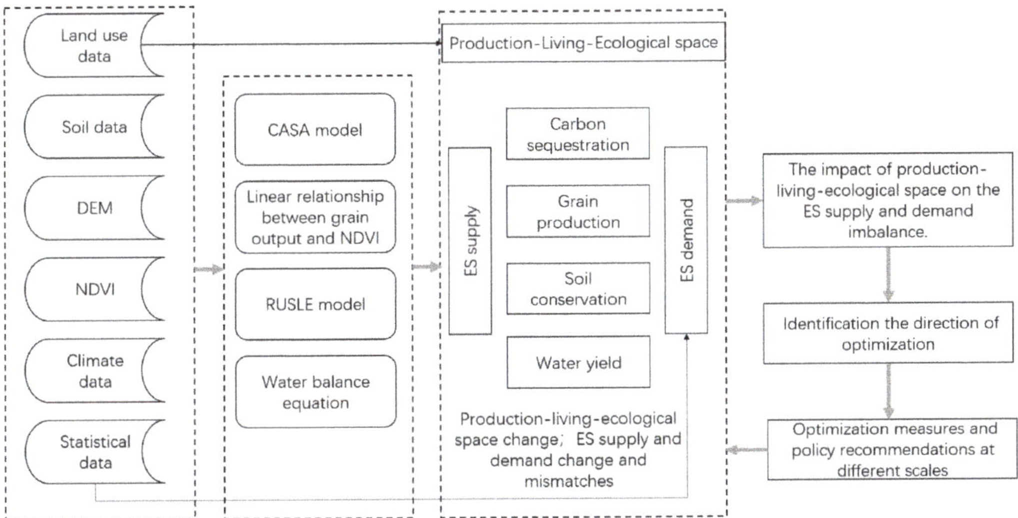

Figure 2. Optimization framework for production-living-ecological space.

Step 2: Quantify ES supply and demand and identify spatial mismatches.

(1) Quantifying ES supply and demand

Water yield service are the ability of an ecosystem to intercept or store water resources from rainfall while mitigating ground runoff [46]. The Yellow River Basin is an important water source in northwest China and assessing the water yield service is of great practical importance for the rational use and conservation of water resources. The water balance model was used to calculate the supply of water yield service [47]. The amount of water consumed per capita in each city in the Yellow River Basin was obtained from data on industrial, agricultural, and domestic water consumption and the resident population of each city and combined with data on population density to obtain the demand for water yield services. The formulas are as follows.

$$WY_{(x)} = P_{(x)} - ET_{(x)} \tag{1}$$

$$ET_{(x)} = \frac{P_{(x)}\left(1 + PET_{(x)}/P_{(x)}\right)}{1 + \omega_{(x)}\left(PET_{(x)}/P_{(x)}\right) + \left(PET_{(x)}/P_{(x)}\right)^{-1}} \tag{2}$$

$$D_w = D_{pw} \times \rho_{pop} \tag{3}$$

where $WY_{(x)}$ represents the annual average yield at pixel x (m^3); $P_{(x)}$ is the average annual precipitation on pixel x (mm); $ET_{(x)}$ is the actual annual evapotranspiration at pixel x (mm); $PET_{(x)}$ is based on the Penman-Monteith formula [48]. D_{wy} is water demand, which in this case equates to water consumption (m^3); D_{pw} is water consumption per capita (m^3/person), which includes water consumption for agricultural, industrial, domestic, and ecological purposes; ρ_{pop} is the local resident population density (person/km^2).

Grain production service, as one of the basic ecological services, plays a vital role in human survival and development [49]. There is a significant linear relationship between crop and livestock production based on the NDVI. The total production of grain was allocated according to the ratio of raster NDVI values to total arable land NDVI values, which in turn characterized the grain production capacity of each raster. Grain demand was estimated by multiplying the per capita grain demand by the population density [17]. The formulas are as follows:

$$GP_{(x)} = GP_{sum} \times \frac{NDVI_x}{NDVI_{sum}} \tag{4}$$

$$D_g = D_{pg} \times \rho_{pop} \tag{5}$$

where $GP_{(x)}$ is the total production of grain for grid x (t/km^2); GP_{sum} is the production of grain products for each province (t); $NDVI_x$ is the normalized difference vegetation index for grid x; $NDVI_{sum}$ is the sum of NDVI of cropland for each province; D_g is the grain demand (t/km^2); D_{pg} is the annual per capita grain consumption (t/person); ρ_{pop} is the resident population density (person/km^2).

Carbon sequestration services are important regulatory services in ecosystems. The CASA model is a common model for calculating NPP due to its high calculation accuracy and easy-to-access data and parameters [50]. The carbon sequestration demand was calculated from the product of population density, per capita energy consumption, and energy carbon conversion rate, where energy consumption was obtained from the statistical yearbooks of the Yellow River Basin provinces. The formulas are as follows:

$$NPP(x, t) = APAR(x, t) \times \varepsilon(x, t) \tag{6}$$

$$D_c = D_{pc} \times \rho_{pop} \tag{7}$$

where NPP is the net primary productivity of the pixel x at time t (gC/m^2·a); $APAR$ is the Absorbed Photosynthetic Active Radiation (MJ/m^2·a), which is estimated from the ratio of total solar radiation (SOL) to absorbed photosynthetically active radiation (FPAR); ε is the efficiency of conversion of photosynthetically active radiation to organic carbon (gC/MJ^2), which is estimated by maximum light energy utilization (0.389 gC/MJ^2), temperature stress (Tε), and water stress (Wε). D_c represents carbon sequestration demand (t/km^2); D_{pc} is the annual per capita carbon consumption (t/person); ρ_{pop} is the resident population density (person/km^2).

Soil conservation service reduces soil erosion and restores soil fertility, which is critical to agricultural production [51]. This study quantified the soil conservation service supply based on the classical revised universal soil loss equation (RUSLE). The ecosystem service demand is the number of ecological goods that humans expect to be able to obtain from an ecosystem. Since actual soil erosion causes unwanted human losses and humans expect to manage these actual amounts of soil erosion, the actual amount of soil erosion is defined as the soil conservation service demand. The formulas are as follows:

$$SC = R \times K \times LS \times (1 - C \times P) \tag{8}$$

$$D_s = R \times K \times LS \times C \times P \tag{9}$$

where SC is soil conservation; D_s is actual soil erosion; R is the precipitation erosion factor; K is the soil erosion factor; LS is slope length factor; P is soil conservation factor; C is vegetation cover factor [52].

(2) ES supply and demand mismatches and shortfalls

The supply and demand of ES are significantly spatially heterogeneous and are reflected in spatial mismatches. The state of ES supply and demand can be characterized by the ecological supply-demand ratio (ESDR), which can be used to reveal the nature of surpluses or deficits [35,53].

$$ESDR = \frac{ESS - ESD}{(ESS_{max} + ESD_{max})/2} \tag{10}$$

where ESS and ESD refer to the ES supply and demand, respectively; ESS_{max} is the maximum value of the ES supply; ESD_{max} is the maximum value of the ES demand. $ESDR$ > 0 indicates a surplus, $ESDR$ = 0 indicates a balanced ES supply and demand, and $ESDR$ < 0 indicates a deficit.

Step 3: The impact of production-life-ecological space on the ES supply and demand imbalance.

The ESDRs for the four major ES in 2000, 2008, and 2018 were calculated by the above method, while the production space ratio/living space ratio/ecological space ratio at the 1 km grid scale was calculated based on 30 m land use data. The data were statistically graded for the years 2000, 2010, and 2018, and then least squares regression analysis was conducted via SPSS to plot the trend line between ESDR and production space ratio/living space ratio/ecological space ratio to indicate negative or positive effects and significance levels. Spatial land management thresholds (i.e., the ratio of production-living-ecological land when there is a deficit in the ES) were then calculated based on the results of the regression analysis.

Step 4: Identification of the direction of optimization and policy recommendations.

A ternary diagram is a type of center of gravity diagram that has three variables but requires the sum of the three to be constant. In an equilateral triangular coordinate system, the position of a point in the diagram represents the proportional relationship between the three variables. In this study, the same ternary was used to visually express the ratio of production-living-ecological space, which was used to identify the optimization area with the main optimization direction, where the ratio occupied by this type of land at the endpoint is 100%. The regions are divided according to the thresholds determined above. The projection of units of different scales is performed, and when the projection falls in the ideal region, it means that the unit does not need to be optimized, and when it falls in other regions, the direction and quantity relationship of optimization can be determined based on the direction and distance from the ideal region.

3. Results

3.1. Structure and Transition of PLE Land Use

The Yellow River basin was mainly dominated by ecological space, with the percentage of ecological space being 55.73%, 55.97%, and 55.99% in 2000, 2010, and 2018, respectively, showing an increasing trend (Figure 3). The northwestern and southern parts of the study area were relatively less densely populated and had a lower level of urbanization and were therefore dominated by ecological space. The percentage of production space was 40.79%, 39.38%, and 38.91% in 2000, 2010, and 2018, respectively, showing a decreasing trend. Production space was mainly located in the eastern coastal areas of the study area, which have better water and heat conditions and are also conducive to crop growth. The region is economically developed, highly urbanized, with a high level of human activity and is a major industrial center and food producer.. The percentage of living space increased from 3.48% in 2000 to 5.10% in 2018. Spatially, living space was mainly distributed around the main cities in the study area, showing a tendency to spread outwards. The chord diagram suggests the scale of transfer of different land uses. The direction shown by the arrow represents the direction of transfer of the land, and the width of the arrow represents the proportion of the area transferred (Figure 3). According to the area conversion of PLES from 2000 to 2018, the largest area of production land was converted outwards, with a total of 35,300 km², of which 10,800 km² was converted to living land and 24,500 km² was converted to ecological land. The smallest area of living land was converted outwards, with a total of 3720 km², of which 3280 km² was converted to production land and 433 km² to ecological land. Ecological land converted mainly into productive land was 21,500 km², while converted into living land was 1940 km².

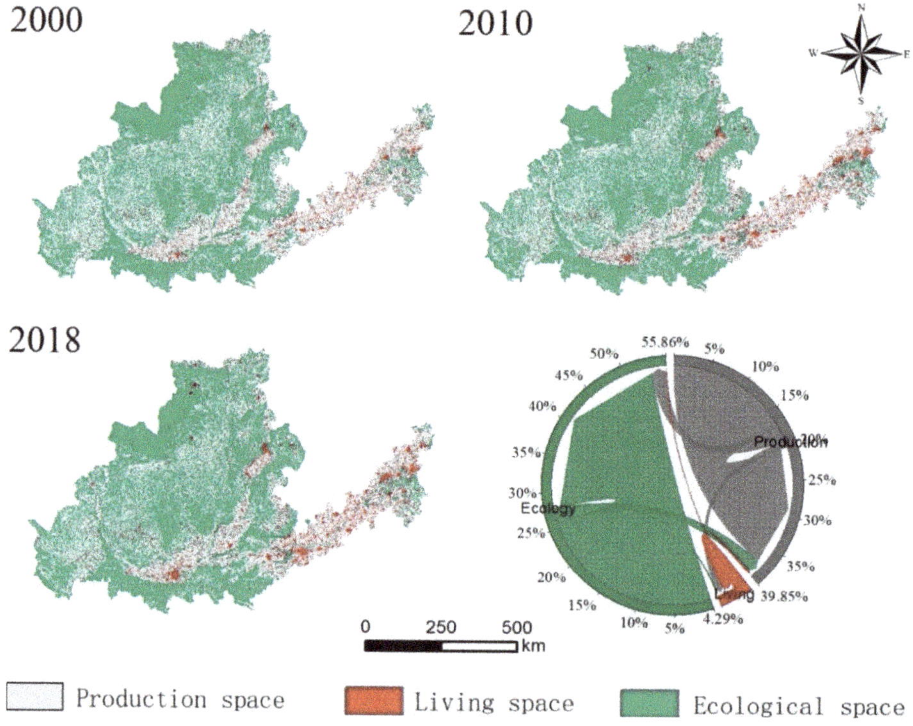

Figure 3. Structure and transition of production-living-ecological space in the Yellow River Basin from 2000 to 2018.

3.2. ES Supply and Demand Change and Mismatches

3.2.1. Water Yield Service

Over the entire period, the total water yield service supply exceeded demand, with surpluses of 47.72 billion m³, 60.31 billion m³, and 41.88 billion m³ in 2000, 2010, and 2018, respectively (Table 2). Water yield service supply in 2000, 2010, and 2018 was 72.32 billion m³, 84.63 billion m³, and 66.90 billion m³, respectively, showing a trend of increase followed by a decrease. The water demand increased significantly from 24.60 billion m³ in 2000 to 25.02 billion m³ in 2018, an increase of 1.69%.

Water yield service supply was strongly influenced by precipitation and evapotranspiration, while water demand was influenced by population density and industrial structure. Precipitation anomalies can increase the uncertainty of the spatial match of water yield service. Although there was an overall surplus of water service, the spatial distribution of water yield service supply and demand also showed a mismatch (Figure 4). The southern and eastern parts of the study area were the main areas of water yield service supply

(Figure 5), but the deficit situation of the water yield service was still significant due to the dense population and agricultural development of the area, which means that there is a huge demand for water resources. Due to the lower water yield service supply in 2018, this has resulted in a significant deficit in water yield service in the South East, with the shortfall areas mainly in the city center.

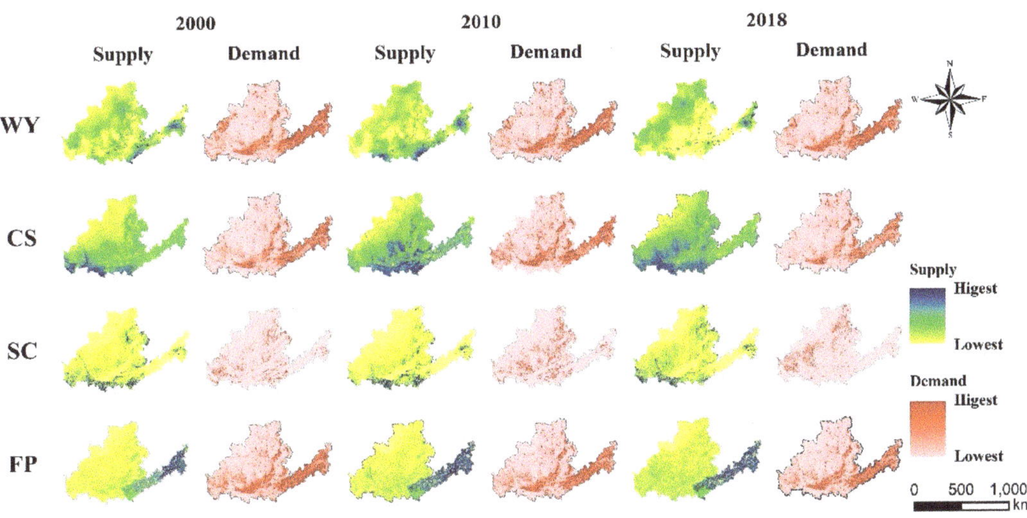

Figure 4. Spatial pattern of supply and demand for individual ecosystem services (ES) in the Yellow River Basin in 2000, 2008, and 2018. ES indicators are: WY—water yield; CS—carbon sequestration; SC—soil conservation; GP—grain production.

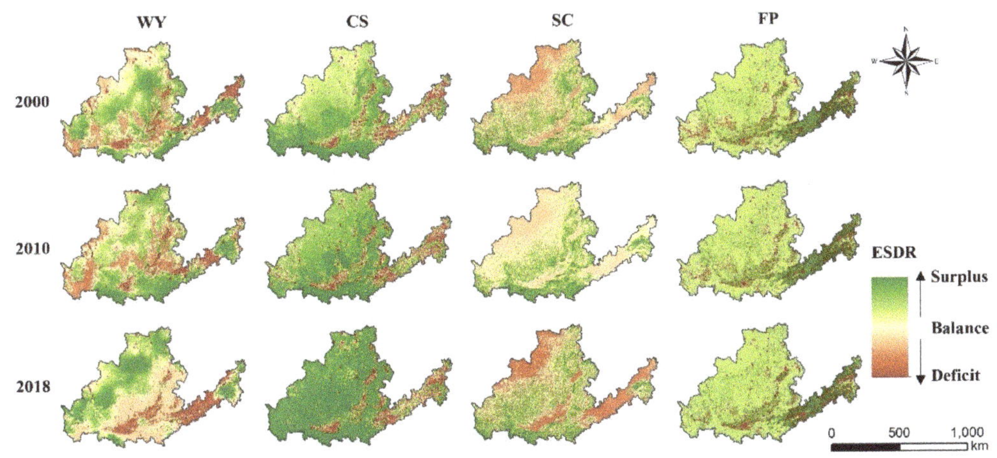

Figure 5. Spatial pattern of the ecosystem service supply-demand ratio in the Yellow River Basin in 2000, 2008, and 2018.

3.2.2. Carbon Sequestration Service

Carbon sequestration service supply has shown an increasing trend, from 134.45 million tons in 2000 to 235.76 million tons in 2018, an increase of 75.35% (Table 2). However, the growth in the carbon sequestration service supply did not cause surpluses. Carbon sequestration demand in 2000, 2010, and 2018 was 134.45 million tons, 210.75 million tons, and 235.76 million tons, respectively, and the carbon sequestration demand in 2010 and

2018 exceeded the carbon sequestration supply with a deficit of 29.66 million tons and 49.48 million tons, respectively, with an upward trend.

Table 2. Ecosystem services supply and demand in the Yellow River Basin in 2000, 2010, and 2018.

Year	Water Yield ($m^3/km^2/a$)		Carbon Sequestration ($tc/km^2/a$)		Soil Conservation ($t/ha/a$)		Grain Production ($t/km^2/a$)	
	Supply	Demand	Supply	Demand	Supply	Demand	Supply	Demand
2000	128,995.41	43,882.20	239.82	134.75	71.80	8.96	223.92	91.52
2010	150,950.80	43,368.23	375.92	428.83	110.53	10.58	326.43	98.28
2018	119,332.21	44,622.76	420.52	508.78	90.39	6.92	382.62	104.55

According to the spatial distribution of carbon sequestration service supply and demand (Figure 4), higher carbon sequestration service supply was mainly concentrated in the south, showing an increasing trend, followed by a decreasing trend. Carbon sequestration supply in the north-western region was relatively low and shows an increasing trend. Higher carbon sequestration service demand was mainly in the main urban area downstream of the study but showed a decreasing trend from 2000 to 2018. There was a clear spatial mismatch in sequestration service (Figure 5), with increased surpluses in the central and western regions of the study area and relatively significant deficits in the eastern regions. The main urban areas around the study area showed a significant deficit in carbon sequestration service, with Zhengzhou showing an increase in the deficit position in 2018.

3.2.3. Soil Conservation Service

From 2000 to 2018, soil conservation service supply exceeded the demand, and both showed an increasing trend (Table 2). The surplus of soil conservation services increased significantly from 3.52 billion tons in 2000 to 5.6 billion tons in 2010 and 4.67 billion tons in 2018. Soil conservation services, as an in-situ service, i.e., one that is generated in situ and benefits in situ, have an aggregate surplus that hardly offsets their spatial mismatch.

In terms of the spatial distribution of the soil conservation services ESDR, the deficit areas were concentrated in the north-central region of the study area and the downtown area in the east (Figure 4). The spatial mismatch in soil conservation services was mainly due to: (1) the north-central region being a loess plateau area, which is very weak for soil and water conservation due to the undulating terrain, loose soil, and poor vegetation cover; (2) the eastern city center area, with strong human activity and high population density in the area, which has led to a reduction in vegetation area. The land-use types are mainly urban land, rural settlements, and other construction lands, which have a poor soil conservation capacity, thus leading to a deficit in soil conservation services.

3.2.4. Grain Production Service

Grain production service increased from 125.54 million tons in 2000 to 214.51 million tons in 2018, an increase of 70.87%. During the same period, grain production demand exhibited an increase of 14.24% from 51.30 million tons in 2000 to 58.62 million tons in 2018 (Table 2). Thus, there was a clear surplus for grain production service, and this surplus showed an increasing trend, from 74.23 million tons in 2000 to 155.90 million tons in 2018.

Despite the overall surplus in grain production service, there were still some spatially mismatched centers (Figure 5). The southeastern part of the study area has relatively good hydrothermal conditions and is a major grain producer, hence the high grain supply. At the same time, the grain demand was relatively high due to the high level of human activity and the relatively high population density in the area. The region's grain production service showed a surplus, indicating that its production capacity was greater than its consumption capacity.

3.3. Influence of PLES Changing on ESDR

3.3.1. Influence of Production Space Changing on ESDR

There was a significant negative impact of production space on the ESDR for carbon sequestration service, soil conservation service, and water yield service during 2000–2018 ($p < 0.01$). The production space explained most of the variance in ESDR for carbon sequestration services, soil conservation services, and water production services, at 87%, 84%, and 48%, respectively (Figure 6a). For the carbon sequestration service, when the production space ratio exceeded 62.83%, the carbon sequestration service swas in deficit. Therefore, to ensure that the carbon sequestration service supply is greater than the demand, it is necessary to ensure that the ratio of production space is less than 62.83%. For the soil conservation service, the range was 0–98.73%. For the water yield service, there was no significant threshold effect due to the limited influence of production space ($k = -0.02$). Production space had a significant positive effect on the ESDR of grain production service and explained most of the variation in the ESDR for the grain supply service, at 89%. When the production space ratio exceeded 18.2%, grain production services supply exceeded the demand. In summary, it is necessary to ensure that the production space ratio in the study area is between 18.2% and 62.83% to ensure that all ecosystem services are in surplus.

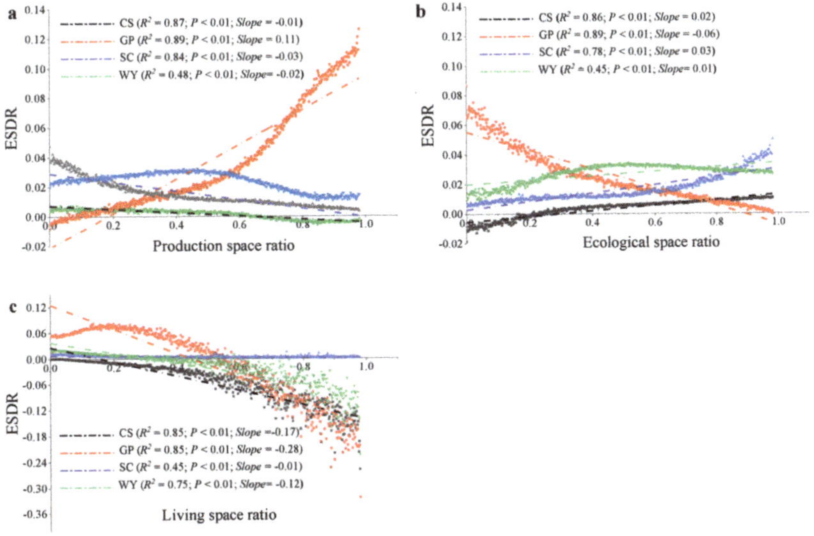

Figure 6. Influence of production space ratio, living space ratio, and ecological space ratio on ESDR in the Yellow River Basin. (**a**). Production space ratio (**b**). Ecological space ratio (**c**). Living space ratio.

3.3.2. Influence of Living Space Changing on ESDR

Living space had a significant negative influence ($p < 0.01$) on water yield service, grain production service, and carbon sequestration service (Figure 6c). For water production services, when the ratio of living space was greater than 31.35%, there was a deficit. This means that the water yield service supply was greater than the demand if the ratio of living space was less than 31.35%. For the carbon sequestration service and grain production service, this threshold was 15.68% and 44.39%, respectively.

Although the ESDR of living space on soil conservation services was negative, the trend was not strong ($k = -0.01$) and explained only part of the variation in soil conservation services ($R^2 = 0.45$). The influence of the living space ratio was not significant, i.e., the soil conservation service was in surplus for any value of the living space ratio between 0–100%. In summary, the living space ratio between 0–15.68% is needed to ensure that all ecosystem services are in surplus in the study area. Due to the limited number of ES selected in this study, this resulted in a significant negative effect of living space on all ES. For some ES, such as landscape aesthetics, there is a dependence on living space, and too little living space will inevitably affect the supply of these ES.

3.3.3. Influence of Ecological Space Changing on ESDR

Ecological space had a significant positive influence on water yield service, soil conservation service, and carbon sequestration service during the period 2000–2018 (Figure 6b). Ecological space explained most of the variation in the ESDR for the soil conservation service and carbon sequestration service, at 78% and 86%, respectively. When the ecological space ratio was less than 29.79%, the carbon sequestration service was a deficit. This threshold did not exist for water yield service or soil conservation service. The ESDR of ecological space on grain production service was negative ($p < 0.01$). When the ecological space ratio was greater than 88.61%, there was a deficit in the grain production service. In summary, an ecological space ratio of 29.79% to 88.61% is needed to ensure that all ecosystem services are in surplus. However, this does not mean that ecological space can be expanded indefinitely, as too much ecological space can squeeze the original production space and lead to a deficit in grain production service.

4. Discussion

4.1. Identification of Optimization Directions in PLES

The ternary phase diagram can visually represent the ratio of PLES in any region, where the endpoints of the triangle indicate a production/living/ecological space ratio of 100% (Figure 7). According to the threshold value of the impact of PLES on the ES supply and demand imbalance, when the ratio of living space is less than 15.68%, the ratio of production space is between 18.2 and 62.83%, and the ratio of ecological space is between 29.79% and 88.61%, while the ES involved in this study are all in surplus. Accordingly, an ideal area, i.e., an area that does not need to be optimized, can be obtained. When the projection of an area falls within the ideal area, it means that this type of area does not need to be optimized, while the rest of the area needs to be optimized to varying degrees, depending on its location. The PLES of a region can be adjusted according to the range in the diagram where the ratio of PLES of any region falls. The greater the distance from the ideal area, the greater the area of change required in the land use pattern of the region.

The size of the ideal area is usually related to the ES selected. The more ES selected, the smaller the ideal area will be in response, meaning that more area will need to be adjusted. The ES can therefore be adjusted in the actual management process to suit the needs of the policymaker accordingly.

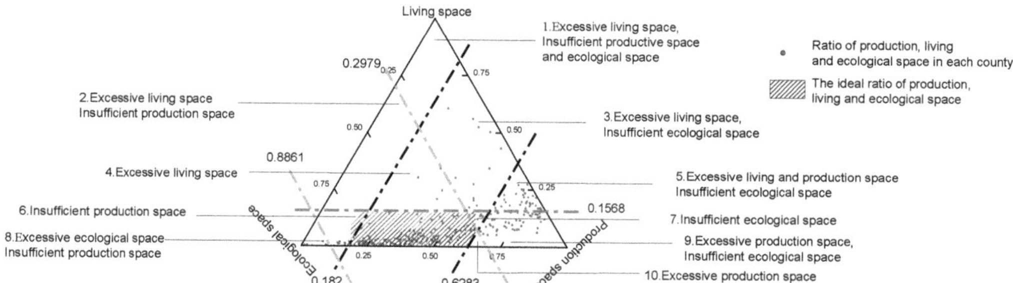

Figure 7. Ternary phase diagram of production-living-ecological space.

4.2. Optimization Measures and Policy Recommendations at Different Scales

The distribution of PLES was highly spatially heterogeneous, which led to a spatial mismatch in ES. Therefore, the PLES optimization should be coupled with optimization measures at different scales to determine the best measures for ecosystem service management. In this study, the ratios of PLES at different scales were counted, and this was used to obtain optimization measures at multiple scales (Figure 8).

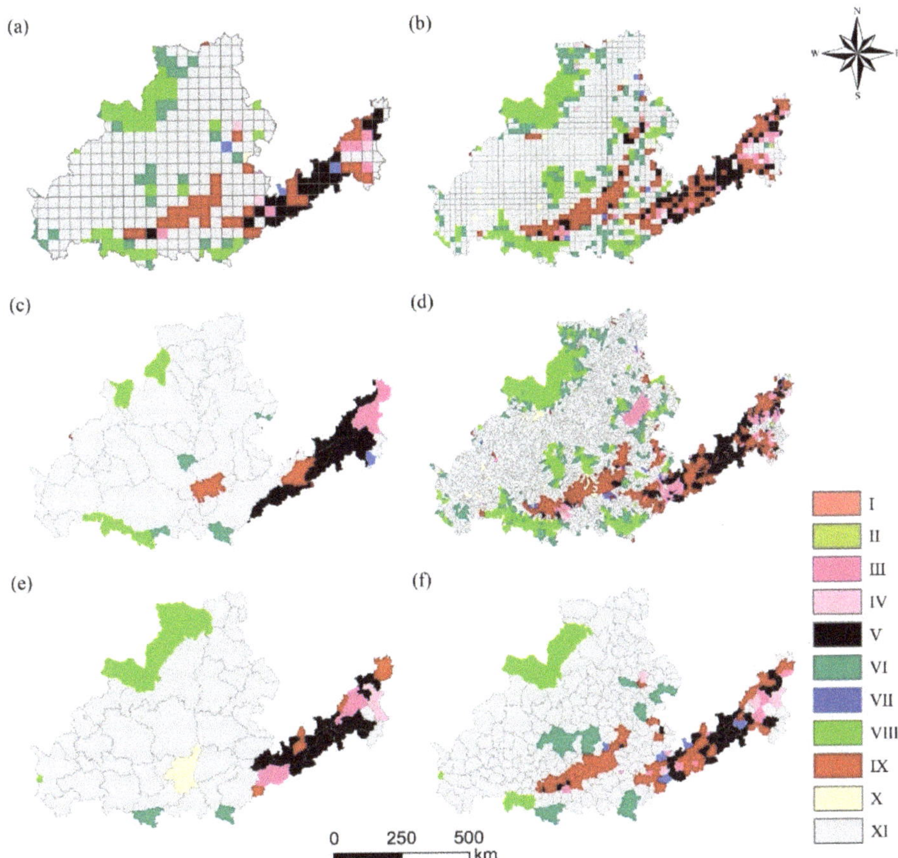

Figure 8. Optimization of production-living-ecological space on multiple scales ((**a**). the 40 × 40 km grid scale; (**b**). the 20 × 20 km grid scale; (**c**). the primary catchment scale; (**d**). the secondary catchment scale; (**e**). the city scale; (**f**). the county scale. TheIto XI corresponds to Figure 7 numbering).

At the grid-scale, there was too much ecological space and not enough production space in the north-western part of the study area, and there is a need for conversion of ecological to production space, such as converting unused land in the area to industrial land. The central part of the study area had too much production space and not enough ecological space, so there is a need to convert the production land in the area to ecological land, such as implementing a system of returning farmland to forest or converting farmland unsuitable for cultivation to forest land. The eastern part of the study area, i.e., the lower reaches of the Yellow River Basin, was mainly characterized by an excess of living space and production space and a shortage of ecological space, so it is necessary to shift the production/living space towards ecological space, increasing the ratio of ecological space and reducing living space. For example, increase the woodland and grassland and reduce the rate of urbanization development. At the primary watershed scale, most of the Midwest was in the ideal mode of PLES, i.e., it did not need to be optimized. There was an excess of productive space in the eastern region and an excess of living space in the coastal region. At the secondary watershed scale, the central and western regions of the study area had more watersheds that need to be optimized, and they behaved in much the same way spatially as at the grid-scale. At the city scale, Erdos had too much ecological space and not enough production space, while Weinan had too much production space. At the county scale, some counties in the central region showed a shortage of production space, while others showed an excess of production space and a shortage of ecological space, implying that production space was not evenly distributed at the county scale in the region.

In conclusion, as the statistical scale increases, there is a general trend towards fewer areas in need of spatial optimization. The main problem in the north-western part of the study area was that there was too much ecological space and not enough production space. Therefore, it is necessary to increase the area of production land in the region, and as the region is also the main area of the Loess Plateau and undertakes important functions of soil and water conservation, the area of regional terraces and industrial land can be increased appropriately. The central part of the study area showed an uneven distribution of production space and a lack of ecological space, so it is necessary to adjust the distribution of production space at several scales to ensure that it is in a reasonable range, while the ratio of ecological land, such as woodland and grassland, can be increased appropriately. The eastern part of the study area had a large ratio of production space and living space area and too little ecological space. Therefore, it is necessary to reduce and harmonize production and living space, while increasing the right amount of ecological space, such as green space and woodland. Here, the direction of the PLES adjustment is mainly explained, which in practice it can be quantified according to the difference between the PLES ratio of a region and the ideal region.

4.3. Limitations and Future Research Directions

This study aims to optimize the PLES with the objective that all ecosystem services can be in surplus as far as possible and proposes corresponding optimizations at multiple scales. This study can provide a basis for decision-making on regional land use management and rational allocation of resources. However, there are some problems with this study. Firstly, in assessing ES, this study used several models, such as the RUSLE model, water balance equation, and CASA model, where differences in data sources and calculation methods can lead to differences in results. Although there are still no effective solutions to these problems, these methods are still widely used [54,55]. Additionally, due to the lack of data and the limitations of ecosystem service models, only the supply and demand of four ES were assessed, which is not comprehensive for the complete management of ES. More ES assessments should be added in future studies. In addition, the use of land-use types for the classification of PLES is a more straightforward method [56]. However, this approach ignores the complex multifunctionality of land. For example, arable land (paddy and dryland) is uniformly classified as production space without taking into account its ecological characteristics. Finally, the issue of scale is also one of the problems studied in this study, with spatial correlation results varying with unit size (grid cell or grain size) [26,57]. In this study, the identification of thresholds was based on the grid-scale using a hierarchical statistical approach. Random points, various grid cell sizes, and basin units should be selected in subsequent studies to explore the differences in the impact of PLES on the ES supply and demand imbalance.

5. Conclusions

Based on various models and methods, this study quantified the mismatch of supply and demand for the four ES in the Yellow River Basin and explores how the spatial pattern of PLES can be adjusted to keep the ES in supply and demand balance. The results show that in 2000, 2010, and 2018, the total supply of the three ecosystem services in the Yellow River Basin was greater than the total demand, except for carbon sequestration services. Along with the implementation of revegetation projects and the establishment of ecological reserves in the region, the supply of many ecosystem services was on the rise. However, increased urbanization and over-concentration of population and economy resulted in a serious spatial mismatch between supply and demand for all four ecosystem services, especially in the major urban centers. The spatial mismatch in ES can be effectively reduced by optimizing the PLES, e.g., increasing the production spatial ratio can effectively increase the supply of grain production service and alleviate the contradiction between supply and demand of grain production service in certain regions. This study provides an optimization objective for the PLES optimization of other regions by providing an ideal region in the ternary phase diagram, i.e., one that can ensure that multiple ES are in surplus at the same time. The direction of optimization of other areas is determined by their relative position to the ideal area, and the amount of adjustment of PLES can be determined by the difference from the ideal area. The PLES optimization framework proposed in this study is very flexible, as reflected in the choice of ES and multi-scale optimization proposals, which can effectively reduce the deficit problem of regional ES in the process of practical application.

Author Contributions: Conceptualization, X.F.; data curation, X.W.; formal analysis, J.M.; funding acquisition, X.W.; investigation, X.F.; methodology, X.F. and J.Z.; resources, X.W.; supervision, X.W.; visualization, X.F., J.Z. and J.M.; writing—original draft, X.F. All authors have read and agreed to the published version of the manuscript.

Funding: This research was funded by the National Key Research and Development Plan of China (2016YFC0501603), the Chinese Academic of Sciences, the Strategic Priority Research Program of the Chinese Academy of Sciences (XDA2002040201).

Conflicts of Interest: The authors declare no conflict of interest.

References

1. Haas, J.; Ban, Y. Urban growth and environmental impacts in Jing-Jin-Ji, the Yangtze, River Delta and the Pearl River Delta. *Int. J. Appl. Earth Obs. Geoinf.* **2014**, *30*, 42–55. [CrossRef]
2. Allington, G.R.; Li, W.; Brown, D.G. Urbanization and environmental policy effects on the future availability of grazing re-sources on the Mongolian Plateau: Modeling socio-environmental system dynamics. *Environ. Sci. Policy* **2017**, *68*, 35–46. [CrossRef]
3. Foley, J.A.; DeFries, R.; Asner, G.P.; Barford, C.; Bonan, G.; Carpenter, S.R.; Chapin, F.S.; Coe, M.T.; Daily, G.C.; Gibbs, H.K.; et al. Global Consequences of Land Use. *Science* **2005**, *309*, 570–574. [CrossRef] [PubMed]
4. Wang, J.; He, T.; Lin, Y. Changes in ecological, agricultural, and urban land space in 1984–2012 in China: Land policies and regional social-economical drivers. *Habitat Int.* **2018**, *71*, 1–13. [CrossRef]
5. Liu, Y.S. Introduction to land use and rural sustainability in China. *Land Use Policy* **2018**, *74*, 1–4. [CrossRef]
6. Ma, W.; Jiang, G.; Li, W.; Zhou, T.; Zhang, R. Multifunctionality assessment of the land use system in rural residential areas: Confronting land use supply with rural sustainability demand. *J. Environ. Manag.* **2019**, *231*, 73–85. [CrossRef]
7. Thorne, J.H.; Santos, M.J.; Bjorkman, J.H. Regional Assessment of Urban Impacts on Landcover and Open Space Finds a Smart Urban Growth Policy Performs Little Better than Business as Usual. *PLoS ONE* **2013**, *8*, e65258. [CrossRef] [PubMed]
8. Li, Y.; Li, Y.; Westlund, H.; Liu, Y. Urban-rural transformation in relation to cultivated land conversion in China: Implications for optimizing land use and balanced regional development. *Land Use Policy* **2015**, *47*, 218–224. [CrossRef]
9. Deines, J.M.; Schipanski, M.E.; Golden, B.; Zipper, S.C.; Nozari, S.; Rottler, C.; Guerrero, B.; Sharda, V. Transitions from irri-gated to dryland agriculture in the Ogallala Aquifer: Land use suitability and regional economic impacts. *Agric. Water Manag.* **2020**, *233*, 106061. [CrossRef]
10. Sommer, W.; Valstar, J.; Leusbrock, I.; Grotenhuis, T.; Rijnaarts, H. Optimization and spatial pattern of large-scale aquifer thermal energy storage. *Appl. Energy* **2015**, *137*, 322–337. [CrossRef]
11. Lin, G.; Jiang, D.; Fu, J.; Cao, C.; Zhang, D. Spatial Conflict of Production-Living-Ecological Space and Sustaina-ble-Development Scenario Simulation in Yangtze River Delta Agglomerations. *Sustainability* **2020**, *12*, 2175. [CrossRef]
12. Tian, F.; Li, M.; Han, X.; Liu, H.; Mo, B. A Production–Living–Ecological Space Model for Land-Use Optimisation: A case study of the core Tumen River region in China. *Ecol. Model.* **2020**, *437*, 109310. [CrossRef]
13. Costanza, R.; De Groot, R.; Braat, L.; Kubiszewski, I.; Fioramonti, L.; Sutton, P.; Farber, S.; Grasso, M. Twenty years of ecosystem services: How far have we come and how far do we still need to go? *Ecosyst. Serv.* **2017**, *28*, 1–16. [CrossRef]
14. Pharo, E.; Daily, G.C. Nature's Services: Societal Dependence on Natural Ecosystems. *Bryology* **1998**, *101*, 475. [CrossRef]
15. Palmer, M.A.; Filoso, S. Restoration of Ecosystem Services for Environmental Markets. *Science* **2009**, *325*, 575–576. [CrossRef]
16. Fu, B.; Wang, S.; Su, C.; Forsius, M. Linking ecosystem processes and ecosystem services. *Curr. Opin. Environ. Sustain.* **2013**, *5*, 4–10. [CrossRef]
17. Wu, X.; Liu, S.; Zhao, S.; Hou, X.; Xu, J.; Dong, S.; Liu, G. Quantification and driving force analysis of ecosystem services supply, demand and balance in China. *Sci. Total Environ.* **2019**, *652*, 1375–1386. [CrossRef]
18. Peng, J.; Wang, X.; Liu, Y.; Zhao, Y.; Xu, Z.; Zhao, M.; Qiu, S.; Wu, J. Urbanization impact on the supply-demand budget of ecosystem services: Decoupling analysis. *Ecosyst. Serv.* **2020**, *44*, 101139. [CrossRef]

19. Zhao, W.; Liu, Y.; Daryanto, S.; Fu, B.; Wang, S.; Liu, Y. Metacoupling supply and demand for soil conservation service. *Curr. Opin. Environ. Sustain.* **2018**, *33*, 136–141. [CrossRef]
20. Fu, B.; Li, Y. Bidirectional coupling between the Earth and human systems is essential for modeling sustainability. *Natl. Sci. Rev.* **2016**, *3*, 397–398. [CrossRef]
21. Bai, Y.; Zhuang, C.; Ouyang, Z.; Zheng, H.; Jiang, B. Spatial characteristics between biodiversity and ecosystem services in a human-dominated watershed. *Ecol. Complex* **2011**, *8*, 177–183. [CrossRef]
22. Gao, Q.; Kang, M.; Xu, H.; Jiang, Y.; Yang, J. Optimization of land use structure and spatial pattern for the semi-arid loess hilly–gully region in China. *Catena* **2010**, *81*, 196–202. [CrossRef]
23. Chen, J.; Jiang, B.; Bai, Y.; Xu, X.; Alatalo, J. Quantifying ecosystem services supply and demand shortfalls and mismatches for management optimisation. *Sci. Total Environ.* **2019**, *650*, 1426–1439. [CrossRef] [PubMed]
24. Sun, Y.X.; Liu, S.L.; Shi, F.N.; An, Y.; Li, M.Q.; Liu, Y.X. Spatio-temporal variations and coupling of human activity intensity and ecosystem services based on the four-quadrant model on the Qinghai-Tibet Plateau. *Sci. Total Environ.* **2020**, *743*, 140721.
25. Santos-Martín, F.; Zorrilla-Miras, P.; Palomo, I.; Montes, C.; Benayas, J.; Maes, J. Protecting nature is necessary but not sufficient for conserving ecosystem services: A comprehensive assessment along a gradient of land-use intensity in Spain. *Ecosyst. Serv.* **2019**, *35*, 43–51. [CrossRef]
26. Peng, J.; Tian, L.; Liu, Y.; Zhao, M.; Hu, Y.; Wu, J. Ecosystem services response to urbanization in metropolitan areas: Thresholds identification. *Sci. Total Environ.* **2017**, *607–608*, 706–714. [CrossRef] [PubMed]
27. Li, D.; Wu, S.; Liu, L.; Liang, Z.; Li, S. Evaluating regional water security through a freshwater ecosystem service flow model: A case study in Beijing-Tianjian-Hebei region, China. *Ecol. Indic.* **2017**, *81*, 159–170. [CrossRef]
28. Ouyang, Z.; Zheng, H.; Xiao, Y.; Polasky, S.; Liu, J.; Xu, W.; Wang, Q.; Zhang, L.; Xiao, Y.; Rao, E.; et al. Improvements in ecosystem services from investments in natural capital. *Science* **2016**, *352*, 1455–1459. [CrossRef]
29. Zhang, Z.; Peng, J.; Xu, Z.; Wang, X.; Meersmans, J. Ecosystem services supply and demand response to urbanization: A case study of the Pearl River Delta, China. *Ecosyst. Serv.* **2021**, *49*, 101274. [CrossRef]
30. Burkhard, B.; Kroll, F.; Nedkov, S.; Müller, F. Mapping ecosystem service supply, demand and budgets. *Ecol. Indic.* **2012**, *21*, 17–29. [CrossRef]
31. Wei, H.; Fan, W.; Wang, X.-C.; Lu, N.; Dong, X.; Zhao, Y.; Ya, X.; Zhao, Y. Integrating supply and social demand in ecosystem services assessment: A review. *Ecosyst. Serv.* **2017**, *25*, 15–27. [CrossRef]
32. Woldeyohannes, A.; Cotter, M.; Biru, W.D.; Kelboro, G. Assessing Changes in Ecosystem Service Values over 1985–2050 in Response to Land Use and Land Cover Dynamics in Abaya-Chamo Basin, Southern Ethiopia. *Land* **2020**, *9*, 37. [CrossRef]
33. Hasan, S.S.; Zhen, L.; Miah, G.; Ahamed, T.; Samie, A. Impact of land use change on ecosystem services: A review. *Environ. Dev.* **2020**, *34*, 100527. [CrossRef]
34. Wilkerson, M.L.; Mitchell, M.G.; Shanahan, D.; Wilson, K.; Ives, C.D.; Lovelock, C.; Rhodes, J. The role of socio-economic factors in planning and managing urban ecosystem services. *Ecosyst. Serv.* **2018**, *31*, 102–110. [CrossRef]
35. Cui, F.; Tang, H.; Zhang, Q.; Wang, B.; Dai, L. Integrating ecosystem services supply and demand into optimized management at different scales: A case study in Hulunbuir, China. *Ecosyst. Serv.* **2019**, *39*, 100984. [CrossRef]
36. Feng, Q.; Zhao, W.; Fu, B.; Ding, J.; Wang, S. Ecosystem service trade-offs and their influencing factors: A case study in the Loess Plateau of China. *Sci. Total Environ.* **2017**, *607–608*, 1250–1263. [CrossRef] [PubMed]
37. Knoke, T.; Paul, C.; Rammig, A.; Gosling, E.; Hildebrandt, P.; Härtl, F.; Peters, T.; Richter, M.; Diertl, K.H.; Castro, L.M. Ac-counting for multiple ecosystem services in a simulation of land-use decisions: Does it reduce tropical deforestation? *Glob. Chang. Biol.* **2020**, *26*, 2403–2420. [CrossRef] [PubMed]
38. Zou, L.; Liu, Y.; Wang, J.; Yang, Y. An analysis of land use conflict potentials based on ecological-production-living function in the southeast coastal area of China. *Ecol. Indic.* **2021**, *122*, 107297. [CrossRef]
39. Yang, Y.; Bao, W.; Liu, Y. Coupling coordination analysis of rural production-living-ecological space in the Beijing-Tianjin-Hebei region. *Ecol. Indic.* **2020**, *117*, 106512. [CrossRef]
40. Lin, G.; Fu, J.; Jiang, D. Production–Living–Ecological Conflict Identification Using a Multiscale Integration Model Based on Spatial Suitability Analysis and Sustainable Development Evaluation: A Case Study of Ningbo, China. *Land* **2021**, *10*, 383. [CrossRef]
41. Świąder, M.; Szewrański, S.; Kazak, J.K. Environmental Carrying Capacity Assessment—The Policy Instrument and Tool for Sustainable Spatial Management. *Front. Environ. Sci.* **2020**, *8*. [CrossRef]
42. Li, Y.; Ye, H.; Sun, X.; Zheng, J.; Meng, D. Coupling Analysis of the Thermal Landscape and Environmental Carrying Capacity of Urban Expansion in Beijing (China) over the Past 35 Years. *Sustainability* **2021**, *13*, 584. [CrossRef]
43. Ding, X.; Zheng, M.; Zheng, X. The Application of Genetic Algorithm in Land Use Optimization Research: A Review. *Land* **2021**, *10*, 526. [CrossRef]
44. Baró, F.; Haase, D.; Gómez-Baggethun, E.; Frantzeskaki, N. Mismatches between ecosystem services supply and demand in urban areas: A quantitative assessment in five European cities. *Ecol. Indic.* **2015**, *55*, 146–158. [CrossRef]
45. Xie, X.; Li, X.; Fan, H.; He, W. Spatial analysis of production-living-ecological functions and zoning method under symbiosis theory of Henan, China. *Environ. Sci. Pollut. Res.* **2021**, 1–18. [CrossRef]
46. Bai, P.; Liu, X.; Zhang, Y.; Liu, C. Assessing the Impacts of Vegetation Greenness Change on Evapotranspiration and Water Yield in China. *Water Resour. Res.* **2020**, *56*, 56. [CrossRef]

47. Jia, X.; Fu, B.; Feng, X.; Hou, G.; Liu, Y.; Wang, X. The tradeoff and synergy between ecosystem services in the Grain-for-Green areas in Northern Shaanxi, China. *Ecol. Indic.* **2014**, *43*, 103–113. [CrossRef]

48. Beven, K. A sensitivity analysis of the Penman-Monteith actual evapotranspiration estimates. *J. Hydrol.* **1979**, *44*, 169–190. [CrossRef]

49. Mehring, M.; Ott, E.; Hummel, D. Ecosystem services supply and demand assessment: Why social-ecological dynamics matter. *Ecosyst. Serv.* **2018**, *30*, 124–125. [CrossRef]

50. Luo, Y.; Lü, Y.; Fu, B.; Zhang, Q.; Li, T.; Hu, W.; Comber, A. Half century change of interactions among ecosystem services driven by ecological restoration: Quantification and policy implications at a watershed scale in the Chinese Loess Plateau. *Sci. Total. Environ.* **2019**, *651*, 2546–2557. [CrossRef]

51. Renard, K.G. *Predicting Soil Erosion by Water: A Guide to Conservation Planning with the Revised Universal Soil Loss Equation (RUSLE)*; United States Government Printing: Washington, DC, USA, 1997.

52. Wischmeier, W.H.; Smith, D.D. *Predicting Rainfall Erosion Losses: A Guide to Conservation Planning*; Department of Agriculture, Science and Education Administration: Washington, DC, USA, 1978.

53. Yu, H.; Xie, W.; Sun, L.; Wang, Y. Identifying the regional disparities of ecosystem services from a supply-demand perspective. *Resour. Conserv. Recycl.* **2021**, *169*, 105557. [CrossRef]

54. Qin, K.; Lin, Y.-P.; Yang, X. Trade-Off and Synergy among Ecosystem Services in the Guanzhong-Tianshui Economic Region of China. *Int. J. Environ. Res. Public Health* **2015**, *12*, 14094–14113. [CrossRef]

55. Abera, W.; Tamene, L.; Tibebe, D.; Adimassu, Z.; Kassa, H.; Hailu, H.; Mekonnen, K.; Desta, G.; Sommer, R.; Verchot, L. Characterizing and evaluating the impacts of national land restoration initiatives on ecosystem services in Ethiopia. *Land Degrad. Dev.* **2020**, *31*, 37–52. [CrossRef]

56. Yu, S.-H.; Deng, W.; Xu, Y.-X.; Zhang, X.; Xiang, H.-L. Evaluation of the production-living-ecology space function suitability of Pingshan County in the Taihang mountainous area, China. *J. Mt. Sci.* **2020**, *17*, 2562–2576. [CrossRef]

57. Chi, Y.; Zhang, Z.; Gao, J.; Xie, Z.; Zhao, M.; Wang, E. Evaluating landscape ecological sensitivity of an estuarine island based on landscape pattern across temporal and spatial scales. *Ecol. Indic.* **2019**, *101*, 221–237. [CrossRef]

Article

Identification and Optimization of Production-Living-Ecological Space in an Ecological Foundation Area in the Upper Reaches of the Yangtze River: A Case Study of Jiangjin District of Chongqing, China

Hongji Chen [1,2,3], Qingyuan Yang [1,2,3,*], Kangchuan Su [3,4], Haozhe Zhang [1,2,3], Dan Lu [1,2,3], Hui Xiang [1,2,3] and Lulu Zhou [1,2,3]

1   School of Geographical Sciences, Southwest University, Chongqing 400045, China;
    chj001292@email.swu.edu.cn (H.C.); ssaijj@email.swu.edu.cn (H.Z.); ludswu@email.swu.edu.cn (D.L.);
    xh982123@email.swu.edu.cn (H.X.); zllzxnlcg@email.swu.edu.cn (L.Z.)
2   Chongqing Jinfo Mountain Kast Ecosystem National Observation and Research Station,
    Chongqing 400715, China
3   Institute of Green Low-Carbon Development, Southwest University, Chongqing 400715, China;
    sukangchuan@swu.edu.cn
4   School of State Governance, Southwest University, Chongqing 400045, China
*   Correspondence: yizyang@swu.edu.cn; Tel.: +86-023-6825-3911

Citation: Chen, H.; Yang, Q.; Su, K.; Zhang, H.; Lu, D.; Xiang, H.; Zhou, L. Identification and Optimization of Production-Living-Ecological Space in an Ecological Foundation Area in the Upper Reaches of the Yangtze River: A Case Study of Jiangjin District of Chongqing, China. *Land* 2021, 10, 863. https://doi.org/10.3390/land10080863

Academic Editor: Dong Jiang

Received: 18 July 2021
Accepted: 14 August 2021
Published: 17 August 2021

Publisher's Note: MDPI stays neutral with regard to jurisdictional claims in published maps and institutional affiliations.

Abstract: The identification of regional production-living-ecological space (PLES) is the basic work for the optimization of territorial space, which can point to the direction for the protection, utilization and restoration of regional territorial space. Identification and optimization of PLES in an ecological foundation area in the upper reaches of the Yangtze River is of great significance for ensuring national ecological security and promoting sustainable social development. In this study, Jiangjin District, located at the tail of the Three Gorges Reservoir area, was selected as a case study. Moreover, based on the land use data of the study area in 2018, the coordination among production, living and ecological functions are analyzed, and the PLES is identified by using the evaluation method of land production-living-ecological function (PLEF) and the coupling coordination degree model. Then, we formulated an optimized zoning scheme of the PLES according to the principles of ecological priority, area advantage and coordinated development. The results show that (1) The living function and production function presented obvious spatial consistency in the study area, while the spatial distribution of ecological function and production function presented significant spatial complementarity. (2) Four categories of spatial combinations can be identified in the study area. Overall, the study area presented a national spatial pattern with production-living-ecological balanced space (PLEBS) and ecological space (ES) as the main body. (3) The PLES in the study area can be divided into four categories. The ecological function should be determined by the ecological conservation area as the primary responsibility, and the comprehensive improvement zone should further improve the coupling and coordination relationship among the PLEF. Moreover, the main production-living and ecological improvement zone and the main production-ecological and living improvement zone should realize the coordinated development of the PLES on the basis of strengthening the leading function.

Keywords: PLES; coupling degree of compatibility; ecological barrier area in the upper reaches of the Yangtze River; Jiangjin District

1. Introduction

Production-living-ecological space (PLES) is an important carrier for the survival and development of human society [1,2]. In recent years, with the development of territorial space planning in China, the PLES has attracted much academic attention, and research on optimizing and coordinating the PLES has become a hot topic for scholars [3]. Along with the rapid development of industrialization and urbanization, the territorial spatial pattern of China has undergone profound changes since the reform and opening up [4–6]. At the same time, many space-related problems have arisen, such as extensive land use, environmental pollution and ecosystem degradation [7–9]. Therefore, the Nineteenth National Congress of the Communist Party of China (CPC) proposed scientifically delineating three control lines (an ecological protection red line, permanent basic farmland, and an urban development boundary) to coordinate the spatial pattern of production, living and ecology and promote the sustainable and balanced development of the economy and environment [10]. In this context, it is of great significance to identify and optimize the PLES to promote the sustainable development of territorial space, ecological civilization and beautiful China construction [11,12].

At present, scholars have carried out much research on the identification and optimization of the PLES and have achieved relatively fruitful results. The research content mainly focuses on the classification system of production-living-ecological land (PLEL) [13–16], the definition and connotation of the concept [17–20], the identification [2,21–23], reconstruction and optimization [24–26] and the pattern evolution of the PLES [27,28]. Qualitative and quantitative identification methods were used to identify and optimize the PLES from the perspectives of land use [21], rural settlements [18] and ecological landscapes [23]. The qualitative identification method mainly refers to the land use type merging method, which is based on the national land classification standard to classify and merge the status of land use, thereby identifying the PLES [29,30]. And the quantitative identification method refers to first constructing an index system and then using the entropy method [31], GIS spatial analysis method [22], coupling coordination degree model [1,22] and other methods or models to carry out functional evaluation and to identify and optimize the PLES. The former lacks consideration of the coordinated development of land use multifunctionality and functional coupling, while the latter has pertinence and comprehensive advantages but has not formed a unified standard system. The coordination of PLES is an effective way to optimize land use, the optimization method of land use in academia provides a good foundation for the study of PLES. At present, most scholars use genetic algorithm [32–34], multi-agent system (MAS) [35], artificial neural networks (ANN) [36] and other metrological methods to build land use optimization models and explore multi-objective optimization schemes for land use. In recent years, more and more scholars have analyzed the problems and the optimization mode of land use from the perspective of ecosystem services, so as to maximize the performance of ecosystem services and land use [37–40]. For example, Herzig [40] demonstrated a land use optimization model to improve ecosystem services by using the multi-objective spatial optimization method. Elliot [39] explored a land use scheme to maximize ecosystem services by optimizing urban land use allocation by using multi-objective integer linear programming (MOILP) model and land use/land cover (LULC) performance score.

Previous studies often used administrative units as the research scale, including macro scales such as the whole country, provinces and cities [25,31,41], meso scales such as cities and counties [42], and microscopic scales such as villages [28]. However, there are also large spatial differences in administrative units, especially in mountainous areas with complex topographical conditions. The spatial information reflected by the research based on the administrative unit scale is not detailed enough. Fortunately, the geographic grid can effectively compensate for this defect at the administrative unit scale [22]. In summary, current research on the identification and optimization of the PLES has achieved phased results, which provide a reference for this article. However, research based on the multifunctional coupling and coordination of land use is still insufficient, and current related research mostly uses the kilometer grid as the basic unit [43]. It is difficult to accurately identify the PLES. In addition, with the fine-grained geographic grid, the identification of the PLES needs to be strengthened.

It is necessary to ensure national ecological security and realize the coordinated development between economy and ecology by accurately identifying the PLES in the ecological barrier area of the upper reaches of the Yangtze River. Located at the upper reaches of the Yangtze River and the end of the Three Gorges Reservoir, Jiangjin District of Chongqing City undertakes the major task of constructing an ecological barrier. Moreover, Jiangjin District is located at the intersection of multiple strategic opportunities such as the construction of the Chengdu-Chongqing Double City Economic Circle, the coordinated development of One District and Two Groups of Chongqing and the construction of the co-urbanization of Chongqing. In the context of the construction of an ecological barrier in the upper reaches of the Yangtze River, and the superposition of multiple strategies, Jiangjin District is facing many development opportunities and challenges [44,45]. Whether socioeconomic, developmental or ecological and concerning environmental protection, it is necessary to use territorial space as a carrier. Therefore, the coordinated development of territorial space can provide a space guarantee for economic development and ecological protection. Therefore, how to optimize the territorial space and promote the sustainable development of the territorial space has become a key issue that urgently needs to be solved for Jiangjin District to consider development opportunities and strengthen ecological protection. In view of this, we selected Jiangjin District as the study area, established the evaluation system of the PLES and used the coupling coordination degree model to quantitatively calculate the coupling coordination degree of the PLEF of the study area. We identified the PLES by using a coupling coordination degree model. Finally, we propose the optimized partition scheme for territorial space based on the identification results. The purpose of this research is to provide a reference for the governance and optimization of territorial space in special areas of multiple strategic intersections.

2. Materials and Methods

2.1. Study Area

Jiangjin District of Chongqing is located in southwestern Chongqing, adjacent to Guizhou Province to the southeast and Sichuan Province to the west and southwest (Figure 1). The geographical coordinates are between 105°49′–106°38′ E and 28°28′–29°28′ N. Jiangjin District has a total area of 3218 km², with five subdistricts and 25 towns. The terrain of Jiangjin District descends gradually from north to south to the Yangtze River Valley, with hills and low mountains in the north and middle and mountainous areas in the south. It has a subtropical humid monsoon climate, with a mild climate and abundant precipitation. The main disasters are high temperature, drought, low temperature and rain, hail and so on. At the end of 2019, the forest area in Jiangjin District reached 16.6 × 104 hm², and the forest coverage rate reached 51.8%. In 2019, the regional GDP of Jiangjin District was 103.67 billion yuan, and the per capita GDP reached 74,452 yuan. At the end of 2019, Jiangjin District had a permanent population of 1.398 million, and the urbanization rate of permanent residents was 69.76%.

As the Chinese government vigorously promotes coordinated regional development, the location advantages and development opportunities of Jiangjin District have become increasingly apparent. The construction of the Chengdu-Chongqing Twin Cities Economic Circle is an important part of the overall national regional development strategy. Jiangjin District is responsible for the construction of the western (Chongqing) Science City Jiangjin area in the construction of the Chengdu-Chongqing Twin Cities Economic Circle. Therefore, it is the vanguard of Sichuan-Chongqing cooperation. In the spatial pattern of the development of the One District and Two Clusters of cities and towns in Chongqing, Jiangjin District is an open new city of science and innovation, a new industrial area, and a cultural and tourist attraction in the main urban area, so it is a leader in the development of the western region of Chongqing. In addition, Jiangjin District is also a pioneering area for the co-urbanization of the main urban area of Chongqing. Obviously, the superposition of many strategies has brought unprecedented development opportunities to Jiangjin District. However, Jiangjin District is located in the upper reaches of the Yangtze River and at the end of the Three Gorges Reservoir, and it is responsible for the construction of the ecological barrier in the Three Gorges Reservoir and the ecological protection of the upper reaches of the Yangtze River. In short, in the context of multiple strategic development opportunities and ecological protection, the question of how to optimize the PLES to provide a great space guarantee both for economic development and to strengthen ecological protection has become a key issue that urgently needs to be resolved in the territorial space planning of Jiangjin District.

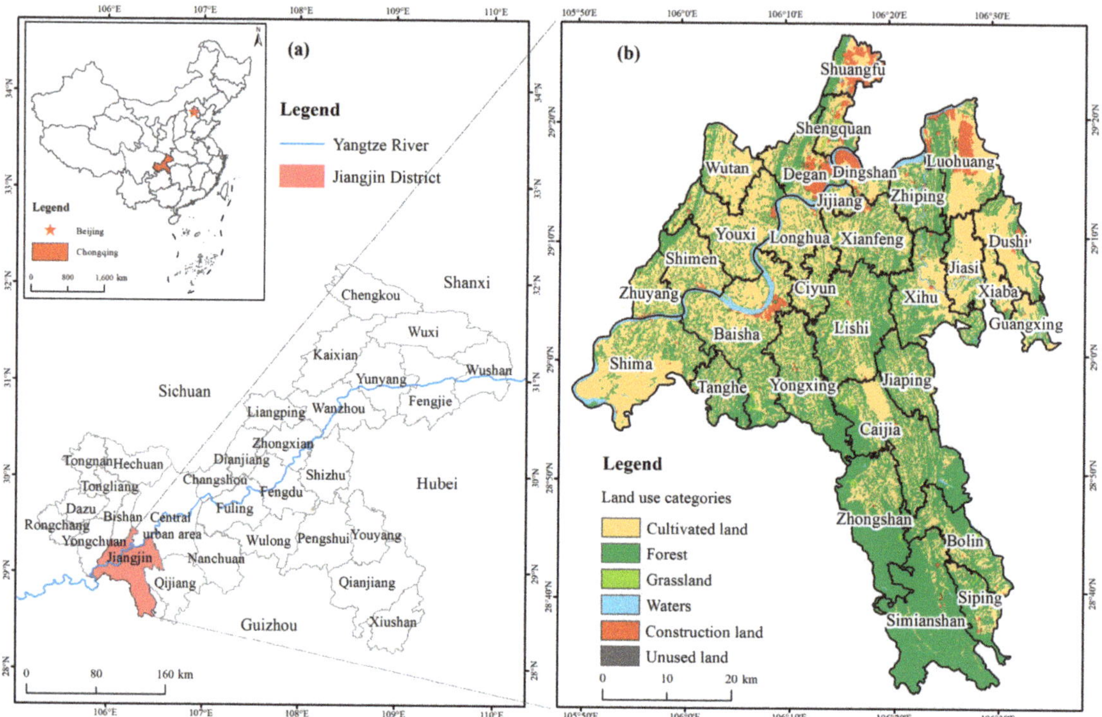

Figure 1. The study area: (**a**) the location of Jiangjin District in Chongqing; (**b**) land use status of Jiangjin District in 2018. Note: Based on the standard map (scale 1:48 million) with the approval number of GS (2019) 1823 on the standard map service website, the base map has not been modified.

2.2. Data Sources and Processing

The data used in this paper mainly includes land use data for 2018 obtained from the Chinese Academy of Sciences Resource and Environmental Science Data Center (https://www.resdc.cn/Default.aspx, accessed on 16 August 2021), with a spatial resolution of 30 m. The data is based on the Landsat 8 remote sensing image of the United States and obtained by manual visual interpretation, and the sampling verification accuracy is above 95%. According to China's land use classification system based on remote sensing monitoring, the land use classification is divided into six categories: cultivated land, forest land, grassland, water area, construction land and unused land. The data processing steps are as follows: firstly, a 300 × 300 m square geographic grid was constructed, superimposed with the current land use data of the study area, and the area of various land types in each geographic grid was calculated. Secondly, according to the evaluation method of the PLEF of land, the production, living and ecological functions of land use types were assigned, respectively, and the values of PLEF of each geographic grid were calculated based on the ArcGIS10.2 software. Thirdly, we used the coupling coordination degree model to measure the coupling coordination degree among the PLEF and the coupling coordination degree between every two of the PLEF, then, the natural break point classification (Jenks) method was used to divide the coupling coordination degree of the PLEF into three types. Based on the results of the above steps, the PLES was identified. Finally, according to the principle of ecological priority, area dominance and coordinated development, the spatial superposition method to calculate and compare the area size of different functional spaces was used, and the similar types of land were merged to get the optimizing zoning of the PLES.

2.3. Methods

2.3.1. Evaluation of the PLEF of Land

The concept of the PLES is proposed based on the land use functions (LUFs). LUFs refer to the products and services provided for human society by land resources through various utilization methods [46,47]. Its concept can be traced back to the European Union's research on agricultural multifunction at the end of the 20th century [48,49]. The PLES is the extension of LUFs that the essence of PLES is a functional space divided according to the products and services it provides on the basis of LUFs [21]. Therefore, the logical connection between land use types and land use functions can be established to construct a scientific classification and evaluation system of the PLES [50]. Based on the research results of scholars and guided by the theory of the PLES, this study divided land use functions into production functions, living functions and ecological functions and divided land use types into production land, living land, and ecological land. According to the difference in the strength of the PLEF of land use, this study introduced the concepts of strong production/living/ecological land, semi-production/living/ecological land, and weak production/living/ecological land. Then, values were assigned to strong production/living/ecological land, semi-production/living/ecological land, and weak production/living/ecological land by using the method of grading assignment.

Taking production land as an example, strong production land means that the production function of land is stronger than other functions, so the assigned function score is up to 5 points. Semi-production land means that the production function of land use is roughly equivalent to other functions, so the assigned function score is up to 3 points. A weak production function means that the production function of land use is weaker than other functions, so 1 point is assigned. In addition, nonproduction land refers to land that does not have a production function and is assigned a value of 0. In this paper, strong/semi/weak production land only reflects the dominant function, not the actual performance of the land's output or the monetary value added per unit of land in terms of productivity. The assignment of living land and ecological land was the same as that of production land (Table 1). First, 300 m × 300 m geographic grids were built based on ArcGIS 10.2 software. Second, the area of each land use type in each geographic grid was counted. Third, the PLEF evaluation method was used to sum the functional scores of production land, living land, and ecological land in each geographic grid. Finally, the spatial distribution map of the production/life/ecological function scores was generated to reveal the distribution pattern of the PLES in the study area. The evaluation method of the PLEF is shown in Formula (1).

$$W_i = \sum_{i=1}^{n} S_i \cdot V_i \qquad (1)$$

In Formula (1), W_i refers to the evaluation score of the production/living/ecological function in each geographic grid. i is the land use type. n is the total number of land use types in each geographic grid. S_i represents the area of each land use type in each geographic grid, and its unit is hm^2. V_i represents the production/living/ecological function value of the unit area of the i-th land use type.

Table 1. Classification and function assignment of the PLEL of Jiangjin District in 2018.

Primary Categories	Secondary Categories and Function Score	Tertiary Categories
Production land	Strong production land (5)	Industrial and mining construction land
	Semi-production land (3)	Paddy field, dry land, urban land, rural settlement
	Weak production land (1)	Forest, grassland, river and canals, lake, reservoir pond
Living land	Strong living land (5)	Urban land, rural settlement
	Semi-living land (3)	Industrial and mining construction land
	Weak living land (1)	Paddy field, dry land, river and canals, lake, reservoir pond
Ecological land	Strong ecological land (5)	Forest, grassland, river and canals, lake, shoaly land, wetland
	Semi-ecological land (3)	Paddy field, dry land
	Weak ecological land (1)	Lake, reservoir pond

2.3.2. Coupling Coordination Degree Model of the PLEF

The concept of coupling is derived from physics and refers to the phenomenon where two or more systems interact and influence each other. The degree of coupling is used to measure the degree of interaction and mutual influence between multiple systems [51]. However, it cannot reflect the synergic relationship between multiple systems [28]. Therefore, we introduced the coupling coordination index to construct a coupling coordination model. First, by evaluating the production, living and ecological function values of each geographic grid, the coupling coordination model is used to measure the degree of coupling and coordination among the PLEF of each geographic grid to reveal the degree of coordinated development of the PLEF. Second, the coupling coordination degree was divided into a coordination zone, break-in zone and incongruous zone by using the natural breakpoint classification method. Coordination zone refers to the coupling among the PLEF is becoming stronger and developing in an orderly direction, and it is in a period of high-level coupling and coordination. Break-in zone means that the PLEF began to develop in coordination and shows the characteristics of benign coupling and coordination. Incongruous zone refers to the interaction of the PLEF is low, and it is in a state of disorderly development. The specific calculation process is shown in Formulas (2) to (4).

$$C = 3 \left\{ \frac{P_i \cdot R_i \cdot E_i}{(P_i + R_i + E_i)^3} \right\}^{1/3} \tag{2}$$

$$T = \alpha P_i + \beta R_i + \gamma E_i \tag{3}$$

$$D = (C \cdot T)^{1/2} \tag{4}$$

In Formula (2), C represents the degree of coupling, with $C \in [0, 1]$; the larger the value of C, the stronger the interaction among the PLEF. P_i, R_i and E_i represent the evaluation scores of productive, living and ecological functions, respectively. In Formula (3), T represents the coordination index among the PLEF. α, β and γ respectively refer to the undetermined coefficients of the production function, living function and ecological function. Due to the lack of unified standards and methods for determining the undetermined coefficients, Wang Cheng et al. [22,28] determined the undetermined coefficients based on the opinions of experienced experts, the main functions of the research area and the contribution of PLEF, and their method is certainly reasonable. Therefore, based on the existing research results [22,28] and experts asked for advice, the undetermined coefficients were determined to be $\alpha = \beta = 0.3$ and $\gamma = 0.4$. In Formula (4), D refers to the degree of coupling and coordination among the PLEF.

Moreover, it is of great significance to further analyze the strengths and weaknesses of the coupling coordination status between every two of the PLEF [28]. Therefore, to explore the coupling coordination status between every two of the PLEF, the coupling coordination degree model is further evolved, as shown in Formulas (5) to (6).

$$C_1 = 2\left\{\frac{P_i \cdot R_i}{(P_i + R_i)^2}\right\}^{1/2} \quad C_2 = 2\left\{\frac{R_i \cdot E_i}{(R_i + E_i)^2}\right\}^{1/2} \quad C_3 = 2\left\{\frac{P_i \cdot E_i}{(P_i + E_i)^2}\right\}^{1/2} \tag{5}$$

$$D = (C \cdot T)^{1/2}, \quad T_1 = \alpha P_i + \beta R_i \text{ or } T_2 = \alpha P_i + \gamma E_i \text{ or } T_3 = \beta R_i + \gamma E_i \tag{6}$$

Based on the existing research results [19,26], when measuring the degree of coupling and coordination between the production function and living function, $\alpha = \beta = 0.5$. When measuring the degree of coupling and coordination between the production function and ecological function, $\alpha = 0.45$ and $\gamma = 0.55$. When measuring the degree of coupling and coordination between the living function and ecological function, $\beta = 0.45$ and $\gamma = 0.55$.

3. Results

3.1. Evaluation Results and Analysis of the PLEF of Land

The evaluation values of the production function, living function and ecological function of Jiangjin District in 2018 were obtained by using the evaluation system of the PLEF of land (Figure 2). The results showed that the evaluation value of the land production function in Jiangjin District presented a spatial differentiation characteristic that was high in the north and low in the south. The high-scoring areas of the land production function were distributed in Wutan Town, Youxi Town, Shima Town and so on. The production function value of land in the southern region of Jiangjin District, such as Caijia Town, Zhongshan Town and Siping town, was relatively low. Areas with high living function values were mainly distributed in areas with high living function values and were mainly distributed in Shuangfu Street, Degan Street and Luohuang Town. In addition, the spatial distribution of living function values and production function values showed obvious consistency. Moreover, the distribution characteristics of ecological function values and the distribution characteristics of production function values showed significant spatial complementarity. Furthermore, areas with high ecological function values were mainly distributed in the south region of Jiangjin District, such as Bolin Town, Zhongshan Town and Caijia Town, which had low production function and living function scores. At the same time, the areas with low ecological function values were distributed in these areas with strong production function and living function.

3.2. Analysis of the Coupling Coordination Degree of the PLEF

3.2.1. Coupling Coordination Degree among the PLEF

The coupling coordination degree and the spatial distribution map of the PLEF of Jiangjin District in 2018 were obtained through the coupling coordination degree model (Figure 3). The natural breakpoint classification method (Jenks) was used to divide the coupling coordination degree of the PLEF of the study area into three categories which can appropriately group similar values and maximize the difference between each grouping [52] (Table 2).

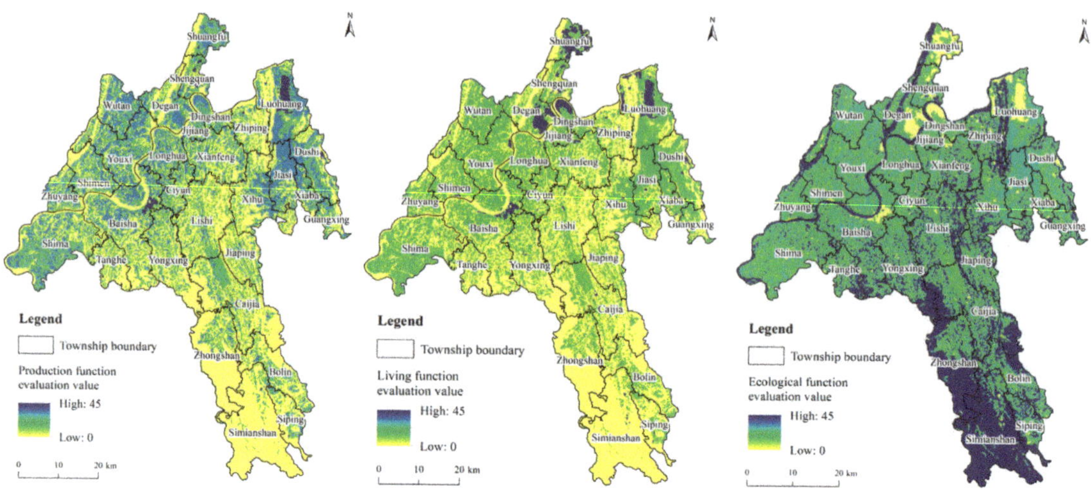

Figure 2. Evaluation value of the PLEF of Jiangjin District in 2018.

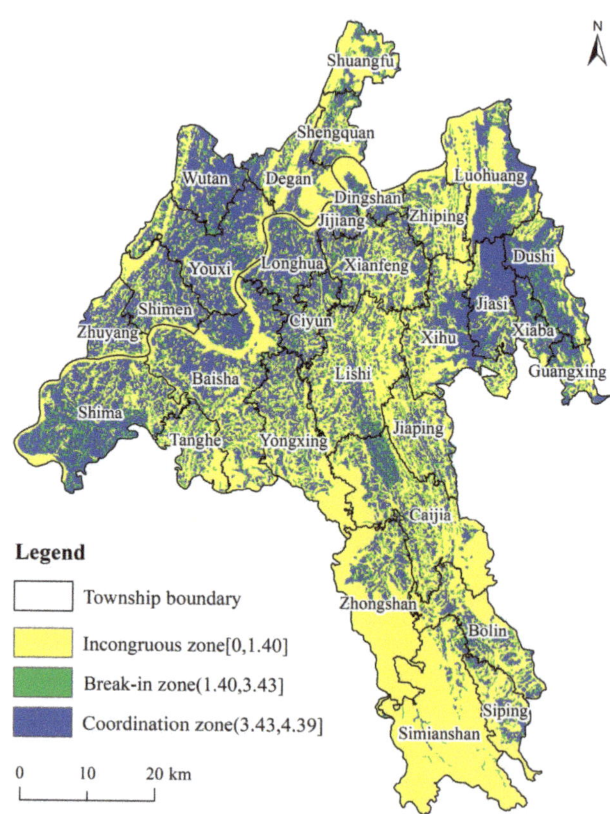

Figure 3. Spatial distribution of the coupling coordination degree of the PLEF of Jiangjin District in 2018.

Table 2. Classification and distribution of the coupling coordination degree of the PLEF.

Function Evaluation Value	Categories	Distribution
[0, 1.40]	Incongruous zone	Townships such as Zhongshan Town, Berlin Town, Degan Street, etc.
(1.40, 3.34]	Break-in zone	Distributed in points between the ncongruous zone and the coordination zone, with a wide range of distribution;
(3.34, 4.39]	Coordination zone	Jiasi Town, Dushi Town, Wutan Town, Youxi Town and other places

3.2.2. Coupling Coordination Degree between Every Two of the PLEF

The degree of coupling and coordination and the spatial distribution map between every two of the PLEF of Jiangjin District in 2018 were obtained according to Formulas (5) and (6) (Figure 4). Moreover, we divided the degree of coupling and coordination between every two of the PLEF into three categories (incongruous zone, break-in zone and coordination zone) by using the natural breakpoint classification method. However, due to the differences in the degree of coordination between every two of the PLEF, the breakpoint value of the division interval was also different (Tables 3–5). In general, the degree of coupling and coordination between every two of the PLEF is generally spatially consistent and high in the north and low in the south.

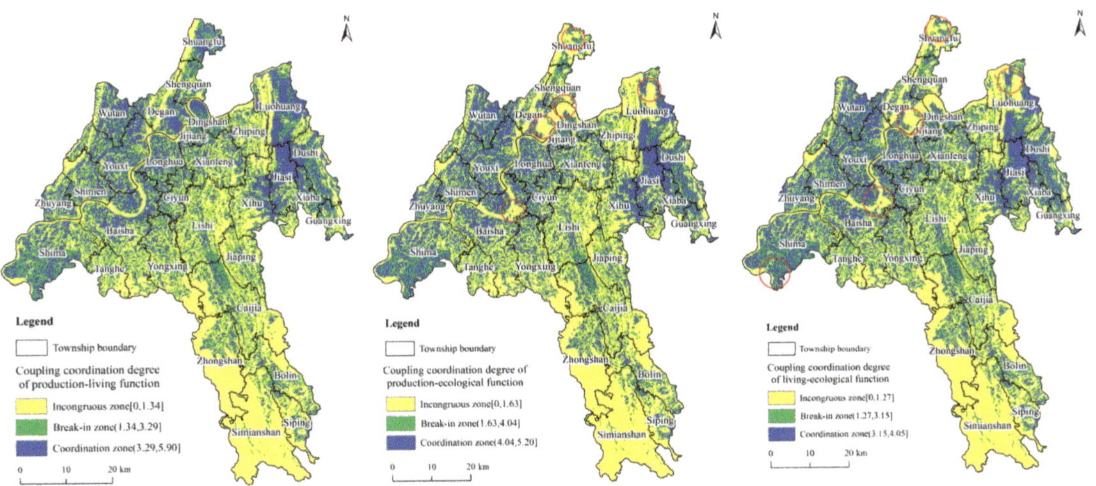

Figure 4. Spatial distribution of the coupling coordination degree between the production-living, the production-ecological and the living-ecological functions of Jiangjin District in 2018.

Table 3. Classification and distribution of the coupling coordination degree of the production-living function (PLF) of Jiangjin District in 2018.

Function Evaluation Value	Categories	Distribution
[0, 1.34]	Incongruous zone	Mainly distributed in Zhongshan Town and Berlin Town in the southern area.
(1.34, 3.29]	Break-in zone	Scattered between the imbalance zone and the coordination zone in the entire study area.
(3.29, 5.90]	Coordination zone	It is mainly distributed in Jiasi Town and Luohuang Town in the east and Wutan Town and Youxi Town in the midwest.

Table 4. Classification and distribution of the coupling coordination degree of the production-ecological function (PEF) of Jiangjin District in 2018.

Function Evaluation Value	Categories	Distribution
[0, 1.63]	Incongruous zone	It is mainly distributed in Shuangfu Street and Degan Street in the north; Luohuang Town and Zhiping Street in the east; and Zhongshan Town and Berlin Town in the south.
(1.63, 4.04]	Break-in zone	Scattered in the whole area of Jiangjin District
(4.04, 5.20]	Coordination zone	They are mainly distributed in Jiasi Town, Dushi Town, and Xiaba Town in the east and Wutan Town, Youxi Town, Shimen Town, and Shima Town in the west and other places.

Table 5. Classification and distribution of the coupling coordination degree of the living-ecological function (LEF) of Jiangjin District in 2018.

Function Evaluation Value	Categories	Distribution
[0, 1.27]	Incongruous zone	It is distributed throughout the whole area, mainly in Shuangfu Street and Degan Street in the north; Luohuang Town in the east and Zhongshan Town and Berlin Town in the south.
(1.27, 3.15]	Break-in zone	The distribution area is small, scattered in the whole area of Jiangjin District.
(3.15, 4.05]	Coordination zone	It is mainly distributed in Jiasi Town, Dushi Town, and Xiaba Town in the east and Wutan Town, Baisha Town, Youxi Town, Shimen Town, and Shima Town and other places in the west.

3.3. Identification of the PLES

According to the spatial distribution results of the coupling coordination degree of the PLEF and the coupling coordination degree between every two of the PLEF, we designated the coordinated area of the PLEF as the balanced space of the PLEF. Moreover, we compared the relative advantages of the coupling and coordination degree between every pair of the PLEF within the scope of incongruous zones and break-in zones of the PLEF. Then, the incongruous zone and break-in zone of the PLEF were identified as production-living space (PLS) and production-ecological space (PES). In addition, according to the value of the PLEF, the incongruous zone between every two of the PLEF was divided into single production, living or ecological space. However, the areas of single production space and single living space identified were very small, and the corresponding land use types have both production and living functions; thus, the two single spaces are classified as production-living compound space. Finally, the PLES pattern of Jiangjin District was obtained. The PLEF can be identified as four categories of space—the production-living-ecological balanced space (PLEBS), the production-living space (PLS), the production-ecological space (PES) and the single ecological space (ES) (Figure 5). The proportions of the four categories of space are 32.04%, 3.41%, 15.83% and 48.72%. In general, the territorial space pattern of Jiangjin District presents the characteristics of the PLEBS and the ES as the main body, supplemented by the PLS and the PES.

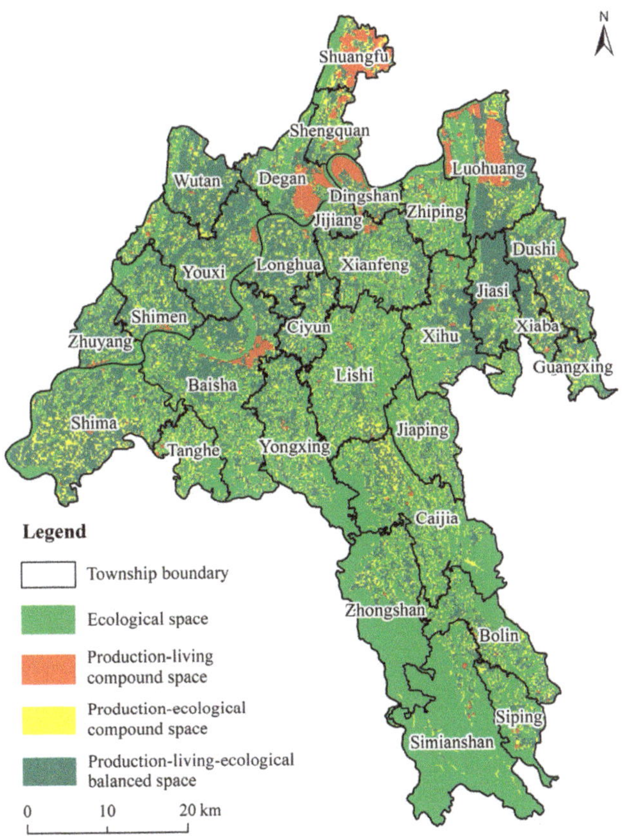

Figure 5. Identification results of the PLES of Jiangjin District in 2018.

4. Optimized Partition of the PLES

4.1. Principles of Optimizing Zoning

(1)   Principle of ecological priority

Jiangjin District is located in the ecological barrier construction area on the upper reaches of the Yangtze River and at the end of the Three Gorges Reservoir. Furthermore, the ecological environment of Jiangjin District is relatively fragile, so ecological protection is a major task of Jiangjin District. Therefore, when zoning for space optimization, all the grids of the ES are classified as ecological conservation areas to protect the ecological environment of Jiangjin District.

(2)   Principle of area dominance

There are bound to be multiple types of spatial combinations within each township. By comparing the area sizes of different spatial types, the spatial type with the largest area is used to determine the optimal spatial division type of each township.

(3)   Principle of coordinated development

The essence of optimizing the territorial spatial pattern is to enhance the capacity of regional sustainable development through the optimization of the structure and distribution of the PLES and to provide a space guarantee for social development, livable life and good ecology. Due to the differences in the strengths and weaknesses of the PLEF in different spatial units, the weak functions should be promoted in the process of space optimization to achieve the goal of the coordinated development of the PLES.

4.2. Scheme of Optimizing Zoning

The geographic grid could describe in detail the spatial heterogeneity of the PLEF, but there was also a problem in that it divided administrative units, which was not conducive to space governance [53,54]. Therefore, to ensure that the optimization results of the PLES have good practical significance, we used townships as the basic unit to optimize the space division and proposed suggestions for different development directions. First, based on the recognition result of the combination categories of PLES, we used the space superposition method to compare the areas of different space categories. Then, the adjacent or similar categories were merged according to the principle of ecological priority, area dominance and coordinated development (Table 6). Finally, the optimal zoning scheme of PLES is proposed, which takes the township as the basic unit (Figure 6).

Table 6. Optimized zoning of the PLES in Jiangjin District.

Space Categories	Scheme of Optimizing Zoning	Goal-Oriented
Ecological space (ES)	The ecological conservation zone	Ecological protection
Production-living space (PLS)	The main production-living and ecological improvement zone	Space coordinated development
Production-ecological space (PES)	The main production-ecological and living improvement zone	Space coordinated development
production-living-ecological balanced space (PLEBS)	The comprehensive promotion zone	Space coordinated development

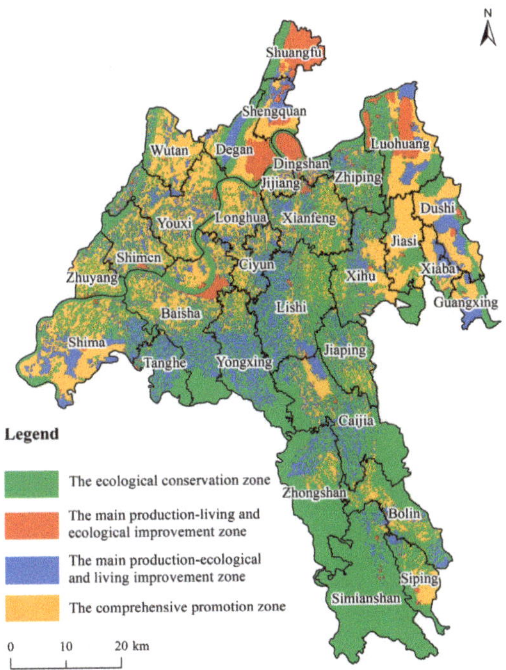

Figure 6. Optimized zoning map of the PLES of Jiangjin District in 2018.

(1) Comprehensive promotion zone. This area has ecological, living and ecological functions, and it mainly distributed in the east of the northwest area of Jiangjin District. The land use types in this area are dominated by paddy fields and dry land, with abundant cultivated land resources but a relative lack of infrastructure. Therefore, this area should make full use of regional resource endowments, vigorously develop high-efficiency agriculture such as flowers, seedlings, and high-quality fruits, establish small and medium-sized entrepreneurship bases, develop supporting industries and leisure tourism, promote the integrated development of agriculture and tourism, and enhance the production function. Second, it should strengthen the harmless treatment of domestic sewage and garbage and the control of agricultural nonpoint source pollution and soil pollution. In addition, protection measures of the ecological environment should be taken to improve the ecological function. Moreover, the construction of public infrastructure services should be strengthened, especially rural roads, medical health and education, to continuously improve the living environment and improve the quality of life.

(2) Ecological conservation zone. This area takes the ecological function as the main function, while living function and production function as the secondary function. The land use types are mainly woodland, grassland and waters. This area covers the Simian mountainous areas in the south of Jiangjin District, the Linfeng mountainous areas in the north of Jiangjin District, the Longmen mountainous areas in the east of Jiangjin District and the waters of the Yangtze River across the north region of Jiangjin District. Undoubtedly, this area bears the responsibility of ecological protection, so it should exploit the advantages of the ecological function and reasonably and appropriately develop ecotourism and eco-industry by protecting the ecological environment. In the southern region, Zhongshan Town, Bolin Town, Caijia Town and other places could moderately develop ecological leisure vacation tourism, seize the advantage of selenium-rich, plant ecological selenium-rich agricultural and sideline products, vigorously cultivate new green kinetic energy and transform the advantages of ecological environment into advantages of green development. The northern region of Jiangjin District should strengthen the comprehensive management

of industrial and agricultural pollution prevention, shoreline protection and restoration and combat illegal fishing to strictly protect the ecological environment and biodiversity of the Yangtze River.

(3) The main production-living and ecological improvement zone. This area is dominated by production and living functions, with weak ecological functions. This area is the core growth pole of the economic development of Jiangjin District. The land use intensity is high, and the economy is relatively developed. However, there were some problems in this area, such as its dense population, production and life pollution, which pose a great threat to ecological security. Moreover, the vegetation coverage rate is low, and the ecological function is weak in this area. Therefore, the area should continue to improve public infrastructure, such as transportation and logistics, strengthen the management of industry and improve the quality of the living environment. By strengthening production and the living function, high-tech industries should be developed, ecological environment management should be strengthened, ecological land areas should be increased, the ecological function should be improved and the coordinated development of the PLEF should be promoted.

(4) The main production-ecological and living improvement zone. This area is characterized by strong production and ecological functions and a weak living function. The region is flat and rich in arable land resources. However, the infrastructure and public services are backward, and the living environment needs to be improved. By protecting arable land resources, this area should strengthen the construction of infrastructure, such as transportation, culture and sports centers, to improve life service functions. In addition, this area should strengthen the treatment of rural domestic garbage, livestock and poultry manure, improve the quality of human settlements and continuously improve regional life features.

5. Discussion

5.1. Advantages of the Method and Comparison of Research Results

Compared with existing research about PLES [55–57], this study conducted research by constructing a fine-grained geographic grid and introducing the coupling coordination degree model of the land the PLEF and combined the coupling and coordination degree of the PLEF between every two of the PLEF for analysis. This overcomes the limitation of the overall coupling and coordination between the administrative unit scale and the single measurement of the PLEF [22]. The research results could provide a valuable reference for clarifying the direction of territorial spatial optimization and enriching the theoretical meaning of the PLES. Moreover, compared with other land use optimization methods [35–38], the evaluation method of PLEF of land was constructed from the perspective of PLES and LUFs in this paper, and the coupling coordination degree model was used to explore the interaction among production, living and ecological functions in the land system. On this basis, we further proposed a space optimization scheme. It has reference value for promoting the application of PLES in the optimization of land use. In addition, the subjective grading method adopted in this article has already referred to the opinions of experienced experts in this field, which to a certain extent can reflect the differences in the strengths and weaknesses of the PLEF of land in the study area.

In addition, it's worth discussing that the research results show that the territorial space of Jiangjin District can be identified as four categories of space: the PLEBS, the PLS, the PES and the ES. Compared with previous studies on the identification of the PLES [21,22], the research results lacked a single production space and a single living space. The reason is that people's production and life are inseparable, and it is difficult to clearly delineate the boundary between the two categories of a single space in terms of spatial expression. In the identification results, the single production space and the living space are relatively small, so they are merged into the PLS.

Moreover, in 2014, the National Development and Reform Commission of China issued "the Thirteenth Five-Year City and County Economic and Social Development Planning Reform and Innovation Guidance Opinions", which required that the proportion of ecological space in key ecological function areas, cities and counties be higher than 50%. In the research results of this study, the ES only accounts for 48.72%. The reason for this is that the PLEBS and the PES both had ecological functions. Therefore, the identification result of this study was in line with the policy requirements, so the research results have reference value.

5.2. Implications for Spatial Management

In recent years, territorial space governance has gradually become the focus of research on human-environment relationship coordination and sustainable development [58]. The upper reaches of the Yangtze River is an important ecological barrier in the western region of China and the most critical area for the ecological environment protection in the Yangtze River Basin. With the rapid development of industrialization and urbanization, the interaction between human and environment in the upper reaches of the Yangtze River has been strengthened, which has produced a series of ecological problems such as water pollution and ecological destruction [12]. Therefore, it is urgent to strengthen regional territorial space governance and strengthen the construction of ecological barrier in the upper reaches of the Yangtze River. Combined with the results of this study, we suggest that the spatial management strategy in the ecological barrier area of the upper Yangtze River should focus on the following two aspects:

(1) At the macro level, the ecological barrier area in the upper reaches of the Yangtze River should adhere to the principle of ecological priority and green development, a more coordinated planning for space utilization and protection should be formulated. Moreover, the main function zoning and the three zones and three lines should be implemented. In addition, the use of PLEL should be strictly controlled to promote the coordinated development of PLES.

(2) At the micro level, more reasonable optimization measures should be taken in view of the main problems facing the management of PLES. In the territorial space with ecological function as the main body, ecological protection and restoration should be further strengthened, and ecological tourism and ecological industry could be reasonably developed under the premise of ecological protection. In addition, on the basis of strengthening the dominant functions, the territorial space with complex functions will enhance the weak functions and promote the coordinated development of the three functions. For example, in the production-living complex space, public service facilities should be improved to strengthen the leading function of production-living. At the same time, the control of production and living pollution in both urban and rural areas should be strengthened. Moreover, it can increase the ecological land area to improve the ecological function.

5.3. Research Limitations

This study also has some limitations. On the one hand, this paper divided the land use functions into production, living and ecological functions, and used the subjective assigning method to assign and grade each function to distinguish the difference between strong and weak functions. However, compared with the current quantitative analysis model and method used by most scholars to study land use optimization, the method adopted in this article has strong subjectivity and cannot reflect the actual situation of the study area more objectively. Therefore, in order to improve the accuracy of land function identification, it is necessary to build a quantitative method or model to distinguish the strength difference of certain land functions in the future. On the other hand, this article only constructed an evaluation system of the PLES from the multifunctional perspective of land use. However, we treated other factors as attributes implicit in land use, such as socioeconomic development and terrain slope, which may lead to certain limitations in the recognition results. Thus, to better adapt to the new needs of land space planning and regional space governance in the future, it is necessary to comprehensively consider the regional natural geographical environment, resource endowments, social and economic development and regional policies and other factors to construct a more comprehensive evaluation system of the PLEF and further explore the formation mechanism of the regional the PLES.

6. Conclusions

(1) The land production function in Jiangjin District presents spatial differentiation characteristics that are high in the north and low in the south. The spatial distribution of the living function and production function showed obvious consistency. In addition, the ecological function of Jiangjin District was relatively dominant, and its spatial distribution showed significant spatial complementarity with the production function and living function distribution characteristics. Areas with high ecological functions are mainly distributed in the waters of the Yangtze River and mountainous areas. In general, the area dominated by living function and production function in the study area showed obvious spatial consistency, and the spatial distribution characteristics of ecological function dominance and production function dominance showed significant spatial complementarity. The spatial distribution pattern of the PLES conforms to the basic characteristics of the ecological barrier zone in the upper reaches of the Yangtze River and the overall requirements of achieving great protection and not engaging in large-scale development.

(2) The degree of coupling and coordination of the PLEF of Jiangjin District in 2018 can be divided into incongruous zone, break-in zone, and coordination zone. The coordination zone of the PLEF is mainly distributed in the northern part of Jiangjin District. The break-in zone of the PLEF is distributed in points between the imbalanced area and the coordination area, with a wide distribution range. The incongruous zone of the PLEF is distributed throughout Jiangjin District and is especially concentrated in the southern region. Moreover, the degree of coupling and coordination between every two of the PLEF generally shows spatial consistency—high in the north and low in the south.

(3) Four space categories can be identified in the study area: the PLEBS, the PLS, the PES, and the ES. The proportions of the four categories of space are 32.04%, 3.41%, 15.83%, and 48.72%. In general, the territorial spatial pattern of Jiangjin District presents the characteristics of the PLEBS and the ES as the main body, supplemented by the PLS and the PES.

(4) The optimized partitions of the PLES in Jiangjin District can be divided into four types: the ecological conservation area, the main production-living and ecological improvement zone, the main production-ecological and living improvement zone and comprehensive promotion zone. The space dominated by production and living functions should enhance the ecological function to realize the coordinated development of the PLES, while the ecological function-oriented space should take the protection of the ecological environment as the primary goal and exploit the advantages of the ecological function to develop ecotourism and ecological industries by protecting the ecological environment.

Author Contributions: The co-authors together contributed to the completion of this article. Specifically, it follows their individual contribution: conceptualization, H.C. and Q.Y.; validation, Q.Y. and K.S.; data curation, H.C., H.Z. and D.L.; formal analysis, Q.Y. and D.L.; methodology, H.C., K.S. and L.Z.; supervision, project administration, Q.Y.; writing—original draft, H.C.; writing—review and editing, H.C., Q.Y. and H.X.; visualization, H.C. All authors have read and agreed to the published version of the manuscript.

Funding: This research was funded by Chongqing Social Science Planning Project, grant number 2020YBZX15.

Institutional Review Board Statement: Not applicable.

Informed Consent Statement: Not applicable.

Data Availability Statement: The data presented in this study are available on request from the author.

Acknowledgments: Thank you to everyone who contributed to this study.

Conflicts of Interest: The authors declare no conflict of interest.

References

1. Wang, D.; Jiang, D.; Fu, J.; Lin, G.; Zhang, J. Comprehensive assessment of production-living-ecological space based on the coupling coordination degree model. *Sustainability* **2020**, *12*, 2009. [CrossRef]
2. Lin, G.; Fu, J.; Jiang, D. Production-living-ecological conflict identification using a multiscale integration model based on spatial suitability analysis and sustainable development evaluation: A case study of Ningbo, China. *Land* **2021**, *10*, 383. [CrossRef]
3. Lin, G.; Jiang, D.; Fu, J.; Cao, C.; Zhang, D. Spatial conflict of production-living-ecological space and sustainable-development scenario simulation in Yangtze River Delta Agglomerations. *Sustainability* **2020**, *12*, 2175. [CrossRef]
4. Yang, Q.; Luo, K.; Lao, X. Evolution and enlightenment of foreign spatial planning: Exploration from the perspective of geography. *Acta Geogr. Sin.* **2020**, *75*, 1223–1236. (In Chinese)
5. Liu, Y.; Fang, F.; Li, Y. Key issues of land use in China and implications for policy making. *Land Use Policy* **2014**, *40*, 6–12. [CrossRef]
6. Lu, D.; Wang, Y.; Yang, Q.; Su, K.; Zhang, H.; Li, Y. Modeling spatiotemporal population changes by integrating DMSP-OLS and NPP-VIIRS nighttime light data in Chongqing, China. *Remote Sens.* **2021**, *13*, 284. [CrossRef]
7. Zou, L.; Liu, Y.; Wang, J.; Yang, Y. An analysis of land use conflict potentials based on ecological-production-living function in the southeast coastal area of China. *Ecol. Indic.* **2021**, *122*, 107297. [CrossRef]
8. Liu, Y.; Li, J.; Yang, Y. Strategic adjustment of land use policy under the economic transformation. *Land Use Policy* **2018**, *74*, 5–14. [CrossRef]
9. Su, K.; Hu, B.; Shi, K.; Zhang, Z.; Yang, Q. The structural and functional evolution of rural homesteads in mountainous areas: A case study of Sujiaying village in Yunnan province, China. *Land Use Policy* **2019**, *88*, 104100. [CrossRef]
10. Li, J.; Sun, W.; Yu, J. Change and regional differences of production-living-ecological space in the Yellow River Basin: Based on comparative analysis of resource-based and non-resource-based cities. *Res. Sci.* **2020**, *42*, 2285–2299. (In Chinese)
11. Duan, Y.; Wang, H.; Huang, A.; Xu, Y.; Lu, L.; Ji, Z. Identification and spatial-temporal evolution of rural "production-living-ecological" space from the perspective of villagers' behavior-A case study of Ertai Town, Zhangjiakou City. *Land Use Policy* **2021**, *106*, 105457. [CrossRef]
12. Zhang, H.; Yang, Q.; Zhang, Z.; Lu, D.; Zhang, H. Spatiotemporal changes of ecosystem service value determined by national land space pattern change: A case study of Fengdu County in The Three Gorges Reservoir Area, China. *Int. J. Environ. Res. Public Health* **2021**, *18*, 5007. [CrossRef]
13. Cao, S.; Hu, D.; Hu, Z.; Zhao, W.; Chen, S.; Yu, C. Comparison of spatial structures of urban agglomerations between the Beijing-Tianjin-Hebei and Boswash based on the subpixel-level impervious surface coverage product. *J. Geogr. Sci.* **2018**, *28*, 306–322. [CrossRef]
14. Sun, X.; Yu, C.; Wang, J.; Wang, M. The intensity analysis of production living ecological land in Shandong Province, China. *Sustainability* **2020**, *12*, 8326. [CrossRef]
15. Wei, C.; Lin, Q.; Yu, L.; Zhang, H.; Ye, S.; Zhang, D. Research on sustainable land use based on production-living-ecological function: A case study of Hubei Province, China. *Sustainability* **2021**, *13*, 996. [CrossRef]
16. Shi, Z.; Deng, W.; Zhang, S. Spatio-temporal pattern changes of land space in Hengduan Mountains during 1990–2015. *J. Geogr. Sci.* **2018**, *28*, 529–542. [CrossRef]
17. Deng, W.; Zhang, J.; Shi, Z.; Wan, J.; Meng, B. Interpretation of mountain territory space and its optimized conceptual model and theoretical framework. *Mt. Res.* **2017**, *35*, 121–128. (In Chinese)
18. Tang, C.; He, Y.; Zhou, G.; Zeng, S.; Xiao, L. Optimizing the spatial organization of rural settlements based on life quality. *J. Geogr. Sci.* **2018**, *28*, 685–704. [CrossRef]
19. Huang, A.; Xu, Y.; Lu, L.; Liu, C.; Zhang, Y.; Hao, J.; Wang, H. Research progress of the identification and optimization of production-living-ecological spaces. *Prog. Geogr.* **2020**, *39*, 503–518. (In Chinese) [CrossRef]
20. Fang, C.L. Scientific cognition and technical paths of urban multiple planning integration in China. *China Land Sci.* **2017**, *31*, 28–36. (In Chinese)
21. Ji, Z.; Liu, C.; Xu, Y.; Huang, A.; Lu, L.; Duan, Y. Identification and optimal regulation of the production-living-ecological space based on quantitative land use functions. *Trans. Chin. Soc. Agric. Eng.* **2020**, *36*, 222–231, 315. (In Chinese)
22. Sui, H.; Song, G.; Zhang, H. Identification of production-living-ecological space at Keshan County level in main grain producing areas in northern Songnen Plain, China. *Trans. Chin. Soc. Agric. Eng.* **2020**, *36*, 264–271, 323. (In Chinese)
23. Li, G.; Fang, C. Quantitative function identification and analysis of urban ecological-production-living spaces. *Acta Geogr. Sin.* **2016**, *71*, 49–65. (In Chinese)
24. Tu, S.; Long, H.; Zhang, Y.; Ge, D.; Qu, Y. Rural restructuring at village level under rapid urbanization in metropolitan suburbs of China and its implications for innovations in land use policy. *Habitat Int.* **2018**, *77*, 143–152. [CrossRef]
25. Tu, S.; Long, H. Rural restructuring in China: Theory, approaches and research prospect. *J. Geogr. Sci.* **2017**, *27*, 1169–1184. [CrossRef]
26. Yu, Z.S.; Chen, Y.Q.; Li, X.J.; Sun, D.Q. Spatial evolution process, motivation and restructuring of "production-living-ecology" in industrial town: A case study on Qugou Town in Henan Province. *Sci. Geogr. Sin.* **2020**, *40*, 646–656. (In Chinese)
27. Tao, Y.; Wang, Q. Quantitative recognition and characteristic analysis of production-living-ecological space evolution for five resource-based cities: Zululand, Xuzhou, Lota, Surf Coast and Ruhr. *Remote Sens.* **2021**, *13*, 1563. [CrossRef]

28. Cheng, W.; Ning, T. Spatio-temporal characteristics and evolution of rural productionliving-ecological space function coupling coordination in Chongqing Municipality. *Geogr. Res.* **2018**, *37*, 1100–1114. (In Chinese)

29. Huang, J.C.; Lin, H.X.; Qi, X.X. A literature review on optimization of spatial development pattern based on ecological-production-living space. *Prog. Geogr.* **2017**, *36*, 378–391. (In Chinese)

30. Liao, G.; He, P.; Gao, X.; Deng, L.; Zhang, H.; Feng, N.; Zhou, W.; Deng, O. The production-living-ecological land classification system and its characteristics in the Hilly Area of Sichuan Province, Southwest China based on identification of the main functions. *Sustainability* **2019**, *11*, 1600. [CrossRef]

31. Wang, X.; Wang, Z.Q.; Jin, G.; Yang, J. Transactions of the Chinese Society of Agricultural Engineering. *Trans. Chin. Soc. Agric. Eng.* **2016**, *32*, 240–248. (In Chinese)

32. Paracchini, M.L.; Pacini, C.; Jones, M.L.M.; Perez-Soba, M. An aggregation framework to link indicators associated with multifunctional land use to the stakeholder evaluation of policy options. *Ecol. Indic.* **2011**, *11*, 71–80. [CrossRef]

33. Porta, J.; Parapar, J.; Doallo, R.; Rivera, F.F.; Sante, I.; Crecente, R. High performance genetic algorithm for land use planning. *Comput. Environ. Urban Syst.* **2013**, *37*, 45–58. [CrossRef]

34. Cao, K.; Huang, B.; Wang, S.; Lin, H. Sustainable land use optimization using Boundary-based Fast Genetic Algorithm. *Comput. Environ. Urban Syst.* **2012**, *36*, 257–269. [CrossRef]

35. Yuan, M.; Liu, Y.; He, J.; Liu, D. Regional land-use allocation using a coupled MAS and GA model: From local simulation to global optimization, a case study in Caidian District, Wuhan, China. *Cartogr. Geogr. Inf. Sci.* **2014**, *41*, 363–378. [CrossRef]

36. Hagenauer, J.; Helbich, M. Mining urban land-use patterns from volunteered geographic information by means of genetic algorithms and artificial neural networks. *Int. J. Geogr. Inf. Sci.* **2012**, *26*, 963–982. [CrossRef]

37. Zheng, W.; Ke, X.; Xiao, B.; Zhou, T. Optimising land use allocation to balance ecosystem services and economic benefits—A case study in Wuhan, China. *J. Environ. Manag.* **2019**, *248*, 109306. [CrossRef] [PubMed]

38. Zhang, L.; Xie, L.; Xiao, Y. Maximising the bene fi ts of regulatory ecosystem services via spatial optimisation. *J. Clean. Prod.* **2021**, *291*, 125272. [CrossRef]

39. Elliot, T.; Bertrand, A.; Almenar, J.B.; Petucco, C.; Proenca, V.; Rugani, B. Spatial optimisation of urban ecosystem services through integrated participatory and multi-objective integer linear programming. *Ecol. Model.* **2019**, *409*, 108774. [CrossRef]

40. Herzig, A.; Ausseil, A.; Dymond, J.R. Spatial optimisation of ecosystem services. In *Ecosystem Services in New Zealand—Conditions and Trends*; Landcare Research: Lincoln, New Zealand, 2013.

41. Zhang, X.; Xu, Z. Functional coupling degree and human activity intensity of production-living-ecological space in underdeveloped regions in China: Case study of Guizhou Province. *Land* **2021**, *10*, 56. [CrossRef]

42. Yang, H.; Lu, X. Study on village type identification based on spatial evolution and simulation of "production-living-ecological space": A case study of Changning City in Hunan Province. *China Land Sci.* **2020**, *34*, 18–27. (In Chinese)

43. Cui, J.; Gu, J.; Sun, J.; Luo, J. The spatial pattern and evolution characteristics of the production, living and ecological space in Hubei Provence. *China Land Sci.* **2018**, *32*, 67–73. (In Chinese)

44. Xiaoyu, C.; Qingyuan, Y.; Guohua, B. Temporal and spatial difference of land resource carrying capacity of Jangjin District in Chongqing. *Res. Environ. Yangtze Basin* **2019**, *28*, 2319–2330. (In Chinese)

45. Wang, C.; Ren, M.; Li, H.; Zhu, Y. Understanding the rural production space system: A case study in Jiangjin, China. *Sustainability* **2019**, *11*, 2811. [CrossRef]

46. Verburg, P.H.; Van De Steeg, J.; Veldkamp, A.; Willemen, L. From land cover change to land function dynamics: A major challenge to improve land characterization. *J. Environ. Manag.* **2009**, *90*, 1327–1335. [CrossRef] [PubMed]

47. Zhang, Y.; Long, H.; Tu, S.; Ge, D.; Ma, L.; Wang, L. Spatial identification of land use functions and their tradeoffs/synergies in China: Implications for sustainable land management. *Ecol. Indic.* **2019**, *107*, 105550. [CrossRef]

48. Huang, J.; Tichit, M.; Poulot, M.; Darly, S.; Li, S.; Petit, C.; Aubry, C. Comparative review of multifunctionality and ecosystem services in sustainable agriculture. *J. Environ. Manag.* **2015**, *149*, 138–147. [CrossRef]

49. Brunstad, R.J.; Gaasland, I.; Vardal, E. Multifunctionality of agriculture: An inquiry into the complementarity between landscape preservation and food security. *Eur. Rev. Agric. Econ.* **2005**, *32*, 469–488. [CrossRef]

50. Liu, J.; Liu, Y.; Li, Y. Classification evaluation and spatial-temporal analysis of "production-living-ecological" spaces in China. *Acta Geogr. Sin.* **2017**, *72*, 1290–1304. (In Chinese)

51. Ma, L.; Jin, F.; Song, Z.; Liu, Y. Spatial coupling analysis of regional economic development and environmental, pollution in China. *J. Geogr. Sci.* **2013**, *23*, 525–537. [CrossRef]

52. Zhao, Z.; Zhao, L.; Wang, Y.; Yuan, T. Analysis of water resources utilization efficiency and water saving potential in Xiong'an New Area under different scenarios. *J. Nat. Res.* **2019**, *34*, 2629–2642. (In Chinese)

53. Tan, M.; Li, X.; Li, S.; Xin, L.; Wang, X.; Li, Q.; Li, W.; Li, Y.; Xiang, W. Modeling population density based on nighttime light images and land use data in China. *Appl. Geogr.* **2018**, *90*, 239–247. [CrossRef]

54. Wang, J.; Sun, J. Some in-depth thoughts on resources and environmental comprehensive scientific expedition using gridding approach in China. *J. Geo-Inf. Sci.* **2015**, *17*, 758–764. (In Chinese)

55. Jin, X.X.; Lu, Y.Q.; Lin, J.H.; Qi, X.H.; Hu, G.J.; Li, X. Research on the evolution of spatiotemporal patterns of production-livingecological space in an urban agglomeration in the Fujian Delta region, China. *Acta Ecol. Sin.* **2018**, *38*, 4286–4295. (In Chinese)

56. Zhu, Y.Y.; Yu, B.; Zeng, J.X.; Han, Y. Spatial optimization from three spaces of production, living and ecology in national restricted zones—A case study of Wufeng County in Hubei Province. *Econ. Geogr.* **2015**, *35*, 26–32. (In Chinese)
57. Cui, S.Q.; Zhu, P.J.; Zhou, G.H.; Zhang, H.H.; Deng, X.Z. Urban spatial function change and regulation path from the perspective of"production-living-ecological": Taking Changsha City as an example. *Res. Environ. Yangtze Basin* **2020**, *29*, 1733–1745. (In Chinese)
58. Gao, L.; Bryan, B.A. Finding pathways to national-scale land-sector sustainability. *Nature* **2017**, *544*, 217–222. [CrossRef]

land

MDPI

Article

Production–Living–Ecological Conflict Identification Using a Multiscale Integration Model Based on Spatial Suitability Analysis and Sustainable Development Evaluation: A Case Study of Ningbo, China

Gang Lin [1,2], **Jingying Fu** [1,2,3,*] **and Dong Jiang** [1,2,3]

[1] Institute of Geographical Sciences and Natural Resources Research, Chinese Academy of Sciences, No. 11A Datun Road, Chaoyang District, Beijing 100101, China; ling@lreis.ac.cn (G.L.); jiangd@igsnrr.ac.cn (D.J.)

[2] College of Resources and Environment, University of Chinese Academy of Sciences, No. 19A Yuquan Road, Haidian District, Beijing 100049, China

[3] Key Laboratory of Carrying Capacity Assessment for Resource and Environment, Ministry of Natural Resources, No. 46 Fuchengmen Road, Xicheng District, Beijing 100812, China

[*] Correspondence: fujy@igsnrr.ac.cn; Tel.: +86-10-6488-9221

Abstract: Production–living–ecological space (PLES) basically covers the scope of spatial activities in people's material production and spiritual life and is the basic carrier of human social development and economic activities. The coordinated development of PLES is an effective method to mitigate land-use conflicts to achieve balanced and coordinated development of the region. However, so far, compared with the single-scale study based on administrative unit, the PLES conflicts between microcosmic grid-scale receives less attention. Considering the important scale problems of the geographical study, this study aims to analyze the synergetic degree of PLES under different scales (administrative-unit, grid, and integrated multiscale) and to scientifically diagnose land use conflicts in Ningbo, China. Results indicated that production land and ecological land in Ningbo were continuously occupied by human activities from 2010 to 2018. The lowest and lower suitability areas of ecological space in Ningbo increased from 2010 to 2018. Land ecological suitability was seriously affected by urban expansion, its ecological value was reduced, and the PLES developed towards the trend of being uncoordinated. Multiscale coupling analysis showed that the PLES in Ningbo was in less conflict on the whole, but with the development of the economy, the coupling coordination degree of PLES was also damaged. This study establishes the different scales of a PLES coupling coordination development degree evaluation index system and enriches the methods of multiscale land use fusion conflict diagnosis and also provides a scientific reference for the optimized and sustainable development of regional territorial space.

Keywords: PLES; multiscale integration; coupling coordination; conflict diagnosis; Ningbo

Citation: Lin, G.; Fu, J.; Jiang, D. Production–Living–Ecological Conflict Identification Using a Multiscale Integration Model Based on Spatial Suitability Analysis and Sustainable Development Evaluation: A Case Study of Ningbo, China. *Land* **2021**, *10*, 383. https://doi.org/ 10.3390/land10040383

Academic Editor: Alexander Khoroshev

Received: 18 February 2021
Accepted: 3 April 2021
Published: 7 April 2021

Publisher's Note: MDPI stays neutral with regard to jurisdictional claims in published maps and institutional affiliations.

1. Introduction

1.1. Motivation and Literature Review

Those who utilize land resources, as well as other stakeholders, are now looking for diversified regional development goals, demand of different phases, and land resource multi-suitability, scarcity, etc. All these pursuits together lead to different sorts of land use conflicts, severely restricting the multi-functional use of land [1,2]. Over the past 40 years of reform and opening up, China's economy has rapidly developed. Especially since the beginning of the 21st century, pressure on resources and the environment caused by population growth and economic development has intensified, and the conflict between humans and land has become increasingly prominent. On the basis of high land expansion and resource consumption, the land-development model overemphasizes economic growth and ignores the overall layout of resources, which leads to great changes in the relationship between humans and land. With the drastic evolution of the land use spatial pattern, the frequency of land use conflict is increasing, and its form and content are more complex and changeable. How we control land use conflict to a manageable proportion has become an important task in the process of land use [3,4]. Land use conflict has a lot to do with the relationship between humans and land. It is a complicated issue involving multiple intertwining factors, such as nature, economy, and society. The only way to manage the conflict zone precisely is to quantitatively identify their functions, and the coupling coordination of land use functions is the point of easing land-use conflict and land resource management. However, because of the complexity and comprehensiveness of land use functions and the different emphasis on the characteristics of the study area (e.g., scale and regional characteristics), there was no unified land use function classification system to scientifically evaluate and diagnose conflict. Production–living–ecological space (PLES), the shortened form for production space (PS), living space (LS), and ecological space (ES), was reclassified considering both land use functions and utilization types [5]. It was formally put forward in the 18th National Congress of the Chinese Communist Party in 2012, and its purpose was to optimize the spatial pattern of land uses by the overall coordination of PS, LS, and ES [6]. PLES basically covers the scope of spatial activities in people's material production and spiritual life. It is the basic carrier of human social development and economic activities. The three are mutually independent and inter-related; they have a symbiotic fusion and restriction effect [7]. Therefore, starting from exploring the differences in the functions of PLES in different regions, as well as the functional relationship between them, it is an effective method to mitigate land-use conflicts to achieve balanced and coordinated development of the region by probing the shortcomings of regional development and clarifying the characteristics of regional spatial patterns. The available research on the methods of quantitative recognition of spatial conflict mainly includes the comprehensive index model, which calculates the spatial conflict index based on the complexity, fragility, and dynamics of the land use system [8]; the competitiveness evaluation model, which ranks the conflicts of the construction, agriculture, and ecology space by the establishment of a competitiveness evaluation index system based on land suitability and driving force [9]; and the suitability evaluation model, which identifies spatial conflicts by evaluating the suitability of the specific land-use types [10]. This kind of method, combined with the geographic information system (GIS), introduced the multi-criteria spatial decision support system [11,12]. Zou et al. developed the conflict identification and intensity diagnosis by this model on the basis of the suitability evaluation [13]. In addition, many scholars have established the spatial-conflict index to identify land conflicts from the perspective of PLES when the PLES concept was introduced [14,15]. However, so far, compared with the single-scale study based on administrative unit, the PLES conflicts between the microcosmic grid-scale receives less attention. Due to the scale dependence of geographical phenomena [16], the spatiotemporal pattern characteristics vary with the geographical scales; there is an urgent need to establish a multiscale integration model to scientifically diagnose land-use conflicts based on the PLES perspective.

1.2. Objective and Contribution

This study aims to analyze the synergetic degree of PLES under different scales (administrative-unit, grid, and integrated multiscale) and to scientifically diagnose land use conflicts in Ningbo, China by synthesizing an evaluation model of sustainable development, a coupling coordination degree model, and a multiscale mathematical model. This is expected to enrich the methods of multiscale land use fusion conflict diagnosis and provide scientific reference for the optimized and sustainable development of regional territorial space.

2. Materials and Methods

2.1. Study Area

The city of Ningbo is an important port on the southeast coast of China, located in the eastern part of Zhejiang province and the southern wing of the Yangtze River Delta. It is adjacent to Qiantang River and Hangzhou Bay in the north, the city of Shaoxing in the west, the city of Taizhou in the south, and the city of Zhoushan in the east across the sea. The terrain is high in the southwest and low in the northeast, and the height difference is about 1000 m. There are two mountain ranges in the city, Siming Mountain and Tiantai Mountain. Hills and mountains above 50 m account for 57% of the total land area. It is one of the 14 coastal cities opened to the rest of the world, and an important city in the Yangtze River Delta urban agglomeration [17]. The total land area of the city was 9816 km^2, the population was approximately 8.5 million, and its GDP was CNY 1240.87 billion in 2020 [18]. Rapid population and economic growth have resulted in high demand for industrial and residential space, leading to unprecedented PLES evolution that poses serious challenges to ecosystems and the natural environment. Regional environmental risk at Ningbo has increased substantially over the past 40 years and will increase over the next several decades with the increasing demands of economic and social development on territorial space resources. The resource bottlenecks and environmental pressure are becoming increasingly prominent, and the contradiction between the supply and demand of land space resources is becoming progressively obvious [19–21]. Ningbo has attracted attention from environmentalists, local authorities, and scientists. Therefore, it is urgent and necessary to construct an ecological civilization. For the past 10 years, while the PS and LS in Ningbo have been expanding, the ES has been shrinking, and PLES contradictions in Ningbo are prominent. How to coordinate the relationship between development, livability, and protection; improve the coordinated development level of PLES in function; and promote the coordinated and sustainable development of the economy, society, and ecology are major issues for the development of Ningbo.

2.2. Research Framework

Based on the characteristics of geographical scale dependence, this study was undertaken to diagnose PLES conflicts from two aspects using a multiscale integration method: administrative-unit scale and the spatial grid scale. The administrative-unit scale highlights the sustainability of PLES, based on the sustainable development goals (SDGs), while the spatial grid scale focuses on the functional suitability of PLES. The administrative-unit scale detects PLES conflicts from the macroscopic perspective on the basis of sustainable utilization of PLES, whereas the spatial grid scale more accurately reflects the microscopic and detailed functional differences of PLES, compensating for the lack of the strong generality of the microscopic scales. Thus, the research framework was developed as follows:

Step 1: explore PLES spatiotemporal variation on the basis of PLES grid data at 1 × 1 km.

Step 2: evaluate PLES suitability on the grid scale and quantify the synergetic degree using the coupling and coordination model.

Step 2: quantify PLES sustainable development evaluation on the administration-cell scale using the coupling and coordination model.

Step 3: diagnose PLES conflicts using the multiscale integration model to combine the evaluation of the synergetic degree of PLES on the grid and administration-cell scales.

2.3. PLES classification

PLES data in Ningbo of 2010 and 2018 evolved from land use data (1 × 1 km) and were provided by Data Center for Resources and Environmental Sciences, Chinese Academy of Sciences (RESDC) (http://www.resdc.cn, accessed on 5 November 2020). Land use classification includes six first-level categories, namely, arable land, woodland, rangeland, water-area and -conservancy facility land, construction land, and unused land and 25 s-level categories. According to the second-level classification standard of the land-use/-cover classification system for remote-sensing monitoring in China, on the basis of the differentiation of land use functions and types, and the linkage table (Table 1), PLES structure and land use types were established.

Table 1. Corresponding table of type between production–living–ecological space (PLES) and land use.

	PLES Classification	Land-Use/-Cover Classification System for Remote-Sensing Monitoring in China [22]
PS	Agricultural-production space	Paddy fields, dry land
	Industrial-production space	Industry and mining, land use for transport construction
LS	Urban-living space	Urban land
	Rural-living space	Rural residential land
	Forestland ecological space	Woodland, shrub land, open forest land, other woodland
	Grassland ecological space	Grassland with high, medium, and low coverage
ES	Water ecological space	Canals, lakes, reservoir pit ponds, permanent glaciers and snowfields, tidal flats, bottomland
	Other ecological space	Sandy land, Gobi, saline alkali land, swamp land, bare land, bare rock land, and other unused land

2.4. Construction of PLES Suitability Evaluation Model on Grid Scale

PLES suitability is to evaluate suitability in different land-utilization types of county territories with high vegetation coverage. In particular, production suitability refers to the suitability of county territories with high vegetation coverage in providing tangible agricultural products or industrial products or intangible products for humans. It mainly reflects the product production level of county territories, which is mainly affected by natural climate, land suitability, and development convenience. Living suitability refers to the suitability of county territories with high vegetation coverage in living conditions such as convenient facilities, housing, and public activities. It investigates the living support level of residents in county territories, which is mainly affected by public facilities, terrain, and the social economy. Ecological suitability refers to the suitability of the research area to provide direct or indirect ecological products and ecological services for humans, which is mainly affected by environmental quality and the social environment. The index system of PLES suitability evaluation in this paper is as follows (see Table 2):

Table 2. Index system of PLES suitability evaluation.

Items	Index	Factor Classification and Score [14,23–27]			
		100	80	60	40
Production suitability	Average annual temperature	≥21	>18–21	>15–18	≤15
	Annual precipitation	≥1800	>1700–1800	>1600–1700	≤1600
	Altitude	<150 m	>150–300 m	>300–500 m	>500 m
	Land use type	Dry land, paddy field, other construction land	Rural land, urban land	Grassland with high coverage	Others
	Gradient	<3°	3°–8°	8°–15°	15°–25°
	Distance from road (m)	500 m	1500 m	3000 m	5000 m
Living suitability	Average annual temperature	≥21	>18–21	>15–18	≤15
	Annual precipitation	≥1800	>1700–1800	>1600–1700	≤1600
	Topographic position index	≤0.54	0.54–0.62	0.62–0.72	≥0.72
	Distance from town center	500 m	1500 m	3000 m	5000 m
	Distance from school and hospital	500 m	1500 m	3000 m	5000 m
	Land use type	Rural land, urban land	Other construction land	N/A	Others

Table 2. Cont.

| Items | Index | Factor Classification and Score [14,23–27] | | | |
		100	80	60	40
Ecological suitability	Land use type	Open forest land, grassland with high coverage, swamp land	shrub land, canals, lakes	Dry land, paddy fields, woodland, shrub land, other woodlands, grassland with medium and low coverage, reservoir pit ponds	Rural land, urban land, other construction land, and others
	Landscape fragmentation	Better regularity	Good regularity	General regularity	Bad regularity
	Normalized difference vegetation index (NDVI)	≥ 0.5	>0.25–0.5	>0.15–0.25	≤ 0.15
	Distance from water body	5000 m	3000 m	1500 m	500 m

Note: weight of each index determined using analytic hierarchy process (AHP) and expert scoring method. The topographic position index, calculated by the slope and the elevation, was used to reflect the influence from the comprehensive geomorphic conditions of Ningbo.

2.5. Coupling and Coordination Model

The coupling coordination degree model is a relationship that can better describe the interaction and influence of two or more systems in the development process. Coupling coordination degree is the degree of mutual promotion and restriction among various systems at high and low levels and reflects the degree of interaction and coordinated development among various systems. This paper mainly adopted this model to calculate the coupling coordination degree of PLES suitability results and regional sustainable development. The coupling coordination degree (CCD) can be obtained according to the following formula. The value of CCD is in the range of 0–1. A higher value indicates a higher degree of coupling coordination [28,29].

$$D = \sqrt{C \times T} \tag{1}$$

$$T = \alpha \times U_1 + \beta \times U_2 + \cdots + \gamma \times U_3 \quad (\alpha = \beta = \gamma = 1/\text{n}) \tag{2}$$

$$C = \sqrt[n]{(U_1 \times U_2 \times \cdots \times U_n) / \left(\frac{U_1 + U_2 + \cdots + U_n}{n} \right)^n} \tag{3}$$

$$U = \sum_{i=1}^{n} (\omega_i \times x_i) \tag{4}$$

where D represents the coupling and coordination degree, T reflects the overall effect and level of each subsystem, C is the coupling degree, U represents subsystem performance, W_i represents the weight value of indicator I, and X_i represents the standard value of indicator i in each subsystem.

There are three subsystems in this study, namely, production, life, and ecology, so C and T are calculated by the following formulas, where U_p, U_l, and U_e are the performance levels of the PS, LS, and ES subsystems, respectively; α, β, and γ represent the contributions of the PS, ES, and LS subsystems, respectively.

$$C = \sqrt[3]{(U_p \times U_l \times U_e) / \left(\frac{U_p \times U_l \times U_e}{3} \right)^3} \tag{5}$$

$$T = \alpha \times U_p + \beta \times U_l + \gamma \times U_e \quad (\alpha = \beta = \gamma = 1/3) \tag{6}$$

2.6. Sustainable Development Evaluation on Administration-Cell Scale

Sustainable development is the fundamental pursuit of territorial space optimization and the starting point of PLES optimization [30]. PLES optimization is consistent with the sustainable development theory of the multi-objective coordination of economy, society, and environment and is the embodiment and implementation of sustainable development theory in urbanization construction. On the basis of SDGs 1–3, 6, 8, 11–13, and 15, this study was guided by "intensive production", "livable life", and "beautiful ecology" to establish the coupling coordination degree evaluation index of PLES, covering the scale of space, structure of space, efficiency of space, quality of space, and other factors (Table 3). It aims to provide a scientific basis for the efficient use of PLES and the sustainable development of economy–society–environment.

Table 3. Index system of spatial coupling coordination degree evaluation for sustainable development goals (SDGs).

Items	SDGs	First-Level Indices	Second-Level Indices
Production space (intensive and efficient)	SDGs 2 and 12	Scale of production space Structure of production space Efficiency of production space	Scale of agricultural land [31] Scale of industrial land [31] Advancement of industrial structure [32] Grain output rate [33] Land output rate [34] Industrial efficiency [34]
Living space (livable life)	SDGs 1 and 11	Scale of living space Quality of living Convenience of living	Size of residential space [31] Green-space coverage ratio [33] Engel's coefficient of urban residents [33] Transportation convenient [35] Traffic accessibility [36]
Ecological space (beautiful ecology)	SDGs 3, 6, 13, and 15	Scale of ecological space Quality of ecological space	Scale of ecological land [31] NDVI [37] Air quality [38] Sewage treatment rate [32] Health level of residents [39]

2.7. PLES Conflict Identification Based on Multiscale Integration Model

In this study, the iterative method was used to establish spatial relations on different scales and perspectives [40]. According to the scores and weights of various functional evaluation factors on different scales, a multiscale mathematical model of PLES conflict weight was established:

$$C_{p,l,e} = (1 - \alpha) \times f_{p,l,e} + \alpha \sum_{i=1}^{m} f_i \times \beta, \tag{7}$$

where $C_{p,l,e}$ represents the comprehensive evaluation value of PLES conflict weight (p, production; l, living; e, ecological) under multiscale integration; α represents the weight of evaluation results at the upper scale (administrative-unit); $f_{p,l,e}$ is the evaluation index of the suitability of PLES on the grid scale; f_i and β represent the evaluation factor of sustainable development of PLES on the administrative-unit scale and the corresponding index weight, respectively. Using the calculated results by the multiscale comprehensive evaluation model to diagnose the PLES conflicts of Ningbo, according to the comprehensive evaluation score of each function of each grid, the level of each function was judged. With ArcGIS, conflict types were classified on average into five levels, namely, no conflict, little conflict, medium conflict, serious conflict, and violent conflict.

3. Results and Analysis

3.1. Spatial–Temporal PLES Characteristics in Ningbo

Figure 1 shows the spatial characteristic of PLES in Ningbo for (a) 2010 and (b) 2018. In general, due to the rapid development of the economy and the acceleration of urbanization in recent years, a great deal of PS and ES in Ningbo was transformed into living space, resulting in a decline of PS and ES, which decreased by 1.68% and 1.89%, respectively, while the area occupied by living space increased by 14.4%. These data show that the rapid development of urbanization and the pursuit of high-quality life in Ningbo from 2010 to

2018 led to the continuous expansion of living land. Production land and ecological land were constantly occupied under the influence of human activities, although the area of PS decreased, and the area of industrial PS slightly increased, which indicates that agricultural PS was greatly affected by human activities. Agricultural PS occupied a large area, and the level of agricultural development was affected.

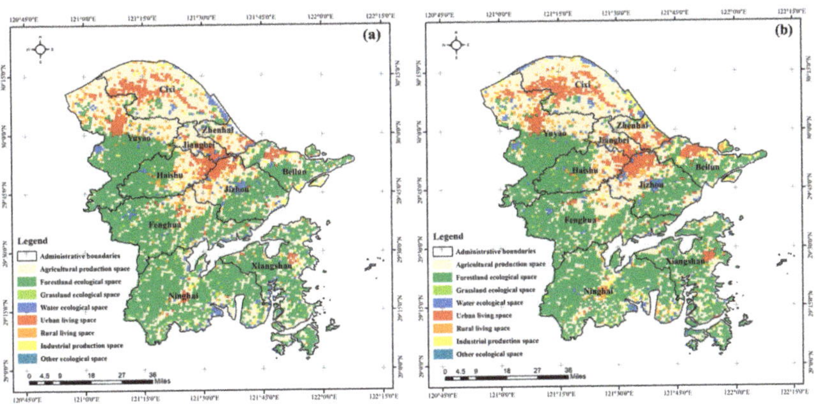

Figure 1. PLES in Ningbo for (**a**) 2010 and (**b**) 2018.

3.2. Coupling and Coordination for PLES Suitability Analysis in Ningbo

Figure 2 shows the suitability evaluation results of PLES in Ningbo. From 2010 to 2018, the areas of lowest and lower suitability of ecological space in Ningbo increased, while areas of general, higher, and highest suitability decreased, indicating that urban expansion seriously affected the ecological suitability of land and reduced its ecological value. The lowest-suitability areas were sporadically distributed in the two years and were mainly in the northern part of Ningbo in 2010. However, in 2018, lowest-suitability areas both spread to the southern part and greatly increased in the northern part, even showing a trend of aggregation. This was mainly due to the acceleration of the urbanization process and the continuous expansion of living space. Increasing amounts of ecological land with flat terrain suitable for urban construction and expansion were used for urban construction. Lower suitability areas also gradually expanded from the north to the south, and the northern area continuously expanded. From 2010 to 2018, lowest, general, and higher suitability areas of living space in Ningbo decreased, while lower and highest suitability areas increased, indicating that higher and general suitability areas were increasingly developed into urban construction areas. With the expansion of urban areas, many lowest-suitability areas were affected by the surrounding cities. Their infrastructure conditions were gradually optimized, and population density gradually increased, making development suitability also gradually increase. From the perspective of space, lowest suitability areas were mainly distributed in the western and northern parts of Ningbo, while highest suitability areas were mainly distributed in urban built-up areas and their surroundings. These areas are densely populated with a long history of development and strong infrastructure conditions to assist in daily life, which are suitable for maintaining the development of living space. Lowest suitability PS areas in Ningbo were mainly distributed in woodland and rangeland in the central and southern parts of Ningbo. These areas are generally at higher elevations, with higher slopes, inconvenient transportation, and high costs for industrial and agricultural production. Higher suitability areas and highest suitability areas of PS were mainly distributed in the northern part of Ningbo, and distribution was similar to that in higher and highest suitability LS areas. Higher suitability areas were mainly distributed in arable land, with a certain basis for agricultural production and development. Highest suitability areas significantly increased from 2010 to

2018, mainly distributed in the urban areas of Ningbo, with dense population, a high land use degree, excellent infrastructure conditions, and a long-term basis for industrial and service production.

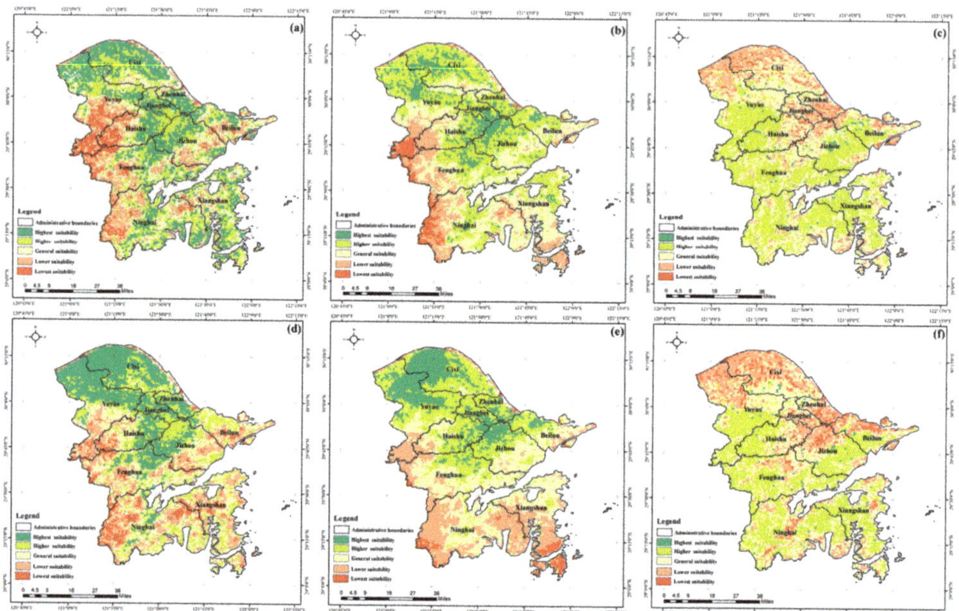

Figure 2. Suitability evaluation of PLES in Ningbo for (**a**) production space, (**b**) living space, and (**c**) ecological space in 2010; and (**d**) production space, (**e**) living space, and (**f**) ecological space in 2018.

Figure 3 shows the results of the coupling coordination degree of PLES suitability in Ningbo. From 2010 to 2018, the uncoordinated areas of PLES in Ningbo remained basically unchanged with small increases. However, near-uncoordinated areas increased by 75.82%. Highest coordinated areas had little change, increasing by only 4.56%. To be specific, the large increase in near-uncoordinated areas indicates that the development of Ningbo in recent years developed PLES towards the uncoordinated direction. From the perspective of space, uncoordinated and near-uncoordinated areas are mainly distributed in the south and west of Ningbo. Those areas are mainly woodland and rangeland. Terrain conditions make it difficult to develop and are merely suitable for maintaining its ecological value, which gradually leads to development conditions between ecological and production–living space being uncoordinated. Higher and highest coordinated areas are mainly located in the south of Ningbo. These are mainly distributed in urban areas, agricultural production space, and their surroundings. These areas have both the local basis for the development of production and living spaces and a higher ability for ecological function development because of their natural geographical factors. The suitability development of PLES in these areas is coordinated.

3.3. Sustainable Development Evaluation of PLES on Administration-Cell Scale in Ningbo

Figure 4 shows the result of the sustainable development evaluation of PLES on the administration-cell scale in Ningbo for 2010 and 2018. Results showed that the state of sustainable development evaluation of PLES in the districts of Jiangbei and Fenghua were uncoordinated, and other counties and districts in Ningbo were near-uncoordinated in 2010. Jiangbei is located in the central urban area of Ningbo, and there, the development of secondary industries promoted economic growth but inhibited the coordinated development of LS and ES. Thus, the development level of PS in Jiangbei was the highest, followed by ES, and the development level of LS was the lowest in 2010. Fenghua is located in the western part of Ningbo; the basic geographical conditions of high altitude and high slope were the main factors restricting the development of LS in this area. Fenghua had the highest development level for ES, PS was second, and the development level of LS was the lowest in 2010. In 2018, the state of sustainable development evaluation of PLES of all counties and districts in Ningbo was higher than that in 2010, and all counties and districts in Ningbo were in a near-uncoordinated state. Among them, Jiangbei and Fenghua paid more attention to the coordinated development of PLES after 2010, the coordination index increased by 0.15 and 0.11, respectively, from 2010 to 2018, and the improvement range of coordination index in other counties and districts was less than 0.10. Although the state of sustainable development evaluation in Ninghai and Xiangshan was higher than that in other regions in 2010, due to the lack of coordinated development of PLES and the influence of geographical conditions, the gap between the development levels of PS, LS, and ES gradually widened, the coordination index only increased by 0.04 and 0.02 from 2010 to 2018, respectively, and their state of sustainable development evaluation of PLES was the lowest in Ningbo.

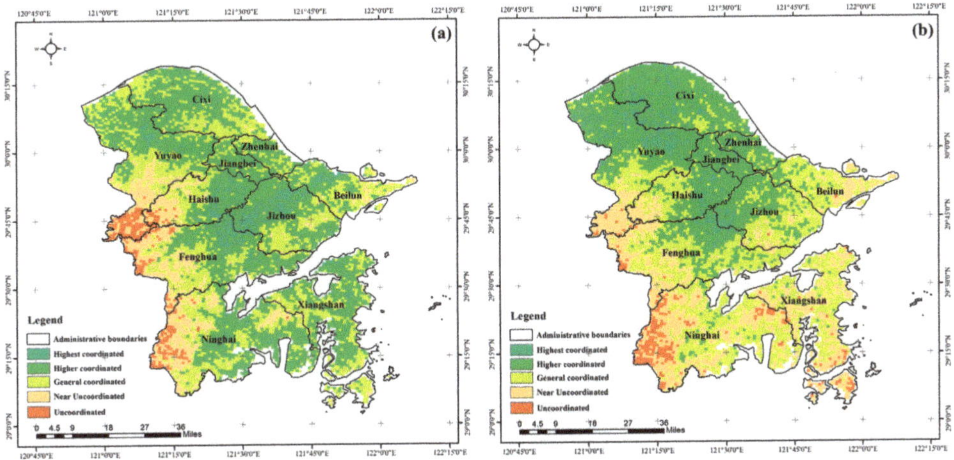

Figure 3. Coupling coordination degree of PLES suitability in Ningbo for (**a**) 2010 and (**b**) 2018.

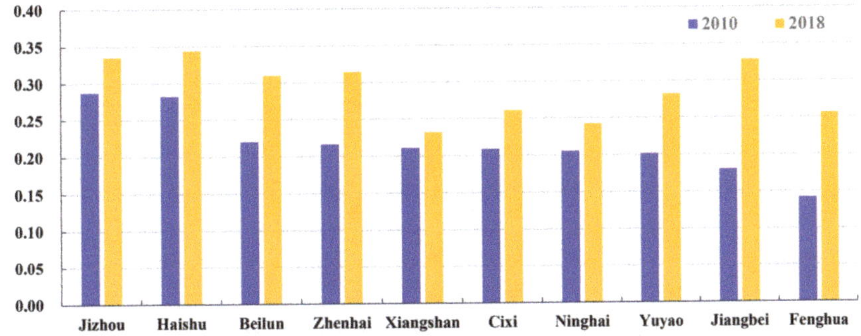

Figure 4. Sustainable development evaluation of PLES in Ningbo for (**a**) 2010 and (**b**) 2018.

3.4. PLES Conflict Identification in Ningbo

Figure 5 shows the multiscale conflict detection results of PLES in Ningbo. In 2010 and 2018, PLES in Ningbo was in a state of coupling coordination. However, with economic development, the coupling coordination degree of PLES in Ningbo was also damaged and developing toward incoordination. The violent conflict area of PLES in Ningbo in 2010 was about 2.42%, mainly concentrated in the district of Zhenhai. In 2018, it increased to 4.01% and shifted to the western part of Ningbo and the southern part of Yuyao. Result analysis of suitability and sustainable development evaluation showed that the main reason for conflicts in Zhenhai and other areas in 2010 was the limited development of living space. The rapid urbanization of Ningbo promoted the development of living space in Zhenhai and other areas, and the increase in space utilization promoted coordinated and balanced PLES development. In the western part of Ningbo and the southern part of Yuyao, due to the effect of terrain conditions, land development was difficult, and the development of production and living space was seriously hindered. The ecological function of land remained unchanged, which led to increasing conflict in the development of PS, LS, and ES. Conflicts in Ninghai, a county in the south of the city, have been continuously increasing. Due to the large scale of ecological land, the relatively small scale of living space, low population density, and poor infrastructure conditions in Ninghai, economic development was limited, resulting in the coordinated development of PS, LS, and ES gradually decreasing, and the level of conflict gradually increasing. By contrast, Jiangbei is located in the center of Ningbo, an excellent geographical location promoting the development of the city. Driven by the great foundation of urban development, and the development opportunity of industrialization and urbanization, the economy, population scale, and environmental quality constantly improved, and the development of the three types of space gradually tended to be coordinated. Moreover, the level of conflict became increasingly lower.

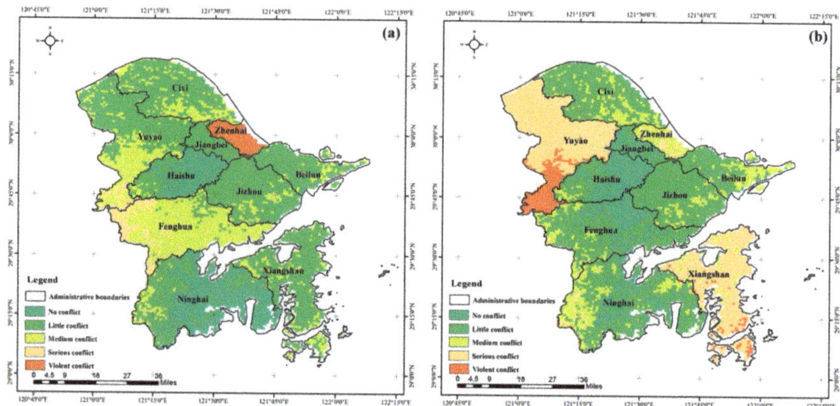

Figure 5. Multiscale PLES conflict identification in Ningbo for (**a**) 2010 and (**b**) 2018.

4. Discussion and Conclusions

There are complex relationships among various geographical scales, and scales and processes interact and influence each other. In this study, PLES coupling coordination degree on the administrative-unit and grid scales were analyzed by multiscale fusion. It contained the administrative-unit scale of PLES function information, reflected the macrobackground of development, and focused on the microlevel of PLES coordination performance, integrating the advantages of the two kinds of scales in order to understand the space–time characteristics of coordinated PLES development. The main conclusions are as follows:

In general, production land and ecological land in Ningbo were continuously occupied by human activities from 2010 to 2018. The industrial production space had a small increase; agricultural-production space occupied a large area, and the level of agricultural development was affected. The PLES coupling coordination degree analysis on the grid scale showed that, from 2010 to 2018, the lowest and lower suitability areas of ecological space in Ningbo increased. Land ecological suitability was seriously affected by urban expansion, and its ecological value was reduced. The near-uncoordinated area of PLES suitability considerably increased by 75.82%. PLES developed towards the trend of being uncoordinated. According to sustainable development on the administrative-unit scale, all counties and districts in Ningbo in 2010 and 2018 were under the near-uncoordinated state, and the sustainable development level of PLES was low. Multiscale coupling analysis showed that the PLES in Ningbo was in less conflict on the whole, but with economic development, the coupling coordination degree of PLES was also damaged and developing towards the uncoordinated direction. From the perspective of space, violent conflict areas of PLES in 2010 were mainly concentrated in Zhenhai. In 2018, this increased to 4.01%, and shifted to the west of Ningbo and the south of Yuyao.

The available research on the methods of quantitative recognition of spatial conflicts more focus on the single scale based on administrative unit, but the PLES conflicts between microcosmic grid scale receives less attention. Considering the important scale problems of the geographical study, PLES coupling coordination degree also has scale dependence. The quantitative research on the grid scale and multi-scale fusion method should be the future of spatial conflicts. It is therefore essential to analyze the spatial conflicts of PLES on different geographical scales. This study established different scales of a PLES coupling coordinated development degree evaluation index system, namely, the administrative-unit scale, grid-unit scale, and multiscale integration, and enriched the methods of multiscale fusion conflict diagnosis. However, the mechanism of multiscale fusion is complex and requires further study.

Author Contributions: G.L. and J.F. contributed to all aspects of this work; D.J. conducted data analysis and G.L. wrote the main manuscript text. All authors reviewed the manuscript. All authors have read and agreed to the published version of the manuscript.

Funding: This work was supported by a grant Strategic Priority Research Program of the Chinese Academy of Sciences (Grant No. XDA19040305), Youth Innovation Promotion Association (Grant No. 2018068) and Institute of Geographical Sciences and Natural Resources Research, Chinese Academy of Sciences (Grant No. E0V00112YZ).

Institutional Review Board Statement: Not applicable.

Informed Consent Statement: Not applicable.

Data Availability Statement: The data presented in this study are available on request from the author.

Acknowledgments: We greatly thank MDPI English service for the editing assistance to the paper.

Conflicts of Interest: The authors declare no conflict of interest.

References

1. Montanari, A.; Londei, A.; Staniscia, B. Can we interpret the evolution of coastal land use conflicts? Using Artificial Neural Networks to model the effects of alternative development policies. *Ocean Coast. Manag.* **2014**, *101*, 114–122. [CrossRef]
2. Tudor, C.A.; Ioja, I.C.; Pătru-Stupariu, I.; Niță, M.R.; Hersperger, A.M. How successful is the resolution of land-use conflicts? A comparison of cases from Switzerland and Romania. *Appl. Geogr.* **2014**, *47*, 125–136. [CrossRef]
3. Chapin, F.S.; Díaz, S. Interactions Between Changing Climate and Biodiversity: Shaping Humanity's Future. *PNAS* **2020**, *117*, 6295–6296. [CrossRef]
4. Shen, Y.; Liu, T.K.; Zhou, P. Theoretical Analysis and Strategies of Natural Ecological Space Use Control. *China Land Sci.* **2017**, *31*, 17–24.
5. Peng, J.; Lv, D.N.; Dong, J.Q.; Liu, Y.X.; Liu, Q.Y.; Li, B. Processes coupling and spatial integration: Characterizing ecological restoration of territorial space in view of landscape ecology. *J. Nat. Resour.* **2020**, *35*, 3–13.

6. Huang, A.; Xu, Y.Q.; Hao, J.M.; Sun, P.L.; Liu, C.; Zheng, W.R. Progress Review on Land Use Functions Evaluation and Its Prospects. *China Land Sci.* **2017**, *31*, 88–97.

7. Adam, Y.O.; Pretzsch, J.; Darr, D. Land use conflicts in central Sudan: Perception and local coping mechanisms. *Land Use Policy* **2015**, *42*, 1–6. [CrossRef]

8. Avriel-Avni, N.; Rofè, Y.; Scheinkman-Shachar, F. Spatial Modeling of Landscape Values: Discovering the Boundaries of Conflicts and Identifying Mutual Benefits as a Basis for Land Management. *Soc. Nat. Resour.* **2020**. [CrossRef]

9. Maleki, J.; Masoumi, Z.; Hakimpour, F.; Coello, C.A.C. A spatial land-use planning support system based on game theory. *Land Use Policy* **2020**, *99*, 105013. [CrossRef]

10. Yang, S.; Dou, S.B.; Li, C.X. Land-use conflict identification in urban fringe areas using the theory of leading functional space partition. *Soc. Sci. J.* **2020**, *10*, 1–16. [CrossRef]

11. Malczewski, J. GIS-based multicriteria decision analysis: A survey of the literature. *Int. J. Geogr. Inf. Sci.* **2006**, *20*, 703–726. [CrossRef]

12. Demesouka, O.E.; Anagnostopoulos, K.P.; Eleftherios, S. Spatial multicriteria decision support for robust land-use suitability: The case of landfill site selection in Northeastern Greece. *Eur. J. Oper. Res.* **2018**, *272*, 574–586. [CrossRef]

13. Zou, L.; Liu, Y.; Wang, J.; Yang, Y.Y.; Wang, Y.S. Land use conflict identification and sustainable development scenario Simulation on China's southeast coast. *J. Clean. Prod.* **2019**, *238*, 117899. [CrossRef]

14. Zou, L.L.; Liu, Y.S.; Wang, Y.Y. An analysis of land use conflict potentials based on ecological-production-living function in the southeast coastal area of China. *Ecol. Indic.* **2021**, *122*, 107297. [CrossRef]

15. Yu, S.H.; Deng, W.; Xu, Y.X.; Zhang, X.; Xiang, H.L. Evaluation of the production-living-ecology space function suitability of Pingshan County in the Taihang mountainous area, China. *J. Mt. Sci.* **2020**, *17*, 2562–2576. [CrossRef]

16. Miller, J.A.; Hanham, R.Q. Spatial nonstationarity and the scale of species-environment relationships in the Mojave Desert, California, USA. *Int. J. Geogr. Inf. Sci.* **2011**, *25*, 423–438. [CrossRef]

17. Liang, H.D.; Guo, J.Z.Y.; Wu, J.P.; Chen, Z.Q. GDP spatialization in Ningbo City based on NPP/VIIRS night-time light and auxiliary data using random forest regression. *Adv. Space Res.* **2020**, *65*, 481–493. [CrossRef]

18. Ningbo Municipal Statistics Bureau. Statistical bulletin of national economic and social development of Ningbo in 2020. *Ningbo Munic. Stat. Bur.* **2021**.

19. Feng, Y.J.; Yang, Q.Q.; Tong, X.H.; Wang, J.F. Long-Term Regional Environmental Risk Assessment and Future Scenario Projection at Ningbo, China Coupling the Impact of Sea Level Rise. *Sustainability* **2019**, *11*, 1560. [CrossRef]

20. Feng, Y.J.; Yang, Q.Q.; Tong, X.H.; Chen, L.J. Evaluating land ecological security and examining its relationships with driving factors using GIS and generalized additive model. *Sci. Total Environ.* **2018**, *633*, 1469–1479. [CrossRef] [PubMed]

21. Li, Y.; Wei, B.Q.; Suo, A.N.; Zhang, Z.F.; Xu, Y.; Liang, Y.H. Spatial and Temporal Coupling Relationships of Coastline Exploitation and Environmental Carrying Safety in Ningbo, China. *J. Coast. Res.* **2020**, *36*, 1292–1301. [CrossRef]

22. Xu, X.L.; Liu, J.Y.; Zhang, S.W.; Li, R.D.; Yan, C.Z.; Wu, S.X. Multi Period Land Use and Land Cover Remote Sensing Monitoring Data Set in China (CNLUCC). 2018. Available online: http://www.resdc.cn/DOI (accessed on 25 January 2021).

23. Cui, J.X.; Kong, X.S.; Chen, J.; Sun, J.W.; Zhu, Y.Y. Spatially Explicit Evaluation and Driving Factor Identification of Land Use Conflict in Yangtze River Economic Belt. *Land* **2021**, *10*, 43. [CrossRef]

24. Zhou, D.; Lin, Z.L.; Lim, S.H. Spatial characteristics and risk factor identification for land use spatial conflicts in a rapid urbanization region in China. *Environ. Monit. Assess.* **2019**, *191*, 677. [CrossRef] [PubMed]

25. Wassenaar, T.; Gerber, P.; Verburg, P.H.; Rosales, M.; Ibrahim, M.; Steinfeld, H. Projecting land use changes in the Neotropics: The geography of pasture expansion into forest. *Glob. Environ. Chang.* **2007**, *17*, 86–104. [CrossRef]

26. Jiang, S.; Meng, J.; Zhu, L. Spatial and temporal analyses of potential land use conflict under the constraints of water resources in the middle reaches of the Heihe River. *Land Use Policy* **2020**, *97*, 104773. [CrossRef]

27. Xiao, L.L.; Liu, Q.Q.; Yu, H.; Lin, M.S. Community regulation in national park based on land use conflict identification: A case study on Qianjiangyuan National Park. *Acta Ecol. Sin.* **2020**, *40*, 7277–7286.

28. Jiang, L.; Bai, L.; Wu, Y.M. Coupling and Coordinating Degrees of Provincial Economy, Resources and Environment in China. *J. Nat. Resour.* **2017**, *32*, 788–799.

29. Lu, H.; Zhou, L.; Chen, Y.; An, Y.; Hou, C. Degree of coupling and coordination of eco-economic system and the influencing factors: A case study in Yanchi County, Ningxia Hui Autonomous Region, China. *J. Arid Land* **2017**, *9*, 446–457. [CrossRef]

30. Zhang, Z.F. Sustainable Land Use Goals, Challenges and Countermeasures in China for SDGs. *China Land Sci.* **2019**, *33*, 48–55.

31. Zhang, X.S.; Xu, Z.J. Functional Coupling Degree and Human Activity Intensity of Production–Living–Ecological Space in Underdeveloped Regions in China: Case Study of Guizhou Province. *Land* **2021**, *10*, 56. [CrossRef]

32. Lu, C.Y.; Li, L.; Lei, Y.F.; Ren, C.Y.; Su, Y.; Huang, Y.F.; Chen, Y.; Lei, S.H.; Fu, W.W. Coupling Coordination Relationship between Urban Sprawl and Urbanization Quality in the West Taiwan Strait Urban Agglomeration, China: Observation and Analysis from DMSP/OLS Nighttime Light Imagery and Panel Data. *Remote Sens.* **2020**, *12*, 32217. [CrossRef]

33. Wei, C.; Lin, Q.W.; Yu, L.; Zhang, H.W.; Ye, S.; Zhang, D. Research on Sustainable Land Use Based on Production–Living–Ecological Function: A Case Study of Hubei Province, China. *Sustainability* **2021**, *13*, 996. [CrossRef]

34. Wang, K.; Tang, Y.K.; Chen, Y.Z.; Shang, L.W.; Ji, X.M.; Yao, M.C.; Wang, P. The Coupling and Coordinated Development from Urban Land Using Benefits and Urbanization Level: Case Study from Fujian Province (China). *Int. J. Environ. Res. Public Health* **2020**, *17*, 5647. [CrossRef] [PubMed]

35. Zhu, S.Y.; Diao, C.L.; Xiao, W.B.; Chen, Q.C.; Wang, H. Convenience Index of Public Transportation for Urban Tourist Attraction: Based on Investigation in Wuhan. *J. Transp. Syst. Eng. Inf. Technol.* **2021**, *21*, 169–175.
36. Wan, J.; Zhang, L.W.; Yan, J.P.; Wang, X.M.; Wang, T. Spatial–Temporal Characteristics and Influencing Factors of Coupled Coordination between Urbanization and Eco-Environment: A Case Study of 13 Urban Agglomerations in China. *Sustainability* **2020**, *12*, 8821. [CrossRef]
37. Wang, D.; Jiang, D.; Fu, J.Y.; Lin, G.; Zhang, J.L. Comprehensive Assessment of Production–Living– Ecological Space Based on the Coupling Coordination Degree Model. *Sustainability* **2020**, *12*, 2009. [CrossRef]
38. United Nations. *Transforming Our World: The 2030 Agenda for Sustainable Development*; United Nations: New York, NY, USA, 2015.
39. Wang, J.; Fan, X.Y. Assessment of Residents' Health Level Based on Fuzzy Comprehensive Evaluation. *J. Guizhou Univ.* **2020**, *37*, 30–34.
40. Wang, Z.; Ma, H. Research on Integration Model of Multi-scale Basic Geographic Information Data: Taking Xianyang City as an Example. *Geomat. Spat. Inf. Technol.* **2019**, *42*, 150–152.

MDPI
St. Alban-Anlage 66
4052 Basel
Switzerland
Tel. +41 61 683 77 34
Fax +41 61 302 89 18
www.mdpi.com

Land Editorial Office
E-mail: land@mdpi.com
www.mdpi.com/journal/land

9 783036 546216